U0358715

全国科学技术名词审定委员会

科学技术名词·工程技术卷（全藏版）

16

海峡两岸航海科技名词

海峡两岸航海科技名词工作委员会

国家自然科学基金资助项目

科 学 出 版 社

北 京

内 容 简 介

  本书是全国科学技术名词审定委员会、台湾中华海运研究协会共同组织海峡两岸的航海界专家会审的海峡两岸航海名词对照本，是在海峡两岸各自公布名词的基础上加以增补修订而成的。内容包括地文航海、天文航海、电子航海、军事航海、航海仪器、航海保证、船艺、船舶操纵与避碰、客货运输、海上作业、航行管理与法规、水上通信、轮机管理等,共收词 7000 余条。供海峡两岸航海界和其他领域的有关人士使用。

图书在版编目（CIP）数据

  科学技术名词. 工程技术卷：全藏版 / 全国科学技术名词审定委员会审定.
—北京：科学出版社，2016.01
  ISBN 978-7-03-046873-4

  I. ①科… II. ①全… III. ①科学技术–名词术语 ②工程技术–名词术语
IV. ①N-61 ②TB-61

  中国版本图书馆 CIP 数据核字（2015）第 307218 号

责任编辑：李玉英 / 责任校对：陈玉凤
责任印制：张 伟 / 封面设计：铭轩堂

科 学 出 版 社 出版
北京东黄城根北街 16 号
邮政编码：100717
http://www.sciencep.com
北京厚诚则铭印刷科技有限公司印刷
科学出版社发行 各地新华书店经销
*
2016 年 1 月第 一 版 开本：787×1092 1/16
2016 年 1 月第一次印刷 印张：30 1/2
字数：713 000
定价：7800.00 元（全 44 册）
（如有印装质量问题，我社负责调换）

海峡两岸航海科技名词工作委员会委员名单

召 集 人: 王逢辰

委　　　员（按姓氏笔画为序）:

　　　　杜荣铭　　李玉英　　杨守仁　　胡正良　　袁安存

召 集 人: 黄正清

委　　　员（按姓氏筆畫為序）:

　　　　朱于益　　林幸蓉　　周和平　　郭長齡　　楊仲箎

　　　　鄭正村

序

  科学技术名词作为科技交流和知识传播的载体,在科技发展和社会进步中起着重要作用。规范和统一科技名词,对于一个国家的科技发展和文化传承是一项重要的基础性工作和长期性任务,是实现科技现代化的一项支撑性系统工程。没有这样一个系统的规范化的基础条件,不仅现代科技的协调发展将遇到困难,而且,在科技广泛渗入人们生活各个方面、各个环节的今天,还将会给教育、传播、交流等方面带来困难。

  科技名词浩如烟海,门类繁多,规范和统一科技名词是一项十分繁复和困难的工作,而海峡两岸的科技名词要想取得一致更需两岸同仁作出坚韧不拔的努力。由于历史的原因,海峡两岸分隔逾50年。这期间正是现代科技大发展时期,两岸对于科技新名词各自按照自己的理解和方式定名,因此,科技名词,尤其是新兴学科的名词,海峡两岸存在着比较严重的不一致。同文同种,却一国两词,一物多名。这里称"软件",那里叫"软体";这里称"导弹",那里叫"飞弹";这里写"空间",那里写"太空";如果这些还可以沟通的话,这里称"等离子体",那里称"电浆";这里称"信息",那里称"资讯",相互间就不知所云而难以交流了。"一国两词"较之"一国两字"造成的后果更为严峻。"一国两字"无非是两岸有用简体字的,有用繁体字的,但读音是一样的,看不懂,还可以听懂。而"一国两词"、"一物多名"就使对方既看不明白,也听不懂了。台湾清华大学的一位教授前几年曾给时任中国科学院院长周光召院士写过一封信,信中说:"1993年底两岸电子显微学专家在台北举办两岸电子显微学研讨会,会上两岸专家是以台湾国语、大陆普通话和英语三种语言进行的。"这说明两岸在汉语科技名词上存在着差异和障碍,不得不借助英语来判断对方所说的概念。这种状况已经影响两岸科技、经贸、文教方面的交流和发展。

  海峡两岸各界对两岸名词不一致所造成的语言障碍有着深刻的认识和感受。具有历史意义的"汪辜会谈"把探讨海峡两岸科技名词的统一列入了共同协议之中,此举顺应两岸民意,尤其反映了科技界的愿望。两岸科技名词要取得统一,首先是需要了解对方。而了解对方的一种好的方式就是编订名词对照本,在编订过程中以及编订后,经过多次的研讨,逐步取得一致。

  全国科学技术名词审定委员会(简称全国科技名词委)根据自己的宗旨和任务,始终把海峡两岸科技名词的对照统一工作作为责无旁贷的历史性任务。近些年一直本着积极推进,增进了解;择优选用,统一为上;求同存异,逐步一致的精神来开展这项工作。先后接待和安排了许多台湾同仁来访,也组织了多批专家赴台参加有关学科的名词对照研讨会。工作中,按照先急后缓、先易后难的精神来安排。对于那些与"三通"

有关的学科，以及名词混乱现象严重的学科和条件成熟、容易开展的学科先行开展名词对照。

在两岸科技名词对照统一工作中，全国科技名词委采取了"老词老办法，新词新办法"，即对于两岸已各自公布、约定俗成的科技名词以对照为主，逐步取得统一，编订两岸名词对照本即属此例。而对于新产生的名词，则争取及早在协商的基础上共同定名，避免以后再行对照。例如101～109号元素，从9个元素的定名到9个汉字的创造，都是在两岸专家的及时沟通、协商的基础上达成共识和一致，两岸同时分别公布的。这是两岸科技名词统一工作的一个很好的范例。

海峡两岸科技名词对照统一是一项长期的工作，只要我们坚持不懈地开展下去，两岸的科技名词必将能够逐步取得一致。这项工作对两岸的科技、经贸、文教的交流与发展，对中华民族的团结和兴旺，对祖国的和平统一与繁荣富强有着不可替代的价值和意义。这里，我代表全国科技名词委，向所有参与这项工作的专家们致以崇高的敬意和衷心的感谢！

值此两岸科技名词对照本问世之际，写了以上这些，权当作序。

2002 年 3 月 6 日

前　　言

  随着海峡两岸间各种交流活动的日益增多,20世纪90年代初在海运界就组织了"海峡两岸通航学术交流"活动,相继开展了有关海运方面的学术交流。由于长达四十多年的分割,海峡两岸各自使用的航海科技名词相异甚多,给两岸的航运交流带来不少困难。为适应即将到来的两岸通航的需要,以确保准确、安全、避免损失,在1995年召开的第三次海峡两岸通航学术交流会上,双方都提出应该及早开展两岸航海科技名词的对照统一。随后,全国科学技术名词审定委员会(以下简称"全国科技名词委")和台湾方面"中华海运研究协会"就开展两岸航海科技名词对照统一进行了积极的联系和组织工作。全国科技名词委委托大连海事大学的专家组成研究组并提出进行此项研究工作的工作计划(草稿)。1996年7月,接受台湾中华海运研究协会的邀请,全国科技名词委副主任潘书祥和海事大学专家赴台座谈,共同商讨,并确定了工作计划。

  座谈会确定:以全国科技名词委公布的《航海科技名词》和台湾有关方面公布的《航海名词》为该项研究的选词基础;在对照研究选词过程中应以航海科技常用词为主,注意收编概念稳定的新词,词条的总收入量争取达到一万条左右;本着求同存异、照顾习惯、逐步一致的原则进行两岸航海科技名词的对照研究。

  1997年7月和2000年8月先后在大连和台北举行研讨会,对双方已提出的词条进行了逐条对照和研讨分析,作了必要的删减和增补。并决定在此基础上再经双方专家组成员审核、校对,订稿后打印成册,送交主办单位印刷出版。

  在研讨中还明确了,随着航海科技的发展,还会不断出现新词条,为使对照研究工作持续进行,双方专家及时沟通,共同商讨,对新出现的航海科技名词及时科学统一定名。

  此次出版的名词共7000余条。工作中疏漏在所难免,望广大读者指出,以利修正。

海峡两岸航海科技名词工作委员会

2001年3月30日

编 排 说 明

一、本书是海峡两岸航海科技名词对照本。

二、本书分正篇和副篇两部分。正篇按汉语拼音顺序编排;副篇按英文名的字母顺序编排。

三、两岸推荐名用黑体字表示。

四、[ ]中的字使用时可以省略。

正篇

五、正名和异名分别排序,在异名处用(=)注明正名和英文名同义词。

六、对应的英文名为多个时(包括缩写词)用","分隔。

副篇

七、英文名所对应相同概念的汉文名用","分隔,不同概念的用① ② ③分别注明。

八、英文名的同义词用(=)注明。

九、英文缩写词排在全称后的( )内。

目　　录

正 篇

A

祖国大陆名	台湾地区名	英 文 名
霭	靄	mist
爱德考克天线	亞德考克天線	Adcook antenna
碍航浮标	危險浮標	danger buoy
碍航物	障礙	obstruction
安放龙骨日期	安放龍骨日期	date of keel laid
安全	安全保證	security
安全棒	安全棒	safety rod
安全报告	安全信文	safety message
安全到达保险	安全到達保險	safety arrival insurance
安全电灯	安全電燈	electric safety lamp
安全吊货钩	安全吊貨鉤	cargo safety hook
安全阀调定压力	安全閥調定壓力	safety valve setting pressure
安全[工作]负荷	安全工作負荷	safe working load, SWL
安全管理手册	安全管理手冊	safety management manual
安全管理证书	安全管理證書	Safety Management Certificate
安全管理制度	安全管理制度	safety management system, SMS
安全航路	安全主航道	safety fairway
安全航速	安全速度	safety speed
安全呼叫	安全呼叫	safety call
安全呼叫格式	安全呼叫格式	safety call format
安全水域标志	安全水域標誌	safety water mark
安全通信	安全通信	safety communication
安全通信程序	安全通信程序	safety communication procedure
安全通信网	安全通信網	safety NET
安全系数	安全因素	safety factor, SF
安全系统	安全系統	safety system
安全信号	安全信號	safety signal
安全业务	安全業務	safety service

祖国大陆名	台湾地区名	英文名
安全优先等级	安全優先順序	safety priority
安全与环保政策	安全與環保政策	safety and environmental protection policy
安全与联锁装置	安全與聯鎖設施	safety and interlock device
安全载流量	安全載流量	safe carrying capacity
安全注射系统	安全噴射系統	safety injection system
岸壁效应	岸壁效應	bank effect
岸冰	岸冰	shore ice
岸电	岸電	shore power
岸电联锁保护	岸電聯鎖保護	interlock protection of shore power connection
岸上管理	岸上管理	shore based management
岸上维修	岸上維修	shore-based maintenance, SBM
岸台费	岸台費	coast station charge
岸吸	岸吸力	bank suction
岸线	海岸線	coastline
暗礁	暗礁	sunken rock
奥米伽	亞米茄	Omega
奥米伽表	亞米茄表	Omega table
奥米伽传播改正量	亞米茄傳播修正值	Omega propagation correction, OPC
奥米伽船位	亞米茄船位	Omega fix
奥米伽导航仪	亞米茄航儀	Omega navigator
奥米伽海图	亞米茄海圖	Omega chart
奥米伽信号格式	亞米茄信號格式	Omega signal format

B

祖国大陆名	台湾地区名	英文名
八字锚泊	[八字]雙錨泊	riding to two anchors
巴拿马运河吨位	巴拿馬運河噸位	Panama Canal Tonnage
巴氏合金	巴比合金	Babbitt metal
巴氏货油舱清洗系统	巴氏貨油艙清洗系統	Butterworth tank cleaning system
罢工条款	罷工保險條款	strike clause
白道	白道	moon's path
白合金轴承	白合金軸承	white metal bearing
白天信号灯	日間信號燈	daylight signaling lamp
白星火箭	白星火箭	white star rocket
白棕绳	馬尼拉繩,白棕繩	Manila rope
百帕	百巴斯噶	hectopascal

祖 国 大 陆 名	台 湾 地 区 名	英 文 名
百万分之几	百萬分之幾	parts per million, ppm
摆幅(=行程)		
摆式罗经	擺式電盤經	pendulous gyrocompass
班轮	定期船	liner
班轮公会	定期船公會	liner conference
班轮公会行动守则公约	定期船同盟行動章程公約	Convention on a Code of Conduct for Liner Conference
班轮提单	定期船載貨證券	liner bill of lading
班轮条款	定期船條件	liner term
班轮运输	定期船業務	liner service
搬运	操作	handling
板材量尺	板材量法	board measure
板式蒸发器	板式蒸發器	plate-type evaporator
半潮	半潮	half tide
半潮礁	半潮礁	half tide rock
半导体	半導體	semiconductor
半导体制冷	半導體冷凍	semiconductor refrigeration
半封闭式制冷压缩机	半封密冷凍壓縮機裝置	semi-hermetic refrigerating compressor unit
半高箱	半高櫃	half height container
半集装箱船	半貨櫃船	semi-container ship
半结	半套結	half hitch
半解析式惯性导航系统	半解析慣性導航系統	semianalytic inertial navigation system
半径差	半徑差	semidiameter, SD
半梁	半梁	half beam
半平衡舵	半平衡舵	semi-balanced rudder
半潜船	半潛船	semisubmerged ship
半日潮	半日週潮	semi-diurnal tide
半双工操作	半雙工作業	semi-duplex operation
半圆法	半圓法	semicircular method
半圆自差	半圓自差	semicircular deviation
半月	半月	half moon
半涨潮	半漲潮	half flood
半致死剂量	半致死劑量	half lethal dose
半致死浓度	半致死濃度	half lethal concentration
半组合曲轴	半組合曲柄軸	half built-up crankshaft
伴流	伴流	wake
伴流横向力	伴流橫向力	transverse force of wake current
伴流系数	跡流因數	wake fraction

祖国大陆名	台湾地区名	英文名
绑扎	捆縛,綁固	lashing
绑扎板	拉緊板	lashing plate
绑扎棒	拉緊桿	lashing bar
绑扎杆	拉緊桿	lashing rod
绑扎钩	拉緊鉤	lashing hook
绑扎环	拉緊環,D 型環	lashing eye
绑扎链	拉緊鏈	lashing chain
绑扎链扣	拉緊鏈扣	chain lashing device
绑扎索	拉緊索	lashing cable
绑扎套筒	拉緊缸	lashing pot
傍靠补给法	傍靠傳遞法	along-side method
傍拖	傍拖	towing alongside
包裹运费	包裹運費	parcel freight
包装	包裝	packing
包装标号	包裝號碼	packaging code number
包装不牢货	包裝不固貨	insufficiently packed cargo
包装材料	包裝材料	packing materials
包装货	包裝貨	packed cargo
包装鉴定	包裝檢查	inspection of package
包装类	包裝分類	packaging group
包装容积	包裝容積	bale capacity
饱和水汽压	平衡蒸汽壓	equilibrium vapor pressure
饱和蒸汽	飽和蒸汽	saturated steam
饱和蒸汽压力	飽和蒸汽壓力	saturated vapor pressure
保函	賠償責任保證書,賠償 保證書	letter of indemnity
保护标志	保護標誌	protective mark
保护海底电缆公约	保護海底電纜公約	Convention for the Protection of Submarine Cables
保护权	保護權	right of protection
保护位置	保護位置	protective location
保留权	拘留權,保留權	right of retention
保赔	防護賠償	protection and indemnity, PI
保赔保险	船舶營運人責任保險	PI insurance
保赔协会	保護及賠償協會	protection and indemnity club, PI club
保赔责任险	防護及賠償責任險	protection and indemnity risk, PI risk
保税库	保稅倉庫	bonded store, bond room
保温箱	保溫[貨]櫃	thermal container

祖 国 大 陆 名	台 湾 地 区 名	英 文 名
保温用具	保溫用具	thermal protective aid
保险价值	保險價值	insured value
保险人	保險人	assurer
保险索赔	保險索賠	insurance claim
保向性	航向保持性	course keeping quality
保修工程师	保證工程師,保固技師	guarantee engineer
保养(=维护)		
保用期	保固期	warranted period
报酬	報酬	reward
报告	報告	report
报告程序	報告程序	reporting procedures
报告点	報告點	reporting point, RP
报关,申报	陳報,申報	declaration
报关行	報關行	custom broker
报警报文	警報信文	alert message
报警打印	警報打印	alarm printer
报警监视系统	警報監視系統	alarm monitoring system
报警数据	警報數據	alert data
报警数据滤除	篩選之警報數據	filtered alert data
报警装置	警報設施	warning device
报头	報頭	preamble
报文	信文	message
暴动和内乱不保	暴動和內亂不保	free of riot and civil commotion
暴风警报	暴風警報	storm warning
暴风雨	暴風雨	heavy rainstorms
爆燃	爆震	detonation
爆炸品	炸藥	explosive
爆炸雾号	音爆霧號	explosive fog signal
爆炸信号	爆炸信號	explosive signal
爆震	爆震	knock
北半球	北半球	northern hemisphere
北冰洋海流	北極海流	arctic current
北大西洋冬季载重线	冬期北大西洋載重線	winter North Atlantic load line
北斗七星	北斗七星	great dipper
北方标	北方標	north mark
北河二(双子 α)	北河二(雙子 α)	Castor
北极	北極	north pole
北极冰	北極流動水田	arctic pack

祖 国 大 陆 名	台 湾 地 区 名	英 文 名
北极光	北極光	northern light, aurora borealis
北极气团	北極氣團	arctic air mass
北极圈	北極圈	arctic circle
北极星	北極星	north star, Polaris
北极星高度改正量	北極星修正	Polaris correction
北极星求纬度	極星求緯法	latitude by pole star
北京坐标系	北京座標系統	Beijing coordinate system
北落师门(南鱼 α)	北落師門(南魚 α)	Fomalhaut
北天极	北天極	north celestial pole
北向上	北向上	north up
北向陀螺	北向迴轉儀	north gyro
备件	備品	spare
备锚	備便拋錨	set the anchor ready for letting go
备用	待命,預備	stand-by
备用发电机	備用發電機	stand-by generator
备用发信机	備用發射機	reserve transmitter
备用锚	備用錨	spare anchor
备用收信机	備用接收機	reserve receiver
备用天线	備用天線	reserve antenna
背景亮光	背景亮光	background light
背离规则	背離規則	departure from these rules
背压式汽轮机	背壓蒸汽渦輪機	back pressure steam turbine
被保险人	被保險人	insured
被呼方	被呼用戶,受話方	called party
被控对象	控制對象	controlled object
被拖船	被拖船	towed vessel
被追越船	被超越船,被追趕船	overtaken vessel
本地差分 GPS	當地差分全球定位系統	local area differential GPS
本地警告	當地警告	local warning
本地用户终端	地面終端台	local user terminal, LUT
本地增强系统	當地增強系統	local area augmentation system
笨重货	笨重貨	awkward cargo
泵舱通海阀	泵室海水閥	pump room sea valve
泵流量	泵能量	pump capacity
泵特性曲线	泵特性曲線	pump characteristic curve
泵吸	抽出	pumping
泵压头	泵壓頭	pump head
泵自动切换装置	泵自動切換設施	pump auto-change over device

祖国大陆名	台湾地区名	英 文 名
比功率	比功率	specific power
比较单元	比較單位	comparing unit
比较器	比測儀	comparator
比例带	比例帶	proportional band
比例调节器	比例調整器	proportioner，proportional regulator
比相	比相	phase comparison
比重环	比重盤	gravity disc
比转数	比速	specific speed
舭(＝舯突出部)		
舭龙骨	舭龍骨	bilge keel
舭肘板	舭腋板	bilge bracket
必要带宽	必須頻帶寬度	necessary bandwidth
毕宿五(金牛 α)	畢宿五(金牛 α)	Aldebaran
闭杯试验	閉杯法試驗	closed cup test
闭环系统	閉環系統	closed-loop system
闭路网络	閉路網路	closed network
闭路用户组	閉路用戶組	closed user group
闭式冷却水系统	閉式冷卻水系統	closed cooling water system
闭式冷却系统	閉式冷卻系統	closed cooling system
闭式喷油器	閉式噴油閥	closed type fuel valve
闭式叶轮	罩筒葉輪	shrouded impeller
闭式液压系统	閉式液壓系統	closed-type hydraulic system
壁,墙	壁,牆	wall
壁式插座	壁式插座	wall plug
壁宿二(仙女 α)	壁宿二(仙女 α)	Alpheratz
避风航路	避風航路	routing for storm avoidance
避风锚地	遮蓋,遮蔽	shelter
避难港	避難港	port of refuge
避难港费用	避難港費用	port of refuge expenses
避碰决策	避碰決策	decision making of collision avoidance
避碰行为	避碰行爲	collision avoidance behavior
避碰专家系统	避碰專家系統	collision avoidance expert system
避碰综合决策	綜合避碰決策	synthetical decision making of collision avoidance
臂距差	曲柄臂距差	difference crank spread
臂距千分表	曲柄軸撓曲針盤量規	crankshaft deflection dial gauge
边舱	翼櫃,翼艙	wing tank
边界润滑	邊界潤滑	boundary lubrication

祖 国 大 陆 名	台 湾 地 区 名	英 文 名
编队(=队列)		
编码位置数据	编碼之位置數據	encoded position data
编码信息	編碼數據	coded information
编码延迟	密碼遲延	coding delay
便利国际海上运输公约	便利國際海上運輸公約	Convention on Facilitation of International Maritime Traffic
变更登记	變更登記	registration of alteration
变工况	可變工作情況	variable working condition
变极调速	變極調速	speed regulation by pole changing
变量泵	變量泵	variable delivery pump, variable capacity pump
变量油马达	變量油馬達	variable-displacement oil motor
变流机组	換流機組	converter set
变频调速	變頻調速	speed regulation by frequency variation
变向泵	可逆泵	reversible pump
变形测量表	變形量規	deformation gauge
变Z形试验	修正式蛇航試驗,修正式之字航行試驗	modified zigzag maneuver test
变压器	變壓器	electric transformer
变质	變質	degenerate
辨识	辨識	identification
辫子(=挠性接头)		
标定断开容量	額定斷電容量	rated breaking capacity
标定工况	額定工作狀況	rated working condition
标定功率	額定功率	rated power, rated output
标定功率修正	定額修正	rating corrections
标定接通容量	額定接續容量	rated making capacity
标定转舵扭矩	額定舵桿扭矩	rated stock torque
标定转速	額定引擎轉速	rated engine speed
标绘航线	標繪航線	plot a course
标绘距离	標繪距離	plot a distance
标记,记号	標記,記號	marking
标牌	標貼	placard
标签	標籤	label
标圈	標圈	aiming circle
标题	標題,名稱	title
标志	標誌	mark
标志船	標誌船	marking vessel

祖 国 大 陆 名	台 湾 地 区 名	英 文 名
标柱测速试验	標柱速率試驗	measured mile trial
标准,规范	標準	standard
标准报告格式	標準報告格式	standard reporting format
标准大气	標準大氣	standard atmosphere
标准定位业务	標準定位業務	standard positioning service, SPS
标准格式	標準格式	standard form
标准航海用语	標準航海用語	standard marine navigational vocabulary, SMNV
标准横向张索补给法	標準強力法輸送設備	standard tensioned replenishment along- side method
标准滑阀图	標準滑閥圖	standard slide valve diagram
标准界面说明	標準介面說明	standard interface description
标准绝缘法兰接头	典型絕緣凸緣接頭	typical insulating flange joint
标准空档深度	標準空隙深度	standard void depth
标准罗航向	標準羅經航向	standard compass course
标准罗经	標準羅經	standard compass
标准排泄接头	標準排洩接頭	standard discharge connection
标准频率和时间信号台	標準頻時信號台	standard frequency and time signal station
标准频率和时间信号业务	標準頻時信號業務	standard frequency and time signal service
标准时	標準時	standard time
标准[误]差	標準誤差	standard error
标准箱(20英尺集装箱)	20呎貨櫃[相]當[數]量	twenty equivalent unit, TEU
飑线	颮線	squall line
表	表	list
表层流	表面流	surface current
表列高度	表列高度	tabulated altitude
表面	表面	surface
表面式空气冷却器	表面空氣冷卻器	surface air cooler
表面硬化	表面硬化	face hardening
表面张力	張力	surface tension
表盘式调速器	針盤調速器	dial type governor
表压力	表壓力	gauge pressure
冰崩	冰崩	ice avalanche
冰舱	冰艙	ice hold
冰冻期	冰凍期	ice period
冰冻条款	冰凍條款	ice clause

祖 国 大 陆 名	台 湾 地 区 名	英 文 名
冰封地带	冰障	ice barrier
冰封区域	冰封區域	ice covered area
冰盖	覆冰［量］	ice cover
冰厚	冰層厚度	ice thickness
冰架	連岸冰, 冰灘	ice shelf
冰间水道	冰間巷道	lead lane
冰壳	脆冰殼	ice rind
冰况报告	冰況報告	ice report
冰况图集	冰況地圖	ice atlas
冰困	冰封	icebound
冰锚	繫冰錨, 冰錨	ice anchor
冰面饱和水汽压	冰面平衡蒸汽壓	equilibrium vapor pressure of ice surface
冰情警报	冰情警報	ice warning
冰情巡逻服务	冰區巡邏服務	ice patrol service
冰丘	冰丘	hummock
冰区航行	冰區航行	ice navigation
冰区界限线	冰區界線	ice boundary
冰区拖航	冰區拖纜航	towing in ice
冰山	冰山	berg
冰山探测	冰山探測	detection of iceberg
冰图	冰圖	ice chart
冰原	冰原, 冰野	ice field
冰缘线	冰緣線	ice edge
冰中操船	冰中操船	shiphandling in ice
冰中护航	冰中護航	convoy in ice
并车电抗器	並聯運轉反應器	parallel operation reactor
并车屏	併俥屏	paralleling panel
并联复式汽轮机	並列複式蒸汽渦輪機	cross-compound steam turbine
并联运行试验	並聯運轉試驗	parallel-running test
并装提单	併裝載貨證券	combined bill of lading
波	波	wave
l-波段紧急无线电示位标	L 頻帶應急指位無線電示標	l-band EPIRB
l-波段紧急无线电示位标系统	頻帶應急指位無線電示標系統	l-band EPIRB system
波动	波動	wave
波峰	波峰	wave ridge, wave crest
波谷	波谷	wave trough, wave hollow

祖 国 大 陆 名	台 湾 地 区 名	英 文 名
波浪补偿器	波浪補償器	swell compensator
波浪谱	波譜	wave spectrum
波浪弯矩	波载彎曲力矩	wave bending moment
波浪要素	波浪要素	wave parameter
波浪周期	波浪週期,波週期	wave period
波纹管	伸縮囊	bellows
波形舱壁	波形艙壁	corrugated bulkhead
波形膨胀管	波形膨脹管	corrugated expansion pipe
H 波型振动	H 型振動	h-mode vibration
波振幅	波幅	wave amplitude
玻璃钢船	玻[璃]纖[維]強化塑 膠船	fiberglass reinforced plastic boat
玻璃钢救生艇	玻璃纖維救生艇	glass-fiber lifeboat
剥皮机	剝皮機	peeler
驳船	駁船	barge
驳船编队系数	編隊係數	formation coefficient
驳船队	駁船隊	barge train
驳船队编组	駁船隊編組	barge train formation
驳船队形图	駁船隊形圖	sketch of barge train formation
驳门	駁門	port
驳运泵	轉駁泵	transfer pump
泊船处	泊地,錨泊處	road stead
泊位条款	碼頭收交貨條件	berth term
泊位租船合同	泊位傭船契約	berth charter party
薄壁	薄壁	thin wall
薄壁结构	薄壁結構	thin-walled structure
薄壁轴承	薄壁軸承	thin wall bearing
薄膜蒸发	薄膜蒸發	thin film evaporation
薄燃料油	稀燃油	thin fuel oil
薄雾(4级)	薄霧(4級)	thin fog
补板	貼補	patching
补偿	補償金	compensation
补偿棒	填隙棒	shim rod
补偿流	補償流	compensation current
补偿调节指针	補償調節指針	compensation adjusting pointer
补偿针阀	補償針閥	compensation needle valve
补充报告	補充報告	supplementary report
补充码	補充碼	complement code

祖 国 大 陆 名	台 湾 地 区 名	英 文 名
补充说明	補充說明	supplemental instruction
补充[天气]预报	補充[天氣]預報	supplementary [weather] forecast
补给船	補給船	delivery ship
补给航速	整補航速	replenishment speed
补给航向	整補航向	replenishment course
补给横距	整補橫距	replenishment distance abeam
补给舰	補給艦	replenishing ship
补给站	補給站	supplying station
补给阵位	整補站	replenishment station
补给直升机	補給直升機	supplying helicopter
补给纵距	整補縱距	replenishment distance astern
补水系统	加水系統	water charging system
捕获	截獲	acquisition
捕获量	漁獲物	catch
捕鲸船	捕鯨船	whaler
捕鲸母船	捕鯨母船	whaling mother ship
捕捞机械	漁撈機械	fishing machinery
捕蟹船	捕蟹船	crabber
捕鱼技术	漁撈學	fishing technology
捕鱼权	捕魚權	right of fishery
不沉性	不沈性	insubmersibility
不定期船运输	不定期船業務	tramp service
不冻港	不凍港	ice free port
不对称浸水	不對稱泛水	unsymmetrical flooding
不发射信标	示標無發送	no beacon emission
不记名提单	不記名載貨證券	blank bill of lading
不洁压舱水	不潔壓艙水	dirty ballast
不可避免的事故	不可避免的事故	inevitable accident
不可拆卸的	不可拆卸者	non-removable
不可倒转柴油机	不可逆轉柴油機	non-reversible diesel engine
不可抗力	不可抗力	force majeure
不可转让提单	不可轉讓載貨證券	non-negotiable bill of lading
不良锚地	不良船席	foul berth
不明过失	不明過失	inscrutable fault
不明阶段	不明階段	uncertainty phase
不平衡舵	不平衡舵	unbalanced rudder
不平衡力	不均衡力	unbalanced force
不清洁提单	不潔載貨證券	foul bill of lading

祖 国 大 陆 名	台 湾 地 区 名	英 文 名
不燃材料	不燃材料	non-combustible material
不完全燃烧	不完全燃燒	imperfect combustion
不在第一港卸的货	留船未卸貨	residue cargo
不知条款	不知條款	unknown clause
不准入境	禁止入境	entrance prohibited
布管船	佈管船	pipeline layer
布缆船	佈纜船	cable layer
布缆机	佈纜機	cable laying machine
布雷航海勤务	佈雷航海勤務	mine-laying navigation service
布氏[硬]度数	勃式硬度數	Brinell figure
步进式(=逐级)		
步桥	窄道,步橋	cat walk
部分封闭救生艇	部分圍蔽救生艇	partially enclosed lifeboat
部分货载	部分貨載	part cargo
部分进汽度	部分進汽度	degree of partial admission
部分装载舱室	部分裝載艙間	partly filled compartment

C

祖 国 大 陆 名	台 湾 地 区 名	英 文 名
财物	財物	property
采样点	抽樣點	sampling point
参数不均匀率	参數不匀率	parameter non-uniform rate
参数设定	参數設定	parameter setting
残水旋塞	排洩旋塞	drain cock
残损鉴定	貨損檢驗	survey on damage to cargo
残碳值	殘留碳	carbon residue
残油标准排放接头	殘油標準排洩接頭	residual oil standard discharge connection
残余边带发射	殘邊帶發射	vestigial-sideband emission
残余废气	殘留氣體	residual gas
残渣油	殘油	residual fuel oil
仓储费	倉儲費	storage charge
仓库交货	倉庫交貨	delivery ex-warehouse
舱壁	艙壁	bulkhead
舱壁防爆填料函	艙壁防爆填料函	anti-explosion bulkhead stuffing box
舱壁甲板	艙壁甲板	bulkhead deck
舱壁图	艙壁圖	bulkhead plan
舱底泵	舭泵	bilge pump

祖 国 大 陆 名	台 湾 地 区 名	英 文 名
舱顶空档测量孔	油面測距孔	ullage port
舱盖	艙蓋	hatch cover
舱盖布	艙口蓋布	hatch tarpaulin
舱盖曳行装置	艙蓋驅動設施	hatch cover driving device
舱口	艙口	hatch, hatchway
舱口吊杆	艙口吊桿	hatch boom
舱口端梁	艙口端梁	hatch end beam
舱口盖绞车	艙口蓋絞車	hatch cover [handling] winch
舱口活动梁	艙口活動梁	hatch shift beam
舱口检验	艙口檢驗	hatch survey
舱口梁	艙口梁	hatch beam
舱口围板	艙口緣圍	hatch coaming
舱口压条	艙口壓條	hatch batten
舱口装卸指挥人	艙口裝卸指揮人	hatch man
舱面货提单	艙面載貨證券	on deck bill of lading
舱内安装设备	裝於艙內設備	internal mounted equipment
舱内货	艙內貨	under-deck cargo
舱内设备	艙內設備	below deck equipment, BDE
舱容图	容量圖	capacity plan
舱容系数	貨艙係數	coefficient of hold
舱室鉴定	艙室檢查	inspection of chamber
舱室通风机	房艙通風機	cabin ventilator
舱外安装设备	裝於艙外之設備	externally mounted equipment
舱外设备	艙面設備	above deck equipment, ADE
舱位登记簿	艙位簿	space book
操舵	操舵	steering
操舵罗航向	駕駛羅經航向	steering compass course
操舵罗经	駕駛羅經	steering compass
操舵试验	操舵試驗	steering test
操舵室	駕駛室	wheel house
操舵遥控传动装置	操舵液壓遙控裝置	steering telemotor
操舵装置	操舵機,操舵裝置	steering gear
操艇	操艇	boating
操纵部位转换	控制位置轉換	transfer of control station
操纵灯号	操縱號燈	maneuvering light signal
操纵杆	操縱桿	control rod
操纵缆	導引線	steering line
操纵能力受限船	操縱能力受限船	vessel restricted in her ability to maneuver

祖 国 大 陆 名	台 湾 地 区 名	英 文 名
操纵失灵	操縱失靈	out of command
操纵失灵号灯	操縱失靈號燈	out of command light
操纵信号	運轉信號,操縱信號	maneuvering signal
操纵性	操縱性能,運轉能力	maneuverability
操纵性衡准	操縱性標準	criteria of maneuverability
操纵性识别	操縱性識別	maneuverability identification
操纵性试验水池	操縱性試驗池	maneuvering tank
操作	操作	handling
操作程序	操作程序	operation sequence
操作级	操作級	operational level
操作说明	操作說明	operating instruction
操作说明书	操作說明書	operation manual
操作污染	操作污染	operational pollution
操作与设备手册	操作與設備手冊	operations and equipment manual
操作指令	作戰指揮	operational command
操作准备	操作準備	operational readiness
槽线	槽線	trough line
侧开门箱	側開貨櫃	open side container
侧面标志	側面標誌	lateral mark
侧扫声呐	水平掃描聲納	side scan sonar
侧视图	側視圖	lateral view
侧推器	側推器	side thruster
侧移式舱盖	側滾式艙蓋	side rolling hatch cover
测爆	爆發限度測計	measuring the explosive limit
测爆仪	測爆儀	explosimeter
测地线	測地線	geodesic
测定值	測定值	measured value
测厚	測厚	thickness measuring
测角器	量角器,測角器,測向器	goniometer
测距仪	測距儀	range finder
测力计,测功器	測功計	dynamometer
测量	測量	surveying
测量船	測量船,測量艦	surveying ship
测量单元	測定單位	measuring unit
测量方法	測計方法	measuring means
测深	測深	sounding
测深潮汐订正	測深訂正潮汐	tidal reduction for soundings
测深尺	測深標尺	sounding scale

祖 国 大 陆 名	台 湾 地 区 名	英 文 名
测深锤	測深錘	sounding lead
测深导航	測深航法	bathymetric navigation
测深管	測深管	sounding pipe
测深线（测深绳）	測深索,測深繩	sounding line
测深仪误差	回聲測深儀誤差	echo sounder error
测速	速率試驗,速率試航	speed trial
测速板	測速板	chip log
测速场	測速場	speed trial ground
测速发电机	測速發電機	tachogenerator
测温传感器	高溫計傳感器	pyrometer probe
测向灵敏度	方探靈敏度	direction finder sensitivity
测向仪	測向儀	direction finder
测氧仪	氧氣分析器	oxygen analyser
测液深标尺	量油尺,量液深尺	dipstick
测者能见地平距离	視程	visible range
[测者]子午圈	天子午線	celestial meridian
层次分析	層次分析	hierarchical analysis
层积云	層積雲	stratus-cumulus
层位	層列	tier
层云	層雲	stratus
插接	疊接(木材)	splicing
查核与管制	查核與管制	verification and control
查询	詢訊	polling
差动油缸	差動氣缸	differential cylinder
差分奥米伽	差分亞米茄	differential Omega
差分测定	差分測定	differential determination
差分测量	差分測量	differential measurement
差分定位	差分定位	differential positioning
差分观测	較差觀測	differential observation
差分全球定位系统	差分式全球定位系統	differential global positioning system, DGPS
差转罗兰 C	差分羅遠 C	differential Loran-C
拆卸检修(=检修)		
柴油机齿轮传动	柴油機齒輪傳動	diesel geared drive
柴油机电力传动	柴油機電力傳動	diesel-electric drive
柴油机电力推进装置	柴油機電力推進裝置	diesel-electric propulsion plant
柴油机负荷图	引擎負載圖	engine load diagram
柴油机和燃气轮机联合	柴油燃氣渦輪機複合動	combined diesel and gas turbine power

祖国大陆名	台湾地区名	英文名
动力装置	力設備	plant
柴油机机油	柴油機潤滑油	diesel engine lubricating oil
柴油机特性	柴油機特性	diesel engine characteristic
柴油机运转范围图	引擎運輸範圍圖	engine layout diagram
柴油-燃气联合动力装置	柴油機及(或)燃氣渦輪機動力裝置	diesel and/or gas turbine power plant
柴油指数	柴油指數	diesel index
柴油主机气动遥控系统	柴油主機氣力遙控系統	pneumatic remote control system for main diesel engine
缠扎	紮縛	seizing
产卵场	産卵場	spawning ground
铲斗	鏟頭	dipper
长[插]接	長接	long splice
长短大圆信号干扰	長短大圓信號干擾	interference between longer and shorter circle path signals
长吨	長噸	long ton
长件货	超長貨	lengthy cargo
长宽比	長寬比	length-breadth ratio
长闪光	長閃光	long flash
长声	長聲	prolonged blast
长式提单	長式載貨證券	long form bill of lading
长行程柴油机	長衝程柴油機	long-stroke diesel engine
长延时	長延時	long time delay
长轴	長軸	major axis
常陈一(猎犬α)	常陳一(獵犬α)	Cor Carole
常规通信信道	正常通信管道	general communication channels
常规无线电通信	一般無線電通信	general radiocommunication
常规无线电业务	一般無線電業務	conventional radio service
常压蒸馏	常壓蒸餾	atmospheric distillation
厂修	廠修	yard repair
场到场	貨櫃場到場	container yard to container yard, CY to CY
场地误差	場地誤差	site error
场站收据	倉庫收貨單	dock receipt
敞顶箱	敞頂貨櫃	open top container
抄关	抄關	searching
抄网	抄網	dip net
超差	時角增量超差	excess of hour angle increment

祖 国 大 陆 名	台 湾 地 区 名	英 文 名
超长货附加费	超長貨附加費	extra-length charge
超长行程柴油机	超長衝程柴油機	super-long stroke diesel engine
超大型发动机	超大型引擎	super large engine
超大型油轮	超級油輪	ultra large crude carrier, ULCC
超导电力推进装置	超導電力推進裝置	superconductor electric propulsion plant
超导电性	超導電性	superconductivity
超导发电机	超導發電機	superconducting generator
超负荷功率限	超載限制	overload limit
超功率事故	超功率事故	super-power accident
超临界速度	超臨界速度	supercritical velocity
超声波	超音波	ultrasonic
超声波探伤	超音波檢查	ultrasonic examination
超声波探伤法	超音波探傷法	supersonic method
超声波探伤器	超音波探傷器	ultrasonic detector
超速保护装置	超速保護設施	overspeed protection device
超速调节器	超速調速器	overspeed governor
超速跳闸机构	超速跳掣機構	overspeed trip mechanism
超越角	超越角	over shoot
超载工况	意外之操作狀況	overriding operational condition
超折射	超折射	super-refraction
超重吊货	吊貨超重	overload of a sling
超重货	笨重貨	heavy-lift cargo
超重货装卸费	逾重貨附加費	heavy-lift charge
潮波	潮波,潮浪	tidal wave
潮差	潮差	tidal range
潮差比	潮距率	range rate
潮高	潮高	height of tide
潮高比	潮高比	height rate
潮高差	潮高差	height difference
潮高基准面	潮汐基準面	tidal datum
潮龄	潮齡	tidal age
潮流	潮流	tidal stream, tidal current
潮流表	潮流表	tidal stream table
潮流图	潮流圖	atlas of tidal stream
潮面	潮面	tide level
潮时差	潮時差	time difference of tide
潮汐	潮汐	tide
潮汐表	潮汐表	tide table

祖国大陆名	台湾地区名	英 文 名
潮汐[海]图	潮汐圖	atlas of tides
潮汐调和常数	潮汐調和常數	tidal harmonic constant
潮汐周期	潮汐週期	tidal period
车令	俥令	engine orders
车令指示器	俥令指示器	engine telegraph order indicator
车上交货	車上交貨	free on truck, free on rail
车钟	俥鐘	engine telegraph
车钟报警	俥鐘警報	engine telegraph alarm
车钟发送器	俥鐘發送器	telegraph transmitter
车钟记录簿	俥鐘紀錄簿	telegraph book, bell book
车钟记录仪	俥鐘紀錄儀	telegraph logger
车钟接收器	俥鐘接收器	telegraph receiver
撤船	撤船	withdrawal of ship
撤销	取消	cancel
尘密	塵密	dust-tight
沉船	沈船,船舶殘骸	wreck
沉船残留物	沈船殘留物	wreck remains
沉船残体	難船漂流物	wreckage
沉船打捞	打撈沈船	raising of a wreck
沉船勘测	沈船勘測	wreck surveying
沉淀	沈澱	sediment
沉淀柜	澄清櫃	settling tank
沉没	沈沒	founder
沉没物	沈沒物	sunk object
沉子	沈錘,沈子	sinker
沉子纲	沈子網	ground rope
晨光始	黎明之開始	beginning of morning twilight
晨雾	晨霧	morning fog
衬垫,垫舱物料	墊材,墊艙,襯材	dunnage
衬料	襯料	lining
称职	稱職	professional competence
撑开	撐開	breast off
撑柱	撐柱	shore
成本加运费价格	成本與運費	cost and freight, C&F
成对频率	成對頻率,頻率對	paired frequencies
成交函	成交書	fixing letter
成批货	整批貨,大宗貨	lot cargo
成品油船	成品油[運載]船	product carrier

祖 国 大 陆 名	台 湾 地 区 名	英 文 名
成套动力装置	整套動力設備	unit power plant
成组	單元化	unitization
成组货	單元化貨物	unitized cargo
承保一切险	全險	against all risk, a. a. r
承磨环	耐磨環	wear ring
承推架	軸承架	bearing beam
承推梁	承推梁	pushed beam
承运船	運送船	carrier
承运人	運送人	carrier
承运人责任期间	運送人責任期	period of responsibility of carrier
承租人	租傭人	charterer
承租人责任终止条款	租傭人責任留置條款	cesser clause of charterer's liability
程序和布置手册	程序和佈置手冊	procedures and arrangement manual
程序控制	程式控制	programmed control, process control
程序信号	程序信號	procedure signal
橙色烟号	橙色煙號	orange smoke signal
吃水	吃水	draft
吃水差	俯仰差	trim
吃水差比尺	俯仰表	trimming table
吃水差曲线图	俯仰差圖	trim diagram
吃水指示系统	吃水指示系統	draft indicating system
持续使用功率	額定連續常用出力	continuous serving rating
持续运转功率	連續運轉功率限	power limit for continuous running
持证艇员	持證救生艇員	certificated lifeboat person
尺度效应	尺度效應	scale effect
齿	齒	tooth
齿轮泵	齒輪泵	gear pump
齿轮传动［装置］	齒輪驅動	gear drive
齿轮油	齒輪油	gear oil
齿圈(＝轮缘)		
齿条–齿轮	齒條與小齒輪	rack and pinion
赤潮	赤潮	red water, red tide
赤道	赤道	equator
赤道潮汐	赤道潮	equatorial tide
赤道地平视差	赤道地平視差	equatorial horizontal parallax
赤道里	赤道里	equatorial mile
赤道无风带	赤道無風帶	doldrums
赤道仪	赤道儀	equatorial telescope

祖 国 大 陆 名	台 湾 地 区 名	英 文 名
赤道坐标系	赤道坐標系統	equatorial system of coordinates
赤经	赤經	right ascension, RA
赤纬	赤緯	declination, Dec
赤纬圈	赤緯平行圈	parallel of declination
充磁开关	激磁開關	pre-exciting switch
充电	充電	charge
充电率	充電率	charging rate
充气系数(=容积效率)		
充塞泡沫塑料打捞	噴以泡沫塑膠浮升	raising by injection plastic foam
充油变压器	充油變壓器	oil-filled electrical transformer
重叠冰	載冰	rafted ice
重发器模式	重發器模式	repeater mode
重复	重複	repeat, RPT
重复起动程序	重複起動程序	repeated starting sequence
重热系数,再热系数	重熱因素	reheat factor
重调(=复位)		
重新定相	重行定相	rephasing
冲程	衝止距	stopping distance
冲动–反动式涡轮机	衝動反動渦輪機	impulse-reaction turbine
冲动级	衝動輪級	impulse stage
冲动式汽轮机	衝動式汽輪機	impulse steam turbine
冲动式涡轮机	衝動式渦輪機	impulse turbine
冲击	衝擊	impact
冲击腐蚀	衝擊腐蝕	impingement corrosion
冲击负载	衝擊負載	impact load
冲击误差	衝擊誤差	ballistic error
冲积岸堤	堤岸	berm
冲角	切水角,攻角	angle of attack
冲水试验	沖水試驗,射水試驗	hose test
冲洗	沖洗	wash
冲洗装置	沖洗裝置	flushing arrangement
冲翼艇	衝翼艇	ram-wing craft
冲桩管线	水沖管路	jetting pipeline
抽气量	抽氣率	bleed air rate
抽汽式汽轮机	抽汽渦輪機	bleeding steam turbine
抽汽系统	抽汽系統,分汽系統	steam bleeding system
稠度	稠度	consistency

祖 国 大 陆 名	台 湾 地 区 名	英 文 名
臭氧发生器	臭氧产生器	ozone generator
出发,离岸	駛離岸	put off
出发港	出發港	port of departure
出港压载水	出港壓艙水	departure ballast
出海	出海	put to sea
出口报告书	結關報告	report of clearance
出口许可	出口許可	export permit
出口许可证	結關出口	port clearance
出链长度	放鏈長度	chain scope
出浅力	再浮力	refloating force
出入口	出入口	access opening
出坞	出塢	undocking
出现,露出	露出量	emergence
出走(**背绳走**)	背著走	walkaway
出租船	出租船	chartered ship
初步运行能力	初始運轉能力	initial operational capability, IOC
初步证据	表面證據	prima facie evidence
初次检验	初次檢驗	initial survey
初次入级	初次入級	initial classification
初给泵	起動泵	priming pump
初见陆地	初見陸地	landfall
初始报告	初步報告	initial report
初始对准	初始校準	initial alignment
初始蒸汽参数	初始蒸汽參變數	initial steam parameter
初稳心	初定傾中心	initial metacenter
初稳心半径	初定傾半徑	initial metacentric radius
初稳心高度	初定傾中心在基線以上高度	initial metacentric height above baseline
初稳性	初穩度	initial stability
初稳性高度	初定傾中心高	initial metacentric height
除冰装置	除冰裝置	deicer
除鳞机	除鱗機	scaler
除湿器	消濕器	dehumidifier
除鼠	滅鼠	deratting
除外条款	豁免條款	exemption clause
除锈	除銹	removing rust
除氧器	除氧器	deaerator
厨房	廚房	galley

祖国大陆名	台湾地区名	英文名
储备	備用	reserve
储备浮力	預[留]浮力	reserve buoyancy
储藏	貯藏室,倉庫	storage
储气瓶	儲氣瓶	air flask
储液缸	貯液櫃	liquid container
储液器	受液器	receiver
储油船	貯油船	oil storage tanker
储油平台	儲油平台	oil storage platform
处理器模式	處理器模式	processor mode
处理设备	處理設備	process unit
处理时间	處理時間	processing time
处置废物	廢棄物處理	disposal of wastes
触发器	觸發器	trigger
触礁	觸礁	strike on a rock
触浅掉头	觸淺短迴轉掉頭	turning short round by one end touch the shoal
触损	接觸損害	contact damage
穿梭油轮	梭運油輪	shuttle tanker
穿越	橫越	crossing
传播路径	傳播路徑	propagation path
传播误差	傳播誤差	propagation error
传递路由	中轉途徑	transit route
传动,传输	傳動,傳輸	transmission
传动轴系	傳動軸系	transmission shafting
传感器	察覺器,測感子,測感器	sensor
传输(=传动)		
传统式输送设备	傳統式輸送設備	conventional transfer rig
传向系统	傳動系統	transmission system
传真	傳真	facsimile, FAX
传真天气图	傳真天氣圖	facsimile weather chart
船岸间距	船岸間距	bank clearance
船边交货	船邊交貨	alongside delivery, free alongside ship
船边交货条款	船邊交貨條款	sous-palan clause
船表	對時錶	hack watch
船舶	船,艦	vessel, ship
船舶安全检查	船舶安全檢查	ship safety inspection
船舶安全学	船舶安全工程	vessel safety engineering
船舶安全营运管理	船舶安全營運管理	management for the safe operation of ships

祖 国 大 陆 名	台 湾 地 区 名	英 文 名
船舶保向宽度	船舶巷道	ship lane
船舶报告系统	船舶報告系統	vessel reporting system
船舶避碰	船舶避碰	ship collision prevention
船舶常数	船舶常數	ship's constant
船舶操纵	船舶操縱	shiphandling
船舶操纵性指数	操縱性能指數	maneuverability indices
船舶代理	船舶代理	ship's agent
船舶登记	船舶登記	ship registration
船舶抵押贷款	船舶押款	bottomry
船舶抵押合同	船舶抵押契約,船舶貸款保證書	bottomry bond
船舶抵押合同持有人	船舶抵押債券持有人	bottomry bondholder
船舶抵押权	船舶抵押權	ship mortgage
船舶地球站	船舶地面台,船舶衛星電台	ship earth station, SES
船舶地球站启用试验	船舶衛星電台啓用試驗	SES commissioning test
船舶地球站识别码	船舶衛星電台識別碼	ship earth station identification, SES ID
船舶电气设备	船舶電力裝置	marine electric installation
船舶电台	船舶電台	ship station
船舶电台表	船舶電台表	list of ship station
船舶电台群呼识别	船舶電台群呼識別	group ship station call identity
船舶电台识别	船舶電台識別	ship station identity
船舶电站	船舶發電所	ship power station
船舶定线制	船舶定航線制	ship's routing
船舶动力装置	船舶動力裝置	marine power plant
船舶动力装置操纵性	船舶動力裝置操縱性	marine power plant maneuverability
船舶动力装置经济性	船舶動力裝置經濟性	marine power plant economy
船舶动力装置可靠性	船舶動力裝置可靠性	marine power plant service reliability
船舶动力装置可维修性	船舶動力裝置可維修性	marine power plant maintainability
船舶动态业务	船舶移動業務	ship movement service
船舶/飞机协作搜寻	船舶/飛機協作搜尋	ship/aircraft coordinated search
船舶辅机	船舶輔機	marine auxiliary machinery
船舶辅助观测	輔助船舶觀測	auxiliary ship observation, ASO
船舶供应商	船舶供應商	shipchandler
船舶共有人	船舶共有人	co-owner of ship
船舶管理人	船舶管理人	ship's husband
船舶管辖权	船舶管轄權	jurisdiction over ship
船舶国籍,船籍	船籍國	nationality of vessel

祖 国 大 陆 名	台 湾 地 区 名	英 文 名
船舶核动力装置	船用核子動力裝置	marine nuclear power plant
船舶呼号	船舶呼號	ship's call sign
[船舶或船货]抵押	船舶或船貨抵押	hypothecation
船舶机械通风	船舶機械通風	ship mechanical ventilation
船舶加速度	船舶加速度	acceleration of ship
船舶减摇装置	船舶穩定裝置	ship stabilizing gear
船舶检验	船舶檢驗	ship survey
船舶交通调查	船舶交通調查	vessel traffic survey, vessel traffic investigation
船舶交通工程	船舶交通工程	vessel traffic engineering
船舶交通管理	船舶交通服務	vessel traffic service, VTS
船舶交通管理站	船舶交通服務站	VTS station
船舶交通管理中心	船舶交通服務中心	vessel traffic service center, VTS center
船舶交通模拟	船舶交通模擬	ship traffic simulation
船舶经纪人	船舶經紀人	shipbroker
船舶经纪人佣金	船舶經紀人傭金	shipbrokerage
船舶经理人	船舶經理人	ship manager
船舶经营人	船舶營運人	ship operator
船舶靠泊系统	船舶靠泊系統	docking system
船舶空气调节	船舶空調	marine air conditioning
船舶领域	船舶領域	ship domain
船舶旅客	船舶旅客	ship's passenger
船舶配备要求	船舶配備要求	ship carriage requirements
船舶碰撞	船舶碰撞	ship collision
船舶碰撞管辖权	船舶碰撞管轄權	jurisdiction of ship collision
船舶碰撞民事管辖权某些规定的国际公约	關於船舶碰撞事件民事管轄國際公約	International Convention on Certain Rules Concerning Civil Jurisdiction in Matters of Collision
船舶碰撞准据法	船舶碰撞適用法	applicable law of ship collision
船舶起货设备检验簿	船舶貨物裝卸設備登記簿	register of cargo handling gear of ship
船舶气象报告	船舶氣象報告	ship weather report
船舶全损险	全損險	total loss only
船舶入级和建造规范	船舶入級與建造規範	Rules for Classification and Construction of Ships
船舶所有人,船东	船舶所有人,船東	shipowner
船舶通风	船舶通風	ship ventilation
船舶[通信]业务	船舶[通信]業務	ship business

祖 国 大 陆 名	台 湾 地 区 名	英 文 名
船舶推进轴系	船舶推進軸系	marine propulsion shafting
船舶无线电导航	船舶無線電導航	marine radio navigation
船舶无线电执照	船舶無線電台執照	ship radio licence
船舶系统	船舶系統	marine system, ship system
船舶消防	船舶滅火	ship's fire fighting
[船舶]一切险	一切險,全險	all risks
船舶优先权	海事優先擔保權,海事留置權	maritime lien
船舶优先权和抵押权国际公约	船舶優先權及抵押權國際公約	International Convention on Maritime Liens and Mortgage
船舶油污险	油污險	oil pollution risk
船舶运动图	運動圖	maneuvering board
船舶战争险	船舶戰爭險	hull war risk
船舶蒸汽锅炉	船用蒸汽鍋爐	marine steam boiler
船舶蒸汽机	船用蒸汽機	marine steam engine
船舶证书	船舶證書	ship certificate
船舶证书检验簿	船舶證書檢驗紀錄簿	ship's certificates surveying record book
船舶制冷装置	船舶冷凍裝置	marine refrigerating plant
船舶轴系	船舶軸系	marine shafting
船舶主机	船舶主機	marine main engine
船舶自动互救系统	自動互助船舶救助系統	automated mutual-assistance vessel rescue system, AMVER
船舶自然通风	船舶自然通風	ship natural ventilation
船舶总流向	船流總向	general direction of traffic flow
船舶租购(**光船租购**)	分期付款之光船租約	bareboat charter with hire purchase
船舶租金	船舶租金	charter hire
船舶租赁	船舶租賃	hiring of ship
船舶阻力特性	船體阻力特性	hull resistance characteristic
船磁	船磁	ship magnetism
船到岸遇险报警	船與岸間遇險警報	ship-to-shore distress alerting
船到船遇险报警	船與船間遇險警報	ship-to-ship distress alerting
船底板	船底殼板	bottom plate
船底漆	船底漆	bottom paint
船底验平	船底校準	ship bottom alignment check
船底纵骨	船底縱肋	bottom longitudinal
船东(=船舶所有人)		
船队	艦隊,船隊	fleet
船队通信网	船隊通信網	fleetNET

祖 国 大 陆 名	台 湾 地 区 名	英 文 名
船筏	船艇,载具	craft
船方不负责装卸积载费	装卸堆装费自理	free in and out stowed
船方不负装卸费用	装卸自理	free in & out, FIO
船架	船架	cradle
船级	船級	class of ship
船级标记	船級標誌	ship class mark
船级符号	船級符號	ship class symbol notation
船级社	船級機構	register of shipping, ship classification society
船籍(=船舶国籍)		
船籍港	船籍港	port of registration, home port
船壳险	船身保險	hull insurance
船宽吃水比	船寬吃水比	beam draft ratio
船楼端	艛端	break
船模	船模	ship model
船模试验	船模試驗	tank experiment
船模系列试验	船模系列試驗	systematic test of ship model
船内驳运	船內轉駁	internal transfer
船内通信设备	船上通信設備	apparatus for on-board communication
船排	船台,滑道(船台),斜道	slipway
船期表	船期表	sailing schedule
船旗国	船旗國	flag state
船旗国法	船旗國法	law of the flag
船旗歧视	船旗歧视	flag discrimination
船上管理	船上管理	shipboard management
船上通信	船上通信	on-board communications
船上通信电台	船上通信電台	on-board communication station
船上训练	船上訓練	on board training
船上训练记录簿	船上訓練紀錄簿	on board training record book
船上油污染应急计划	船上油污染應急計劃	shipboard oil pollution emergency plan
船首倍角法	艏倍角定位法	doubling angle on the bow
船速	船速	ship speed
船速分布	航速分配	ship speed distribution
船台	造船台	ship building berth
船体保养	船體保養	hull maintenance
船体结构	船身構造	hull construction
船体密性试验	船體密性試驗	tightness test for hull

祖 国 大 陆 名	台 湾 地 区 名	英 文 名
船体下坐	艉坐	squat
船体以上高度	[號燈]距船身高度	height above the hull
船外除泥	船外除泥	removing mud around wreck
船位	船位	ship position, fix
GPS 船位	全球定位系統船位	GPS fix
船位报告	船位報告	position report
船位差	船位差	position difference
船位精度	船位精度	accuracy of position
[船位]误差平行四边形	誤差平行四邊形	error parallelogram
[船位]误差三角形	誤差三角形	cocked hat
[船位]误差椭圆	船位誤差橢圓	error ellipse of position
[船位]误差圆	未定的船位圈	circle of uncertainty
船位圆	位置圈	circle of position
船吸效应	船間相互作用	interaction between ships
船行风速,视风速	船行風速	velocity of ship wind
船形灯标	船形燈標	light-float
船医	船醫	ship's doctor
船艺	船藝	seamanship
船用泵	船用泵	marine pump
船用柴油	船用柴油	marine diesel oil
船用柴油机	船用柴油機	marine diesel engine
船用齿轮箱	船用齒輪箱	marine gear box
船用雷达	船用雷達	marine radar
船用离合器	船用離合器	marine clutch
船用联轴器	船用聯軸節	marine coupling
船用汽轮机	船用蒸汽渦輪機	marine steam turbine
船用轻柴油	船用輕柴油	marine gas oil
船用全功能焚烧炉	船用多功能焚化爐	marine multifunction incinerator
船用燃气轮机	船用燃氣渦輪機	marine gas turbine
船用物料	船用物料	marine store
船员	船員	crew
船员定员	船員配額	complement
船员法	海員法	law of mariner
船员名单	船員名單,船員名冊	crew list
船员证书	船員證書,航海人員證書	certificate of seafarer
船员自用物品报关	船員自用物品申請單	crew's customs declaration

祖国大陆名	台湾地区名	英 文 名
船长	船長,艦長	captain
船长借支	船長借支	advance to captain
船钟	船鐘	clock
喘振	顫動,激度	surge
喘振限	喘振限,波振限	surging limit
串级调速	梯列控制調速	speed regulation by cascade control
串联式推进装置	串聯推進系統	tandem propulsion system
串联往复泵	串聯往復泵	reciprocating pump in series
串联系泊装油系统	串列裝卸系統	tandem loading system
串列螺旋桨	串聯螺槳	tandem propeller
窗式空气调节器	窗型空調	window type air conditioner
窗楣	窗楣	brow
床铺	床[鋪]	bunk
吹灰器	吹灰器	soot blower
吹开风	離岸風	offshore wind
吹拢风	向岸風	on shore wind
垂荡	[船身]起伏	heaving
垂线间长	垂標間距	length between perpendiculars, LBP
垂线偏角	垂線偏差	deviation of the vertical
垂直	垂直	vertical
垂直波束宽度	垂直波束寬度	vertical beam width
垂直补给	垂直整補	vertical replenishment
垂直磁棒	上升用磁鐵	vertical magnet
垂直工作范围	垂直工作範圍	vertical working range
垂直光弧	垂直弧區	vertical sector
垂直角定位	垂直角定位	fixing by vertical angle
垂直角距离	垂直角距離	distance by vertical angle
垂直角位置线	垂直角位置線	position line by vertical angle
垂直圈	垂直大圓	vertical circle
垂直天线	垂直天線	vertical antenna
垂直危险角	垂直危險角	vertical danger angle
垂直线	垂直線,豎線	vertical line
垂直轴阻尼法	垂直軸阻尼法	damped method of vertical axis
春分点	春分點	vernal equinox
磁暴	磁暴	magnetic storm
磁北	磁北	magnetic north
磁差	地磁差	[magnetic] variation, Var
磁场调速	磁場控制調速	speed regulation by field control

祖国大陆名	台湾地区名	英 文 名
磁赤道	地磁赤道,磁赤道	magnetic equator
磁方位	磁方位	magnetic bearing, MB
磁航向	磁航向	magnetic course, MC
[磁]罗差	羅經誤差,羅經差	compass error
磁罗经	磁羅經,磁羅盤	magnetic compass
磁罗经校正	磁羅經校正	magnetic compass adjustment
磁罗经指向误差	磁羅經指向誤差	directive error of magnetic compass
磁偏角	磁偏轉	magnetic deflection
磁倾角	地磁傾角,磁傾角	magnetic dip
磁倾角针	地磁俯角針	dip needle
磁倾仪	地磁俯角儀	dip circle
磁通门罗经	磁通門羅經	flux gate compass
磁针	磁針	magnetic needle
磁致伸缩效应	磁效伸縮效應	magnetostrictive effect
磁子午线	磁子午線	magnetic meridian
次级天文钟	次級天文鐘	hack chronometer
刺网	刺網	gill net
从动齿轮	從動齒輪	driven gear
从价	從價	ad valorem
从价运费	從價運費	ad valorem rate
从签约地法	從簽約地法	lex loci contractus
从事捕鱼的船舶	從事捕魚中船舶	vessel engaged in fishing
粗糙度	粗糙度	roughness
粗糙货	粗貨	rough cargo
粗差	人爲誤差	gross error
粗对准	粗校準	coarse alignment
[粗]过滤器	濾器	strainer
粗缆	粗纜	cable
粗同步法	粗同步法	coarse synchronizing method
催化热裂	觸媒裂解	catalytic cracking
淬火	淬火	quenching
存储转发	存儲轉發	store and forward
存储转发单元	儲存前管器	store and forward unit
错平结,祖母结	假平結	granny knot

D

祖 国 大 陆 名	台 湾 地 区 名	英 文 名
搭边	搭邊	lap
搭接	搭接	lap joint
搭载费	［搭载］船費	embarkation charge
打横	船身突橫(順風駛帆時)	broach to
打水砣	打水砣	heave the lead
打印结束信号	打印結束信號	printing finished signal
打住,稍松	打住,稍鬆	check
打桩船	打椿船	floating pile driver
大潮	大潮	spring tide
大潮升	大潮升	spring rise, SR
大地测量	大地測量	geodetic survey
大地水准面	重力球體	geoid
大地水准面高度图	大地水平面高度圖	geoidal height map
大定位误差	定位顯著誤差	large location error
大风警报	大風警報	gale warning, GW
大风浪航行工况管理	惡劣氣候航行作業形式管理	heavy weather navigation operating mode management
大风浪中船舶操纵	大風浪中操船	shiphandling in heavy weather
大副	大副	chief officer, chief-mate
大管轮,二车	二管輪	second engineer
大角(牧夫 α)	大角(牧夫 α)	Arcturus
大浪(5 级)	洶濤	rough sea
大量生产	大量生產	quantity production
大陆架	大陸架,陸棚	continental shelf
大陆架界线	大陸架界線	continental shelf boundary
大陆坡	大陸斜坡	continental slope
大气	大氣	atmosphere
大气环流	大氣環流	general atmospheric circulation
大气透射率	大氣傳達量	atmospheric transmissivity
大倾角稳性	大傾側角時之穩度	stability at large angle of inclination
大桶(合 63 加仑)	液量單位(合 63 加侖)	hogshead
大雾(0 级)	濃霧(0 級)	dense fog
大西洋东区	大西洋東區	Atlantic Ocean Region East

祖国大陆名	台湾地区名	英文名
大西洋极锋	大西洋極鋒	Atlantic polar front
大西洋区	大西洋區域	Atlantic Ocean Region, AOR
大西洋西区	大西洋西區	Atlantic Ocean Region West
大型浮标(=高灯芯浮标)		
大型油船	巨型油輪	very large crude carrier, VLCC
大修	翻修,大修	over haul
大洋航路	大洋航路	ocean passage
大洋航行	大洋航行	ocean navigation
大洋水深图	大洋水深圖	ocean sounding chart
大圆	大圈	great circle
大圆顶点	頂點	vertex
大圆方位	大圓方位	great circle bearing, GCB
大圆分点	大圓中分點	intermediate point of great circle
大圆改正量	半輻合角	half-convergency
大圆海图	大圓海圖	great circle chart
大圆航线算法	大圓航法	great circle sailing
大圆航向	大圓航向	great circle course, GCC
大圆距离	大圓距離	great circle distance
大抓力锚	高抓著力錨	high holding power anchor
代理人	代理人	agent
代旗	代旗	substitute flag
代位	代位	subrogation
代位求偿权	代位求償權	right of subrogation
带顶标浮标	桿形浮標	beacon buoy
带缆口令	帶纜口令	mooring orders
带外发射	頻帶外發射	out-of-band emission
带罩叶轮	帶罩葉輪	propeller impeller
袋装货	袋裝貨	bagged cargo
单板舵	單板舵,平板舵	single plate rudder
单边带发射	單邊帶發射	single sideband emission, SSB emission
单边带无线电话	單邊帶無線電話	single sideband radiotelephone, SSB RT
单编结	魯班扣,魯班單扣	sheetbend
单差	單差	single difference
单船围网	單船圍網	single boat purse seine
单点定位	單點定位	point positioning
单点系泊	單點繫泊	single point mooring, SPM
单独海损	單獨海損	particular average, P/A

祖国大陆名	台湾地区名	英文名
单独仲裁员	單獨仲裁人	sole arbitrator
单杆作业	單吊桿系統	single boom system
单工	單工,單式	simplex
单工操作	單工作業	simplex operation
单管设备	單管設備	single-hose rigs
单横队	單橫隊	single line abreast
单级废气涡轮增压	單級渦輪增壓	single-stage turbocharging
单级汽轮机	單級蒸汽輪機	single-stage steam turbine
单级闪发	單級閃蒸發	single stage flash evaporation
单级压缩机	單級壓縮機	single stage compressor
单绞辘	單滑車	single whip
单卷筒绞车	單筒絞車	single drum winch
单联滤器	單過濾器	single strainer
单流式蒸汽机	單流蒸汽機	uniflow steam engine
单锚泊	單錨泊	riding to single anchor
单锚腿系泊	單錨腿繫泊	single anchor leg mooring
单排球轴承	單排滾珠軸承	single-row ball bearing
单人坐板	工作吊板	bosun's chair
单套结	單套結	bowline
单体船	單[胴]體船	mono-hull ship
单头缆	單頭纜	single rope
单凸轮换向	單凸輪換向	single cam reversing
单拖	單拖	otter trawling
单拖网船	單桅網漁船	otter trawler
单位	單位	unit
单位容积制冷量	單位氣缸容積冷凍效果	refrigerating effect per unit swept volume
单位轴马力制冷量	單位制動馬力冷凍效果	refrigerating effect per brake horse power
单相运行保护	單相保護	protection against single-phasing
单向航路	單向航路	one-way route
单向旋回法	單迴旋法	single turn
单向止回阀(**止回阀**)	止回閥	check valve
单效蒸发	單效蒸發	single effect evaporation
单油头	單油頭	single probe
单胀式蒸汽机	單脹式蒸汽機	single expansion steam engine
单轴系	單軸系	single shafting
单转子摆式罗经	單轉子擺式電羅經	single gyro pendulous gyrocompass
单字母信号码	單字母信號碼	single letter signal code
单纵队	單縱隊	single line ahead, single column

祖 国 大 陆 名	台 湾 地 区 名	英 文 名
单作用式柴油机	單動式柴油機	single-acting diesel engine
单作用油缸	單動式氣缸	single acting cylinder
担架	擔架	litter
弹药舱	彈藥艙	magazine
淡化器具	去鹽設備	desalting kit
淡化设备	除鹽器	de salting apparatus
淡水泵	淡水泵	fresh water pump
淡水系统	淡水系統	fresh water system
淡水循环泵	淡水循環泵	fresh water circulating pump
淡水载重线	淡水載重線	fresh water load line
淡水注入管	淡水注入管	fresh water filling pipe
当地水位	當地水位	local water level
挡浪板	擋浪板	breakwater
导板	導板	guide plate
导边	導緣(螺槳)	leading edge
导标	引導示標	leading beacon
导程	導程	lead
导出包络	導出包絡	derived envelope
导电液体	導電液體	conducting liquid
导阀,控制阀	[嚮]導閥	pilot valve
导管推进器	導罩螺槳	ducted propeller
导航参数	導航參變數	navigation parameter
导航雷达	導航雷達	navigation radar
导航设备	助航設施	navigation aids
导航声呐	導航聲納	navigation sonar
导航卫星	導航衛星	navigational satellite
导航线	導航線	leading line
导口	導槽入口	entry guide
导缆孔	[繫船]道索管	mooring pipe
导缆器	索導	fairlead
导链轮	導鏈器	chain cable fairlead
导流罩舵	球形舵	bulb rudder
导线测量	折航測量	traverse survey
导向叶片	導葉片	guide vane
导星	導星	guiding star
岛架	島架	island shelf
岛礁区航行	多礁水域航行	navigating in rocky water
倒车	倒俥	back

祖 国 大 陆 名	台 湾 地 区 名	英 文 名
倒车操纵阀	倒俥操縱閥	astern maneuving valve
倒车冲程	倒俥停俥距離	reverse stopping distance
倒车舵	倒俥舵	flanking rudder
倒车隔离阀	倒俥護閥	astern guarding valve
倒车功率	倒俥動力	backing power, astern power
倒车排汽室喷雾器	倒俥排氣室噴霧器	astern exhaust chest sprayer
倒车汽轮机	倒俥蒸汽渦輪機	astern steam turbine
倒车燃气轮机	倒俥燃氣渦輪機	astern gas turbine
倒车试验	倒俥試航	astern trial
倒车装置	倒俥裝置	means of going astern
倒航工况管理	倒航操作形式管理	astern running operating mode management
倒缆	倒纜	spring
倒签提单	倒簽載貨證券	anti-dated bill of lading
倒拖	倒拖	reverse towing
倒装	順序不當裝載	overstowing arrangement
到岸价格	含保險費與運費之貨價	cost insurance and freight, CIF
到达船	到港船	arrived ship
到达点	到達點	arrival point
到达港	到達港	port of arrival
到付运费	到付運費	freight to collect
到港日期	到港日期	date of arrival
到港压载水	抵港壓艙水	arrival ballast
到货通知	到貨通知	arrival notice
德国船级社	德國驗船協會	German Lloyd's
灯标	燈標	lighted mark
灯船	燈船	light-vessel
灯浮标	燈浮標	lighted buoy
灯高	燈[距水面]高	elevation of light
灯光船	集魚燈船	fishing light boat
灯光射程	燈光射程	light range
灯光通信	燈號通信	flashing light signaling
灯光信号	燈光信號	light signal
灯光诱鱼	燈光誘魚	lamp attracting
灯具间	燈具室	lamp room
灯塔	燈塔	lighthouse
灯塔供应船	燈塔補給船	lighthouse tender
灯桩	燈標,燈浮標	light beacon

祖国大陆名	台湾地区名	英 文 名
登岸证	登岸證	landing permit, shore pass
登船梯	乘載梯	embarkation ladder
登记长度	登記長度	registered length
登记宽度	登記寬度	registered breadth
登记深度	登記深度	registered depth
登陆航海勤务	登陸海勤	navigation service for landing
登陆舰	登陸艦艇	landing ship
登轮检查	登輪檢查	inspection by boarding
蹬索	鐙,繫索	stirrup
等潮差线	等潮差線	corange line
等潮时	等潮時	cotidal hour
等潮时图	等潮圖	cotidal chart
等潮时线	等潮線	cotidal line
等磁差图	等偏差線,等角線圖	isogonic chart
等待	等候	waiting
等高差	等高差	equation of equal altitude
等高法	等高法	equal altitude method
等高圈	等高圈	circle of equal altitude
等高线	等高線	contour lines
等级航道	分級航道	graded fairway
等角投影	等角投影	equiangular projection
等平均有效压力限	平均有效壓力限制	mean effective pressure limit
等深线	等深線	depth contour
等深线图	水深圖,海底地形圖,等深線圖	bathymetric map
等温线	等溫,等溫線	isotherm
等效	當量	equivalent
等效全向辐射功率	等效等向輻射功率	equivalent isotropically radiated power, EIRP
等压面	等壓面	isobaric surface
等压面图	等值圖	contour chart
等压线	等壓線	isobar
等值	同等設備	equivalent
等转矩限	轉矩轉速限	torque/speed limit
低标准船	次標準船	substandard ship
低播焰性	低度火焰蔓延	low flame spread
低潮	低潮	low water, LW
低潮时	低潮時	low water time

祖国大陆名	台湾地区名	英 文 名
低低潮	較低低潮(一日中)	lower low water, LLW
低地球轨道	低空地球軌道(低空軌道)	low earth orbit
低电压保护	低電壓保護	low-voltage protection
低电压释放	低電壓釋放	low-voltage release
低高潮	較低高潮(一日中)	lower high water, LHW
低轨道搜救卫星系统	低軌道搜救衛星系統	low Earth orbit SAR satellite system
低轨道搜救卫星系统本地用户终端	低軌道衛星搜救系統地面終端台	LUT in a LEOSAR system
低滑油压力保护装置	低潤滑油壓跳脱設施	low-lubricating oil pressure trip device
低极轨道卫星搜救系统	衛星輔助搜救系統	COSPAs-SARSAT system
低极轨道卫星搜救系统信文	衛星輔助搜救系統信文	COSPAs-SARSAT message
低极轨道卫星搜救组织理事会	國際衛星輔助搜救組織理事會	COSPAs-SARSAT council
低频压力	低頻應力	low-frequency stress
低热值	低熱值	lower calorific value
低速柴油机	低速柴油機	low speed diesel engine
低温腐蚀	低溫腐蝕	low temperature corrosion
低压	低壓,低氣壓	low [pressure]
低压槽	槽形低壓	trough
低压蒸汽发生器	低壓蒸汽發生器	low pressure steam generator
低音雾笛	霧號器	diaphone
低真空保护装置	低真空保護設施	low-vacuum protective device
堤坝	堤	dyke
滴点	落點	drop point
滴油润滑	滴油潤滑	drop lubrication
狄塞尔循环	狄賽爾循環	diesel cycle
笛号	笛號	whistle signal
底舱	底[貨]艙	lower hold
底层流	底層流	bottom current
底迹	底跡	bottom tracking
底盘	車底盤,底盤	chassis
底拖网	底曳網	bottom trawl
底质	底質	quality of the bottom
抵押登记	抵押登記	mortgage registration
地波	地波(無線電)	ground wave
地磁赤道	地磁赤道	geomagnetic equator

祖 国 大 陆 名	台 湾 地 区 名	英 文 名
地磁极	地磁極	geomagnetic pole
地方恒星时	當地恒星時	local sidereal time, LST
地方[平]时	地方平均時,地方平時	local mean time, LMT
地方时	地方時	local time
地方时角	當地時角	local hour angle, LHA
地极	地極	earth pole, terrestrial pole
地脚螺栓	壓緊螺栓	holding down bolt
地理分类	地區劃分	geographic sorting
[地理]经度	地理經度	[geographic] longitude
地理区域群呼	地理區域群呼	geographical area group call
[地理]纬度	地理緯度	[geographic] latitude
地理信息系统	地理資訊系統	geographic information system, GIS
地理坐标	地理坐標	geographic coordinate
地面部分	地面部分	ground segment
地面部分操作员	地面部分運作者	ground segment operator
地面电台	地面電台	terrestrial station
地面搜救处理器	地面搜救處理器	ground search and rescue processor
地面台	地面台	ground station
地面[天气]图	地面天氣圖	surface [weather] chart
地面通信网络	地面通信網絡	terrestrial communication network
地面通信系统	地面通信系統	terrestrial communication system
地面无线电通信	地面無線電通信	terrestrial radiocommunication
地面真地平	感觀水平面	sensible horizon
地平	地平	horizon
地平经度	地平经度	azimuth
地平视差	地平視差	horizontal parallax
地平坐标系	水平坐標制	horizontal coordinate system
地球	地球,地	Earth
地球半径	地球半徑	radius of the earth
地球扁率	地球扁率	flattening of earth
地球长半轴	赤道半徑	equatorial radius
地球-电离层波导	地球-電離層波導	earth-ionospheric waveguide
地球偏心率	地球偏心率	eccentricity of earth
地球椭球体	地球橢球體	spheroid of earth
地球椭圆体	地球橢圓體	earth ellipsoid
地球形状	地球形狀	earth shape
地球圆球体	地球球體	terrestrial sphere
地球站	地球台	earth station

祖国大陆名	台湾地区名	英 文 名
地区条款	地區條款	local clause
地天波改正量	地波對天波修正值	ground wave to sky wave correction
地文航海	地文航海術	geo-navigation
地心纬度	地心緯度	geocentric latitude
地心纬度改正量	地心緯度修正	correction of geocentric latitude
地形岸线测量	地形岸線測量	topographic and coastal survey
地中海系泊法	地中海繫泊法	Mediterranean mooring
地轴	地軸	earth axis
地转风	地轉風	geostrophic wind
地转风速	地轉風速	velocity of geostrophic
帝(小熊β)	帝,北極二(小熊β)	Kochab
第三代中速柴油机	第三代柴油機	third generation engine
缔约国政府	締約國政府	contract government
点燃式发动机	火花點火引擎	spark ignition engine
点蚀	斑蝕	pitting
点图	點圖型	dot pattern
电报线立标	電報線標桿	telegraph cable beacon
电报学	電報術	telegraphy
电传	電傳	teleprinter exchange
电磁摆	電磁擺	electromagnetic pendulum
电磁波测距仪	電磁波測距儀	electromagnetic wave distance measuring instrument
电磁阀	電磁閥	solenoid valve
电磁换向阀	電磁方向控制閥	solenoid directional control valve
电磁计程仪	電磁計程儀	electromagnetic log, EM log
电磁控制罗经	電磁控制電羅經	electromagnetically controlled gyrocompass
电磁自差	電磁自差	electromagnetic deviation
电动舵机	電舵機	electric steering engine
[电动机]起动阻塞控制	起動阻塞控制	start blocking control
电动[力矩]平衡器	電平衡器	electrical balancer
电动起货机	電動起貨機	electric cargo winch
电动气动式	電動氣動式	electro pneumatic type
电动液压传动	電動液壓驅動	electro hydraulic drive
电动液压舵机	電動液力舵機	electro-hydraulic steering engine
电镀	電鍍	galvanization
电弧焊	電[弧]焊[接]	arc welding
电[化腐]蚀	電流侵蝕	galvanic corrosion

祖 国 大 陆 名	台 湾 地 区 名	英 文 名
电化学腐蚀	電化腐蝕	electrochemical corrosion
电话	電話	telephony
电话分组交换网	電話分組交換網	packet switching telephone network
电机员	電機工程師	electrical engineer
电机转速误差	馬達轉速誤差	motor revolution error
电解液	電解質,電解液	electrolyte
电缆	電纜	cable
电缆浮标	電纜浮標	cable buoy
电缆松紧指示器	電纜鬆緊指示器	cable slack meter
电离层	游離層	ionosphere
电离层折射改正	游離層折射修正	ionospheric refraction correction
电力推进	電力推進	electric propulsion
电力推进船	電力推進船	electric propulsion ship
电力推进电机间	電力推進馬達室	electrical propulsion motor room
电力拖动	電力驅動	electric drive
电力拖动装置	電力驅動設備	electric drive apparatus
电力执行机构	電力引動器	electric actuator
电起动	電力起動	electrical starting
电−气变换器	電動氣力轉變器	electro-pneumatic transducer
电气捕鱼	電氣捕魚	electrical fishing
电气联锁	電聯鎖	electrical interlocking
电气日志	電機日誌	electrical log book
电−气式主机遥控系统	電動氣力遙控系統	electric-pneumatic remote control system for main engine
电热融霜	電除霜	electric defrost
电热融霜定时器	電除霜定時器	electric defrost timer
电渗析法	電透析法	electrodialysis method
电枢反应	電樞反應	armature reaction
电台	電台	station
电文格式	電文格式	message format
电信	電信	telecommunication
电压电流变换器	電壓電流轉變器	voltage-current transducer
电液换向阀	電動液力方向控制閥	electro-hydraulic directional control valve
电液伺服阀	電動液力伺服閥	electro-hydraulic servo valve
电液伺服机构	電動液力伺服致動器	electric-hydraulic servo actuator
电致伸缩效应	電伸縮效應	electrostrictive effect
电子方位线	電子方位線	electronic bearing line, EBL
电子管	真空管	valve

祖 国 大 陆 名	台 湾 地 区 名	英 文 名
电子管振荡器	真空管振盪器	valve oscillator
电子海图	電子海圖	electronic chart
电子海图数据库	電子海圖資料庫	electronic chart data base
电子海图显示与信息系统	電子海圖顯示與資訊系統	electronic chart display and information system, ECDIS
电子航海	電子儀航行術	electronic navigation
电子[设备]海上维修	海上電子維修	electronic maintenance at sea
电子[式]调速器	電子調速器	electronic governor
电子式主机遥控系统	主機電子遙控系統	electronic remote control system for main engine
电子数据交换	電子數據交換	electronic data interchange
电子提单	電子載貨證券	electronic bill of lading
电子调节器	電子調整器	electronic regulator
电子邮件	電子郵件	electronic mail
电子邮件业务	電子郵件業務	electronic mail service
垫舱货	填墊貨	dunnage cargo
垫舱物料(＝衬垫)		
垫底货	底艙貨	base cargo
垫木	墊木	chock
垫水	艙底水	tank bottom water
殿后舰	殿後船艦	rear ship
吊车费	起重機使用費	cranage
吊放式救生筏	吊桿下水救生筏	davit launching liftraft
吊杆	吊桿	derrick
吊杆叉头	吊桿根	derrick heel
[吊杆]跟部滑车,艉龙骨墩	[吊桿]跟部滑車	heel block
吊杆架	吊桿架	boom cradle
吊杆跨距	吊桿伸出舷外距離	boom outstretch
吊杆偏角	吊桿偏角	slewing angle
吊杆索具	吊桿索具	derrick rigging
吊杆托架	吊桿承座	derrick rest
吊杆仰角	吊桿俯仰角	boom topping angle
吊杆柱	主柱	king post
吊杆柱平台	主柱突出枭	kingpost outrigger
吊杆转轴	鵝頸	goose neck
吊杆座	鵝頸形吊桿座	gooseneck bracket
吊缸	吊缸	lift out piston

祖 国 大 陆 名	台 湾 地 区 名	英 文 名
吊货杆	吊貨桿	cargo boom
吊货钩	吊貨鉤	cargo hook
吊货索	吊貨索	cargo runner, cargofall
吊货索环	起吊索具	hoisting sling
吊货索卷筒	吊貨索捲軸	cargo drop reel
吊货网	吊貨網	cargo net
吊货眼板	吊貨眼板	cargo purchase eye
吊梁	吊樑	lifting beam
吊锚	起錨	cat
吊锚杆	吊錨桿	cat davit, anchor davit
吊锚滑车	吊錨滑車	cat block
吊升信号	吊升信號	hoisting signal
吊艇钩	吊艇鉤	boat lifting hook
吊艇机	小艇吊機	boat winch
吊艇架	小艇吊架	boat davit
吊艇索	小艇吊索	boat fall
吊艇柱座	凸式套座	pedestal socket
吊艇装置	小艇吊放裝置	boat handling gear
吊装	吊上吊下船	lift on/lift off
钓船	釣船	line fishing boat
钓饵	釣餌	bait
钓竿	釣竿	angle rod
钓竿箱	釣竿盒	fishing rod box
调查庭	調查庭	court of survey
调查委员会	調查委員會	board of investigation, board of inquiry
掉抢	掉餁	come about
掉头	短迴轉	turning short round
掉头区	迴旋區	swinging area
跌落水	落水	by the board
叠标	疊標桿	transit beacon
蝶阀	蝶形閥	butterfly valve
丁香结	丁香結	clove hitch
顶边舱	舷側氣櫃	top side tank
顶标	顶[上]標[誌]	topmark
顶潮流航行	頂潮航行	stem the tide
顶点纬度	頂點緯度	latitude of vertex
顶风停船	頂風停船	in iron
顶风停住	頂風停住	in stays

祖 国 大 陆 名	台 湾 地 区 名	英 文 名
顶头风	頂頭風,逆風	head wind
顶头浪	頂頭浪,逆浪	head sea
顶推	推頂	pushing
顶推操纵缆	推頂操舵纜	pushing steering line
顶推船队	推駁船隊	pusher train
顶推架	推頂架	pushing frame
顶推柱	推頂柱	pushing post
顶推装置	推頂裝置	pushing gear
顶置气阀式发动机	頂置閥式引擎	valve-in head type engine
顶装法	上層積載	load on top, LOT
订舱	訂載	booking
订舱单	訂運單	berth note
订租确认书	成交書(論程傭船需用)	fixture note
定边	定向	sense determination
定光	定光	fixed light
定镜差	水平鏡誤差	side error
定量油马达	固定排量油馬達	fixed-displacement oil motor
定期保险	定期保險	time insurance
定期检验	定期檢驗	periodical survey
定期维修	定期保養	regular maintenance
定期预防维修	預防保養	preventive maintenance
定期租船	計時雇船,計時租賃	time charter
定日镜	定日鏡	heliostat
定容循环	定容循環	constant volume cycle
定时广播	定時廣播	for scheduled broadcast
定位	定位	position fix
定位格架	定位格	location grid
定位锚	定位錨	positioning anchor
定相	定相(傳真、真蹟用)	phasing
定向导航灯	指導向燈	directional light
定向天线	定向天線	directional antenna
定向无线电信标	定向無線電示標	directional radio beacon
定压涡轮增压	定壓渦輪增壓	constant pressure turbo-charging
定压循环	定壓循環	constant pressure cycle
定约承运人	契約運送人	contract of carrier
定值控制	穩定控制	stabilization control
定制的	訂製者	made-to-order

祖国大陆名	台湾地区名	英文名
定置网	定置漁網	stationary fishing net
定置渔具	定置漁具	stationary fishing gear
定轴性	迴轉慣性	gyroscopic inertia
东大距(行星)	東大距(行星)	greatest eastern elongation
东方标	東方標	east mark
东风波	東風波	easterly wave
东西距	東西距	Departure, Dep
冬季季节区带	季節性冬期地帶	winter seasonal zone
冬季季节区域	季節性冬期區域	winter seasonal area
冬季载重线	冬期載重線	winter load line
冬至点	冬至	winter solstice
动航向稳定性	動向穩度	dynamic course stability
动横倾角	動横傾角	dynamical heeling angle
动横倾力臂	動横傾力	dynamical heeling lever
动横倾力矩	動横傾力矩	dynamical heeling moment
动界	界	arena
动镜差	垂直誤差	perpendicular error
动力活塞	動力活塞	power piston
动力黏度	動力黏度	dynamical viscosity
动力涡轮	動力輪機	power turbine
动力相似	動態相似性	dynamic similarity
动力行程	動力衝程	power stroke
动力装置单位质量	動力裝置單位質量	power plant specific mass
动力装置燃油消耗率	動力裝置燃油消耗率	power plant effective specific fuel oil consumption
动力装置生命力	動力裝置壽命	power plant viability
动力装置[有效]热效率	動力裝置[有效]熱效率	power plant [effective] thermal efficiency
动量理论	動量理論	momentum theory
动能	動能	kinetic energy
动配合	動配合	movable fit
动平衡	動力平衡	dynamical balance
动索	動索	running rigging
动态	動態	dynamic state
动态定位	動態定位	dynamic positioning, kinematic positioning
动态模拟	動態模擬	dynamic simulation
动稳性	動穩度	dynamical stability
动稳性力臂	動穩度力臂	dynamical stability lever

祖 国 大 陆 名	台 湾 地 区 名	英 文 名
动物箱(**牲口箱**)	牲口櫃	cattle container
动压头	動落差	dynamic head
动叶片	轉動葉片	moving blade
动植物检疫	動植物檢疫	animal or plant quarantine
斗式挖泥船	抓斗挖泥船	grab dredger
独立波	孤立波	solitary wave
堵漏	堵漏	leak stopping
堵漏器材	堵漏器材	leak stopper
堵漏水泥箱	水泥堵	cement box
堵漏毯	堵漏毯,防水墊	collision mat
渡船	渡船	ferry
镀铬	鍍鉻	chrome plating
镀铬缸套	鍍鉻缸套	chrome plated liner
镀铬环	鍍鉻環	chrome-plate ring
镀锡铁皮(=马口铁)		
镀锌	鍍鋅	galvanization
端开门箱	端開貨櫃	open end container
短[插]接	短編接	short splice
短程国际航行	短程國際航程	short international voyage
短吨	短噸	short ton
短缆系结	短纜繫結	short-line connection
短路	短路	short circuit
短路电流	短路電流	short circuit current
短声	短[笛]聲	short blast
短卸	短卸	short-landed, short-delivery
短卸货	短卸貨	shortlanded cargo
短周期事故	短週期事故	short period accident
短轴	短軸	minor axis
短装	短裝	short-shipped
段同步	段同步	segment synchronization
段信号	段信號	segment signal
断裂区	破裂區,破裂帶	zone of fracture
断裂应变	裂斷應變	breaking strain
断续海岸	斷續海岸	broken coast
锻接	鍛接	forge weld
锻造型曲轴	鍛造曲柄軸	forged type crankshaft
堆垛机	承座接	stacker
堆积冰	圓丘冰	hummocked ice

祖国大陆名	台湾地区名	英 文 名
堆积锥	承接錐	stacking cone
堆码	裝載	stowage
堆码件	堆積裝具	stacking fitting
堆热功率	反應器熱功率	heat output of reactor
堆芯	索芯	core
堆装试验	堆積試驗	stacking test
队列,编队	編隊	formation
队列方位	編隊方位	formation bearing
队列角	編隊角	formation angle
队列线	編隊線	formation line
队列轴	編隊軸	formation axis
队形变换	隊形變換	changing formation
队形标志灯箱	信號箱	station marker light box
队形长度	隊形長度	length of formation
队形灯	部位燈	station light
队形宽度	隊形寬度	width of formation
对称灌水	平衡泛水	counter flood
对称进水	對稱浸水	cross flood
对第三者负责的保险	責任保險	liability insurance
对讲电话机	電話對講機	intercommunication telephone set
对接	對接	butt joint
对景图	視圖	view
对流层	對流層	troposphere
对流层折射改正	對流層折射修正	tropospheric refraction correction
对水移动	對水移動	making way through water
对拖	雙拖	pair trawling, twin trawling
对遇	正對	end on
对遇局面	迎艏正遇情況	head-on situation
对置活塞式柴油机	對衝活塞柴油機	opposed piston diesel engine
吨税	噸稅	tonnage dues
吨位	噸位	tonnage
趸船	駁船	pontoon
多波束测深系统	多波束測深系統	multi-beam sounding system
多船避碰	多船避碰	multi-ship collision avoidance
多次反射回波	多次反射回波	multiple reflection echo
多点系泊系统	多點繫泊系統	multi-point mooring system
多浮筒系泊	多浮筒繫泊	multi-buoy mooring
多级闪发	多級閃蒸發	multiple-stage flash evaporation

祖 国 大 陆 名	台 湾 地 区 名	英 文 名
多级压缩机	多級壓縮機	multi-stage compressor
多进口脉冲转换器	多進口脈波變換器	multi-entry pulse converter
多进口脉冲转换器系统	多進口脈衝轉換器系統	multi-entry pulse converter system
多径	多[途]徑	multipath
多径传播	多徑傳播	multipath propagation
多径误差	多徑誤差	multipath error
多能打桩架	多功能打樁架	multiple pile driven tower
多普勒定位	都卜勒定位	Doppler location
多普勒计程仪	都卜勒計程儀	Doppler log
多普勒计数	都卜勒計數	Doppler count
多普勒频移	都卜勒頻移	Doppler shift
多普勒位置信息	都卜勒位置資訊	Doppler position information
多式联运	多式聯運	multimodal transportation
多式联运经营人	多式聯運經營人	multimodal transport operator
多式联运提单	多式聯運載貨證券	multimodal transport bill of lading
多体船	多[胴]體船	multi-hull ship
多桅帆船	多桅帆船	barkentine
多卫星链路	多衛星鏈路	multi-satellite link
多效添加剂	多功能添加劑	multipurpose additive
多效蒸发	多效蒸發	multiple-effect evaporation
多用途货船	多用途貨船	multipurpose cargo vessel
多用途拖船	多用途拖船	multipurpose towing ship
多用途渔船	多用途漁船	multipurpose fishing boat
多支承舵	多舵針舵	multi-pintle rudder
多值性(**模糊度**)	模糊度	ambiguity
多专长高级船员	多專長甲級船員	polyvalent officer
舵	舵	rudder
舵板	舵板	rudder plate
舵柄	舵柄	tiller
舵掣	舵韌,舵制動器	rudder brake
舵承	舵承	rudder carrier
舵杆	舵桿	rudder stock
舵杆接头	舵頸接頭	rudder coupling
舵机舱	舵機艙	steering engine room
舵机间	舵機室	steering gear room
舵机追随机构	舵機從動裝置	steering hunting gear
舵角	舵角	helm angle, rudder angle
舵角指示器	舵角指示器	rudder angle indicator

祖 国 大 陆 名	台 湾 地 区 名	英 文 名
舵力	舵力	rudder force
舵令	舵令	helm order, steering orders
舵轮	舵輪	steering wheel
舵面积	舵面積	area of rudder
舵面积比	舵面積比	rudder area ratio
舵扇	扇形舵柄	rudder quadrant
舵效	舵效	rudder effect
舵压力	舵壓力	rudder pressure
舵压力中心	舵壓力中心	rudder pressure center
舵叶	舵葉	rudder blade
舵轴	舵軸	rudder axle
舵柱	舵柱	rudder post
惰轮,中间轮	惰輪	idling gear
惰锚	下風錨,不著力錨	lee anchor
惰性气体	惰性氣體	inert gas, IG
惰性气体发生器	惰氣産生器	inert gas generator
惰性气体风机	惰氣鼓風機	inert gas blower
惰性气体系统	惰[性]氣[體]系統	inert gas system, IGS
惰性气体窒息灭火系统	惰氣窒火系統	inert gas smothering system
惰转时间	空轉時間	idle time

E

祖 国 大 陆 名	台 湾 地 区 名	英 文 名
额定光力射程	公稱光程	nominal range
额定起重量	額定載量	rated load weight
恶臭货	惡臭貨	malodorous cargo
恶劣天气	惡劣氣候	heavy weather
恩氏黏度	恩氏黏度	Engler viscosity
二层甲板	中甲板	tween deck
二车(＝大管轮)		
二冲程柴油机	二衝程柴油機	two stroke diesel engine
二冲程发动机	二衝程引擎	two-cycle engine
二次场	二次場	secondary field
二次辐射	二次輻射	re-radiation
二次力矩补偿器	二次力矩補償器	the second order moment compensator
二次配电系统	二次配電系統	secondary distribution system
二次喷射	二次噴射	secondary injection

祖 国 大 陆 名	台 湾 地 区 名	英 文 名
二次相位因子	二次相位因數	secondary phase factor, SPF
二次行程回波	二次回波	second-trace echo
二副	二副	second officer, second mote
二管轮	三管輪	third engineer
二回路	副電路,二次電路,二次迴路	secondary circuit
二级水手	普通水手	ordinary seaman, OS
二级无线电电子证书	第二級無線電電子員證書	second-class radioelectronic certificate
二级增压	二級增壓	two-stage supercharging
二列冲动叶轮	雙排衝動葉輪	two-row impulse wheel
二时更	暮更	dog watch
二桅帆船	雙桅帆船	brig, brigantine
二位三通换向阀	二位三通方向控制閥	two-position three way directional control valve
二自由度陀螺仪	二自由度迴轉儀	two-degree of freedom gyroscope

F

祖 国 大 陆 名	台 湾 地 区 名	英 文 名
发报局	發報局	office of origin, O/O
发报台	發信台	station of origin
发出	發出	forward
发电机[控制]屏	發電機控制屏	generator control panel
发电机励磁系统	發電機激磁系統	generator excited system
发电机组	發電機組	generating set
发火顺序	點火順序	firing order
发火性能	點火性	ignition quality
发货人	發貨人	consignor
发裂纹	[毛]細裂痕	hair crack
发射	發射	emission
发射静止搜救卫星无效报警	發射無效之定置搜救衛星警報	GEOSAR invalid alert transmitted
发射类别	發射等級	class of emission
发射天线	發射天線	transmitting antenna
发信台识别符	發射台識別符	transmitter identification character
发展	隊形變換	evolution
阀	閥	valve

祖国大陆名	台湾地区名	英 文 名
阀杆导承	閥桿導件	valve spindle guide
阀箱	閥櫃,閥箱	valve chest
法定检验	法定檢驗	statutory survey
法定时	法定時	legal time
法国船级社	法國驗船協會	Bureau Veritas, BV
法兰键槽	凸緣上之槽孔	slot in flange
法院地法	審判地法,法院地法[庭]	lex fori
帆布	帆布	canvas
帆船	帆船	sailing ship, sailer
帆脚杆	帆桁	boom
帆缆间	帆纜庫	boatswain's store
帆缆箱	帆纜箱	boatswain's chest
帆缆作业	帆纜作業	canvas and rope work
帆下角	縱帆踵,吊鋪攀	clew
帆缘索	帆[帳]緣索	bolt rope
繁殖场	繁殖場	breeding ground, nursery ground
繁殖季节	繁殖期	breeding season
反动度(**反应度**)	反應程度	degree of reaction
反动级	反應級	reaction stage
反动式汽轮机	反動式汽輪機	reaction steam turbine
反舵角	反舵角	counter rudder angle
反光材料	反光材料	retro-reflective material
反接制动	逆流制動	counter-current braking
反馈	反饋	feedback
反气旋	反氣旋	anticyclone
反射罗经	反射羅經	reflector compass
反射器	反射器	reflector
反渗透法	反滲透法	reverse osmosis method
反向	反向,回動	reverse
反应堆周期	反應器週期	reactor period
反作用力	反作用力	counter-acting force
返航	返航	homeward voyage
泛滥标(=涨潮标志)		
方便旗	權宜船籍	flag of convenience
方块图	方塊圖	block diagram
方位	方位	bearing
方位标志	方位標誌	cardinal mark

祖国大陆名	台湾地区名	英文名
方位等距投影	方位等距投影	azimuthal equidistant projection
方位定位	交叉方位定位	fixing by cross bearings
方位分辨力	方位分析度	bearing resolution
方位角差	方位角差	azimuth difference
方位距离定位	方位距離定位	fixing by bearing and distance
方位圈	方位圈	bearing circle
方位投影	方位投影	azimuthal projection
方位投影图	天頂投影圖	zenithal chart
方位陀螺	方位迴轉儀	azimuth gyro
方位[陀螺]仪	定向迴轉儀	directional gyroscope
方位位置线	方位位置線	position line by bearing
方位线	方位線	bearing line
方向控制阀	方向控制閥	directional control valve
方向稳定性	方向穩定性	directional stability
方向效应	方向效應	direction effect
方形系数	方塊係數	block coefficient
方[型]艉	方型艉	square cut stern
防爆	防爆	explosion prevention
防爆式风机	防爆風扇	explosion proof fan
防爆型	防爆型	explosion-proof type
防冰装置	防冰設備	anti-icing equipment
防波堤	防波堤	breakwater
防喘系统	防波動系統	surge-preventing system
防滴型	防滴型	drip proof type
防风雨的	耐候	weather-proof
防腐处理	防蝕處理	anti-corrosion treatment
防腐锌阳极	鋅陽極防蝕	zinc-anode for protection
防护货	防護貨	protecting cargo
防护衣	防護衣	protective clothing
防滑甲板涂料	防滑甲板漆	antiskid deck paint
防火舱壁	防火艙壁	fire proof bulkhead
防火控制图	火災控制圖	fire control plan
防火门	防火門	fire door
防火网	防焰網	flame screen
防溅挡板	防濺板	splash plate
防空编队与部署	防空編隊與序列	air defense formation and disposition
防浪阀	止浪閥	storm valve
防磨损换位	索位掉頭	freshen the nip

祖 国 大 陆 名	台 湾 地 区 名	英 文 名
防倾装置	抗搖裝置	anti-toppling device
防热漆	防熱漆	heat-resistant paint
防鼠挡	防鼠板,防鼠罩	rat guard
防水的	不透水	water-proof
防水胶带	防水膠帶	watertight adhesive tape
防水手电筒	防水手電筒	waterproof electric torch
防水型	防水型	water proof type
防松装置	抗鬆裝置	anti-slack device
防台锚地	防颱錨地	typhoon anchorage
防污漆	防污漆	anti-fouling paint
防锈	防鏽	anti-rust, rust proof
防锈漆	防鏽漆	antirust paint
防锈油	防鏽油	rust preventive oil
防烟面具	防煙面具	smoke mask
防淤堤	攔沙壩	sand-blocking dam
防止垃圾污染	防止垃圾污染	prevention of pollution by garbage
防止倾倒废物及其他物质污染海洋公约	防止傾倒廢棄物及其他物質污染海洋公約	Convention on the Prevention of Marine Pollution by Dumping of Wastes and other Matters
防止生活污水污染	防止污水污染	prevention of pollution by sewage
防止油污染	防止油污染	prevention of pollution by oil
防撞舱壁	防撞艙壁,防碰艙壁,碰撞隔堵	collision bulkhead
房舱(=居住舱)		
房舱布置图	房艙佈置圖	cabin plan
仿真器(=模拟器)		
仿真天线	假天線	artificial antenna
纺锤形浮标	紡錘形浮標	spindle buoy
放电	放電	discharge
放电率	放電率	discharging rate
放电特性曲线	放電特性曲線	discharge characteristic curve
放射性废水箱	輻射性廢水箱櫃	radioactive waste water tank
放射性废物箱	固體輻射性廢棄物儲存箱櫃	radioactive solid waste storage tank
放射性物质	放射物質,放射性物質	radioactive substance
放松	鬆弛	slack away
放网	投網	shooting net
飞车	螺槳空轉	propeller racing

祖 国 大 陆 名	台 湾 地 区 名	英 文 名
飞剪[型]艏	飛剪式艏	clipper stem, clipper bow
飞轮	飛輪	fly wheel
飞行信息区	飛航情報區	flight information region
飞行中	飛行中	on the fly
飞重式调速器	配重調速器	weight governor
非常征用权	擄用破壞中立國船舶非 常徵用權	right of angary
非规则波	不規則波浪	irregular wave
非静止卫星通信容量	非定置衛星通訊之容量	volume of non-GEOSAR traffic
非排水船舶	非排水型船	non-displacement craft
非生物资源	非生物資源	non-living resources
非收放型减摇鳍装置	非收放型鰭板穩定器	non-retractable fin stabilizer
非铁金属、有色金属	非鐵金屬	non-ferrous metal
非危险区域	非危險區域	non-hazardous areas
非增压的	非增壓的	unsupercharged
非增压发动机	吸氣式發動機	aspirated engine
非周期过渡条件	非週期過渡條件	aperiodic transitional condition
非周期罗经	無週期羅經,立復羅經, 安定羅經	aperiodic compass
废气锅炉	廢氣熱交換器	exhaust gas heat exchanger
废气净化	排氣淨化	exhaust purification
废气涡轮发电机组	廢氣渦輪發電機組	exhaust turbine generating set
废气涡轮复合系统	廢氣渦輪複合系統	exhaust turbo compound system
废气涡轮增压	渦輪增壓	turbocharging
废弃物	廢棄物	waste
废热回收	廢熱回收	waste heat recovery
废热回收装置	廢熱回收裝置	waste-heat recovery plant
废物处理系统	廢物處理系統	waste disposal system
沸腾蒸发	沸騰蒸發	boiling evaporation
费率本,运价本	費率規章,收費制	tariff
分舱因数	隔艙因數	factor of subdivision
分舱载重线	艙間吃水線	subdivision loadline
分道通航制	分道通航制	traffic separation schemes, TSS
分点	春秋分點	equinox
分点潮	春秋分點潮,二分潮	equinoctial tide
分点大潮	春秋分大潮	equinoctial spring tides
分段航行	分段航行	sectional navigation
分段引航	分段引水	sectional pilotage

祖 国 大 陆 名	台 湾 地 区 名	英 文 名
分隔带	分道區	separation zone
分隔线	分道線	separation line
分货种运费	商品運費	commodity freight
分级控制	分級控制	step control
分节驳船	組合駁船	integrated barge
分节驳船队	組合駁船隊	integrated barge train
分解	分解	decomposition
分开式燃烧室	分離式燃燒室	divided combustion chamber
分离盘	分離盤	separating disc
分离盆	分離盆	separating bowl
分罗经	羅經複示儀	compass repeater
分配	配送	distribution
分区协调人	分區協調人	sub-area coordinator
分水机	淨油機	purifier
分线盒	分配箱	distribution box
分杂机	淨油機	clarifier
分站	配電所,變電所	substation
分子结构	分子結構	molecular structure
分组交换	分組交換	packet switching
焚烧炉	焚化爐	incinerator
粉碎设备	粉碎設備	comminuter
风暴潮	暴風潮	storm tide
风暴中航行	惡劣氣候航行	navigating in heavy weather
风洞	風洞	wind tunnel
风斗	風斗	wind scooper
风花	風花圖	wind rose
风级	風級	wind scale
风浪	風成浪	wind wave
风浪高度	風[成]浪高	wind wave height
风冷式冷凝器	氣冷式冷凝器	air-cooled condenser
风流压差(=漂流角)		
风门	擋板	damper
风区	受風區	fetch
风生流	風生流	wind driven current
风时	風時	wind duration
风速	風速	wind velocity, wind speed
风速表	風速計	anemometer
风速计	風速紀錄計	anemograph

祖 国 大 陆 名	台 湾 地 区 名	英 文 名
风箱	風箱,皮老虎	bellows
风向	風向	wind direction
风向标	風向標	wind vane
风压	風壓	wind pressure
风压差	風壓差角	leeway angle
风压差系数	風壓差係數	leeway coefficient
风压横倾力臂	風壓傾側力臂	wind heeling lever
风压横倾力矩	風壓傾側力矩	wind heeling moment
风雨密	風雨密	weathertight
风阻力	風阻力	wind resistance
风阻力矩	風阻力矩	moment of wind resistance
风阻力系数	風阻力係數	wind resistance coefficient
封闭箱(**封闭集装箱**)	封閉貨櫃	closed container
封舱抽水打捞	封艙抽水浮升	raising by sealing patching and pumping
封舱设备	艙口壓緊裝置	hatch battening arrangement
封缸	封缸	closing cylinder
封港	封港	embargo
封锁	封鎖	block
峰包功率	峰包功率	peak envelope power
峰值	尖峰值	peak value
锋	鋒	front
锋面过境	鋒過境	frontal passage
缝帆工具	縫帆工具	sailmaker's tool
缝隙腐蚀	間隙腐蝕	crevice corrosion
缝隙天线	槽孔天線	slot antenna
佛氏铁	校磁鐵棒	Flinders' bar
弗劳德数	佛勞數	Froude's number
伏尔肯弹性联轴器(一种橡胶弹性元件联轴器)	福爾幹撓性聯軸節	Vulkan flexible coupling
扶强材	加強肋	stiffener
扶手	扶手	hand rail
服务标准	航務標準	criterion of service
服务处所	服務空間	service space
服务区	服務區	service area
氟利昂	氟氯烷冷凍劑	freon
浮标	浮標,浮筒	buoy
浮冰	浮冰	floe ice

祖国大陆名	台湾地区名	英 文 名
浮冰群	块冰	pack ice
浮船坞	浮坞,浮船坞	floating dock
浮桨	浮檠	buoyant paddle
浮力	浮力	buoyancy
浮力调节系统	浮力調節系統	buoyancy regulating system
浮码头	浮箱	pontoon
浮锚	浮錨,海錨	floating anchor
浮桥	浮橋	pontoon bridge
浮式采油生产平台(浮式采油平台)	浮式産油平台	floating oil production platform
浮式储油装置	浮動型貯油裝置	floating storage unit, FU
浮式生产储油及卸载设施	浮動型石油生産貯藏及卸載設施	floating production storage and offloading facility, FASO
浮式生产储油装置	浮式生産儲存裝置	floating production storage unit, FPSU
浮式输油软管	浮式輸油軟管	floating oil loading hose
浮水杓	浮水杓	buoyant bailer
浮筒打捞	用救難浮箱浮升	raising with salvage pontoons
浮筒系钩	浮筒繫鉤	buoy hook
浮拖网	浮曳網	floating trawl
浮心	浮[力中]心	center of buoyancy
浮心高度	浮心高	height of center of buoyancy
浮心距中距离	縱向浮心與舯距離	longitudinal distance of center of buoyancy from midship
浮油层取样器	浮油層取樣器	float oil layer sampler
浮油回收船	去油沫器,撇油器(撈油船)	oil skimmer
浮装	浮載	float on/float off
浮子	浮子	float
浮子纲	浮繩	float line
幅移键控	移幅按鍵	amplitude shift keying, ASK
辐合	輻合	convergence
辐合线	輻合線	convergence line
辐散	輻散	divergence
辐散线	輻射線	divergence line
辐射热	輻射熱	radiant heat
辐照监督管	照射檢測管	irradiation inspection tube
斧头	斧頭	axe
俯极	下天極	depressed pole

祖国大陆名	台湾地区名	英 文 名
俯角	俯角	dip
辅锅炉	副鍋爐	donkey boiler
辅锅炉自动控制系统	輔鍋爐自動控制系統	auxiliary boiler automatic control system
辅冷凝器循环泵	輔機冷凝器循環泵	auxiliary condenser circulating pump
辅汽轮机	輔蒸汽渦輪機	auxiliary steam turbine
辅助操舵装置	輔助操舵裝置	auxiliary steering gear
辅助柴油机	輔助柴油機	auxiliary machinery diesel engine
辅助发电柴油机	輔助柴油發電機	auxiliary generator diesel engine
辅助软管设备	輔助軟管設備	auxiliary hose rigs
辅助天气图	輔助天氣圖	auxiliary weather chart
辅助装置	輔助設施	auxiliary device
腐蚀	腐蝕	corrosion
腐蚀磨损	腐蝕耗損	corrosion wear
腐蚀疲劳	銹蝕疲勞,腐蝕疲勞	corrosion fatigue
腐蚀性物质	腐蝕性物質	corrosives
负反馈	負反饋	negative feedback
负反馈控制系统	負反饋控制系統	negative feed back control system
负荷分配	負載分配	load-sharing
负荷特性	負載特性	load characteristic
负荷限制旋钮	負荷限制旋鈕	load limit knob
负荷指示器	負荷指示器	load indicator
负载调节器	負載調節器	load governor
负责操作人员的职务	負責作業人員之職位	responsible operator position
附加二次相位因子	附加二次相位因數	additional secondary phase factor, ASF
附加费用	附收費用	accessorial charge
附加检验	額外檢驗	additional survey
附加运费(**附加费**)	附加費	additional charge
附体阻力	附屬物阻力	appendage resistance
复合应力	複應力	compound stress, composite stress
复合增压	複合增壓	compound supercharging
复励发电机	複激發電機	compound generator
复励阻抗	複激阻抗	compounding impedance
复示磁罗经	電導羅經	transmitting compass
复式冲动涡轮机	複式衝動渦輪機	compound impulse turbine
复式压力级涡轮机	複壓渦輪機	pressure-compounded turbine
复位,重调	重置,復歸	reset
复原力臂	扶正力臂	righting arm, righting lever
复原力臂曲线	扶正力臂曲線	righting lever curve, righting arm curve

祖 国 大 陆 名	台 湾 地 区 名	英 文 名
复原力矩	扶正力矩	righting moment
复原力偶	扶正力偶	righting couple
复原稳性力臂	剩餘扶正力臂	residual righting lever
复杂循环燃气轮机	複合循環燃氣渦輪機	complex cycle gas turbine
副标志	副標誌	counter mark
副车钟	副俥鐘	sub-telegraph
副航道	副航道	sub-channel
副机日志	副機日誌	auxiliary engine log book
副热带高压	副熱帶高壓	subtropical high
副热带无风带	馬緯度	horse latitude
副艏材	艏護木, 艏牆	apron
副台	副台	secondary station
副台信号	副台信號	slave signal
副拖缆	副拖纜	auxiliary towing line
副陀螺	輔迴轉儀	auxiliary gyro
富余水深	餘裕水深	under keel clearance, UKC
覆板	複板, 加力板	doubling plate
覆盖区	覆蓋區	coverage
覆盖区边缘	涵蓋邊緣	edge of coverage

G

祖 国 大 陆 名	台 湾 地 区 名	英 文 名
改版图	大改正海圖	large correction chart
改向性	變向能力	course changing ability
改向性试验	變向能力試驗	course changing ability test
钙基润滑脂	鈣基滑脂	calcium grease
盖斯林格弹性联轴器	蓋氏可撓聯結器	Geislinger flexible coupling
概率航迹区	或然航跡區	probable track area
概率误差	或然差	probable error
概率误差圆	圓形概差	circular error probable
概位	概略位置	position approximate, PA
干冰	乾冰	dry ice
干舱证书	[貨艙]乾燥證明	dry certificate
干出高度	出水高度	drying height
干出礁	出水礁石	drying rock
干出滩	潮灘	tidal flat
干船坞	乾[船]塢	dry dock

祖 国 大 陆 名	台 湾 地 区 名	英 文 名
干底润滑	乾油槽潤滑	dry sump lubrication
干粉灭火系统	乾粉滅火系統	dry powder fire extinguishing system
干货	乾貨	dry cargo
干货船	乾貨船	dry cargo ship
干绝热直减率	乾絕熱線直減率	dry adiabatic lapse rate
干罗经	乾羅經	dry compass
干扰	干擾	trouble
干散货箱	乾散貨櫃	dry bulk container
干湿表	空氣濕度計,乾濕表	psychrometer
干湿计	乾濕計	psychrograph
干式缸套	乾式缸套	dry cylinder liner
干式蒸发器	乾式蒸發器	dry-type evaporator
干舷	乾舷	freeboard
干舷甲板	乾舷甲板	freeboard deck
干舷漆	乾舷漆	top-side paint
干预公海非油类物质污染议定书	油以外物質污染事故在公海行使干涉議定書	Protocol Relating to Intervention on the High Seas in Case of Pollution by Substances other than Oil
干燥剂	乾燥劑,催乾劑	drier
干燥器	乾燥器	drier
干燥室	乾燥室	drying room
干支流交汇水域	主支流匯流區	convergent area of main and branch
杆式液压调速器	槓桿式液壓調速器	lever-type hydraulic governor
杆状浮标	樁標	spar buoy
竿钓	桿	rod
感潮河	潮河	tidal river
感潮河段	潮達區	tide reaching zone
感染性物质	傳染性物質	infectious substance
感应船磁	船體感應磁	ship induced magnetism
干流	幹流	trunk stream
干线	幹繩	main line
干线导管	幹繩導管	main line guide pipe
干线放线机	放繩機	line casting machine
干线理线机	理繩機	line arrangement machine
干线起线机	捲繩機	line hauler
刚度	剛性,剛度	rigidity
刚性	剛性,抗撓性,勁度	stiffness
刚性结构	剛性結構	rigid construction

祖国大陆名	台湾地区名	英 文 名
刚性救生筏	硬式救生筏	rigid liferaft
刚性联轴器	剛性聯結器	fast coupling
刚性轴系	剛性軸系	stiff shafting
刚性转子	堅固轉子	rigid rotor
缸径最大磨损	缸徑最大磨損	bore maximum wear
缸套冷却水泵	缸套冷卻水泵	jacket cooling water pump
[缸套]磨损率	缸套磨損率	[liner] wear rate
钢	鋼	steel
钢船	鋼船	steel ship
钢筋水泥船	鋼筋混凝土船	ferro-concrete vessel, reinforced concrete vessel
钢麻绳	鋼麻合燃索	spring lay rope
钢丝绳	鋼絲索	wire rope
钢丝绳剪	切繩器	wire rope cutter
钢芯铜线	鋼心銅線	steel-cored copper wire
港泊图	港圖	harbor plan
港池	港池	harbor basin
港界	港界	harbor boundary, harbor limit
港口	港	harbor
港口电台	港埠電台	port station
港口服务处	港務接洽處	harbor service kiosk
港口管理	港口管理	port management
港口国	港口國	port state
港口国管理	港口國管制	port state control, PSC
港口国管理检查	港口國管制檢查	inspection of port state control
港口雷达	港口雷達	harbor radar
港口使费	貨櫃碼頭費用	terminal charge
港口通过能力	港口能量	port capacity
港口吞吐量	港口貨物吞吐量	port's cargo throughput
港口习惯	港口慣例	custom of port
港口营运业务	港埠營運業務	port operation service
港口租船合同	港口租船契約	port charter party
港内速度	港內船速	harbor speed
港区	港區	harbor area
港湾测量	港灣測量	harbor survey
港务费	港工捐	harbor dues
港务监督	港務監理	harbor superintendence administration
[港务]杂费	[港務]雜費	petty average

祖 国 大 陆 名	台 湾 地 区 名	英 文 名
港章	港口規章	port regulations
港作船	港勤船	harbor launch
高层云	高層雲	alto-starts
高潮	高潮	high water, HW
高潮时	高潮時	high water time
高程	標高,仰角	elevation
高处	在上,在高處	aloft
高灯芯浮标,大型浮标		high focal plane buoy
高低潮	較高低潮(一日中)	higher low water, HLW
高低压继电器	高低壓繼電器	high and low pressure relay
高度差	高度差	altitude difference
高度差法	高度差法	altitude difference method
高度圈	等高圈	almucantar
高度视差	高位視差	parallax in altitude
高费率货	工資加成貨	penalty cargo
高高潮	較高高潮(一日中)	higher high water, HHW
高积云	高積雲	alto-cumulus
高架桥	高架道路,陸橋	viaduct
高架索补给装置	高架索補給裝置	highline rig
高架索绞车	高線絞機	highline winch
高架纤维绳传递装置	纜索高線傳遞設備	fiber rope highline rig
高精度定位系统	高精準定位系統	high precision positioning system
高空[天气]图	高空[天氣]圖	upper-level [weather] chart
高空作业	高空作業	aloft work
高肋板	深肋板	depth floor
高频通信	高頻通信	HF communication
高速柴油机	高速柴油機	high speed diesel engine
高速艇筏	高速艇筏	high speed craft
高速诱导空气调节系统	高速誘導空調系統	high velocity induction air conditioning system
高温腐蚀	高溫腐蝕	high temperature corrosion
高温计	高溫計	pyrometer
高压	高氣壓	high [pressure]
高压脊,脊	隆起緣,脊	ridge
高压水冲洗	高壓水冲洗	hydro-blasting
告警阶段	警戒階段	alert phase
搁浅	擱淺	aground
搁浅船	擱淺船	vessel aground

祖 国 大 陆 名	台 湾 地 区 名	英 文 名
格林尼治恒星时	格林[威治]恒星時	Greenwich sidereal time, GST
格林尼治平时	格林威治平均時	Greenwich mean time
格林尼治时角	格林[威治]時角	Greenwich hour angle, GHA
格林尼治子午线	格林[威治]子午線	Greenwich meridian
格网航法	方格航法	grid navigation
格网偏差	方格偏差	grid variation
格子舱盖	格子窗口	grating hatch
隔舱填料函	艙壁填料函	bulkhead stuffing box
隔离	隔離	separated from
隔离表	隔離表	segregation table
隔离病房	隔離病院	lazaret
隔离阀	隔離閥	isolating valve
隔离法	隔離法	isolating method
隔离开关	隔離開關	isolating switch
隔片	間隔物	spacer
隔票	分離	separation
隔热箱	絕熱櫃	insulating container
隔日潮	隔日潮	double day tide
隔绳	間索	tackline
隔音室	隔音室	sound-proof chamber
个人示位标	個人示位標	personal locator beacon, PLB
铬镍合金	克鏴美	chromel
给定值	調定值	set value
给水倍率	給水率	feed water ratio
跟踪	目標追蹤	tracking
工厂交货	工廠交貨	ex works
工程船	工程船	engineering ship
工况报警	狀況警報	condition alarm
工况监视器	狀況偵測器	condition monitor
工况显示器	狀況指示器	condition indicator
工作日	工作日	working day, WD
工作时间	工作時間	hours of service
公差	公差	tolerance
公吨	公噸	metric ton
公共呼叫频道	公共呼叫頻道	common calling channel
公海	公海	high seas
公海海上安全信息	公海海上安全資訊	high seas maritime safety information
公海自由	公海自由	freedom of the open seas

祖 国 大 陆 名	台 湾 地 区 名	英 文 名
公路罐车	公路液罐車	road tank vehicle
公司旗	公司旗	house flag
公务电报	公務電報	service telegram
公用处所	公用空間	public space
公用电话交换网络	公用交換電話網路	public switched telephone network
公用数据交换网络	公衆交換數據網路	public switched data network
公证检验	公證檢驗	notarial survey
公证鉴定	公證檢定	inspection by notary public
公证权	公證權	right of notary
公制粗牙螺纹	公制粗螺紋	metric coarse thread
公众通信	公衆通信	public correspondence
公众通信业务	公衆通信業務	public correspondence service
功率储备	功率餘裕	power margin, power reserve
功率换算	功率換算	power conversion
功率密度	輸出密度(原子爐)	power density
功率容积比	功率容積比	power-to-volume ratio
功率输入传动装置	功率輸入傳動裝置	power take-in drive
功能	功能	function
功能试验	功能試驗	function test
供电网	供電網絡	supply network
供应船	勤務艦,補給船	tender
供油提前角	供油提前角	fuel supply advance angle
共轭赤经	恒星時角	sidereal hour angle, SHA
共同安全	共同安全	common safety
共同海损	共同海損	general average, GA
共同海损保险	共同海損支付保險	general average disbursement insurance
共同海损担保	共同海損擔保	general average security
共同海损费用	共同海損費用	general average expenditure
共同海损分摊	共同海損分攤	general average contribution
共同海损分摊保证金	共同海損保證金	general average deposit
共同海损分摊价值	共同海損分攤價值	contributory value of general average
共同海损理算	共同海損理算	general average adjustment
共同海损理算书	共同海損理算書	general average adjustment statement
共同海损时限	共同海損時限	time limit of general average
共同海损损失	共同海損損失	general average loss or damage
共同海损条款	共同海損條款	general average clause
共同海损牺牲	共同海損犧牲	general average sacrifice
共同海损行为	共同海損行爲	general average act

祖国大陆名	台湾地区名	英 文 名
共同海损总额	共同海損總額	total amount of general average
共同危险	共同危險	common danger, common peril
共振,谐振	共振,諧振	resonance
沟通	溝通	communication
钩	鉤	hook
钩吊周期	吊鉤週期	hook cycle
钩篙	鉤篙	boat hook
构件	構件	structural member
孤立危险物标志	孤立危險物標誌	isolated danger mark
谷密	穀密	grain-tight
谷物	穀類	grain
谷物防动装置	穀類防動裝置	grain fitting
谷物横倾体积矩	穀類體積橫傾力矩	grain transverse volumetric upsetting moment
谷物倾侧力臂	穀類傾側力臂	grain upsetting arm
谷物移动角	穀類移動角	shifting angle of grain
鼓风机	鼓風機	blower
鼓轮	鼓輪	drum
鼓形控制器	圓筒控制器,鼓形控制器	drum controller
鼓形转子	鼓形轉子,鼓形輪子	drum rotor
固定冰	堅冰	fast ice
固定的	固定的	stationary
固定负荷	靜載負荷	dead load
固定高膨胀泡沫灭火系统	固定高脹力泡沫滅火系統	fixed high expansion forth fire-extinguishing system
固定环形天线	固定環形天線	fixed loop antenna
固定件	固定裝具	securing fitting
固定距标	距離指標	range marker
固定螺距桨	[固]定[螺]距螺槳	fixed pitch propeller, FPP
固定泡沫灭火系统	固定泡沫滅火系統	fixed forth fire-extinguishing system
固定式采油平台	固定式産油平台	fixed oil production platform
固定式甲板泡沫系统	固定甲板泡沫系統	fixed deck foam system
固定式气体灭火系统	固定氣體滅火系統	fixed gas fire extinguishing system
固定压力喷水灭火系统	固定壓力噴水滅火系統	fixed pressure water-spraying forth fire-extinguishing system
固定渔网	定骨網	fixed net
固定自差	固定自差	constant deviation

祖 国 大 陆 名	台 湾 地 区 名	英 文 名
固定作业程序	现行作業程序	standing operating procedure
固态电路	固態電路	solid state circuit
固体货站	乾貨整補站	solid cargo station
固体散货	散裝固體貨物	solid bulk cargo
固体散装货物安全操作 规则	散裝固體貨物安全實務 章程	Code of Safe Practice for Solid Bulk Car- goes
固体	固體,固態	solid
固有频率	固有頻率	inherent frequency
固有缺陷	固有缺陷	inherent vice
故障灯	故障號燈	breakdown lights
故障	故障	trouble
故障间隔平均时间	平均故障間隔[時間]	mean time between failures
故障检测与辨识	故障檢測與識別	fault detection and identification, FDI
故障检测与排除	故障檢測與排除	fault detection and exclusion, FDE
故障探测器	故障探測器	fault detector
故障信号	故障信號	fault signal
故障诊断	故障診斷	fault diagnosis
锢囚锋	包圍鋒	occluded front
刮刀	刮刀	scraper
刮油环	刮油張圈	scraper ring
挂靠港	寄泊港	port of call
挂满旗	掛滿旗	full dress
关闭[水密门窗]	關閉水密[門窗]	dog down
关封	海關封條	customs seal
关税	關稅	customs duties
关于在领海和港内使用 国际海事卫星船舶地 面站的国际协议	在領海及港内使用國際 海事衛星船舶地球台 之國際協約	International Agreement on the Use of IN- MARSAT ship Earth Station within the Territorial Sea and Ports
观测	觀測,觀察	observation
观测船位	觀測船位	observed position, OP
观测船位误差	觀測船位誤差	error of observed position
观测高度	觀測高度	observed altitude
观测高度改正	觀測高度修正	observed altitude correction
观测柜	窺測油櫃	observation tank
观测经度	觀測經度	observed longitude
观测纬度	觀測緯度	observed latitude
观察孔	窺孔	sighting port
管节	管套節	union

祖 国 大 陆 名	台 湾 地 区 名	英 文 名
管理船舶过失	管理船舶過失	default in management of the ship
管理级	管理級	management level
管路附件	管路配件	pipeline fittings
管路特性曲线	管路特性曲線	pipeline characteristic curve
管隧	管道	pipe tunnel
管辖权条款	管轄權條款	jurisdiction clause
管线标	管線標	pipeline mark
管形燃烧室	管形燃燒室	tubular combustor
管制值机员	管制值機員	controlling operator
贯穿螺栓	貫穿螺栓	through bolt
贯索四(北冕 α)	貫索四(北冕 α)	Alphecca
惯性导航系统	慣性導航系統	inertial navigation system, INS
惯性力	慣性力	inertia force
惯性试验	慣性試驗	inertial trial
惯性转头角	慣性轉頭角	overshoot
盥洗室	盥洗室	lavatory, toilet
灌水试验	注水試驗	water filling test
灌注法	灌注法	flooding method
罐式箱	槽[貨]櫃	tank container
罐头加工船	罐頭工作船	canning factory ship
罐形浮标	罐形浮標	can buoy
光船	空船	bare boat
光船租赁	光船租賃	bareboat charter
光船租赁登记	光船租賃登記	bareboat charter registration
光电燃气探测器	光電煙道氣探測器	photo electric flue gas detector
光电效应	光電效應	photoeffect
光弧	號燈光弧	sector of light
光弧界限	光弧界限	limit of sector
光力射程,照距	光照距,光強度視程	luminous range
光行差	光行差	aberration
光学经纬仪	光學經緯儀	optical theodolite
光诱围网	燈誘圍網	light-purse seine
广播星历	廣播天文曆	broadcast ephemeris
广播业务	廣播業務	broadcasting service
广域增强系统	廣域系統	wide area augmentation system, WAAS
归位,收紧	歸位	bring home
规避操舵	迴避操舵法	evasive steering
规避航向	迴避航向	evasive course

祖 国 大 陆 名	台 湾 地 区 名	英 文 名
规避机动	迴避操縱	evasion maneuvre
规范(=标准)		
规则波	規則波浪	regular wave
规则库	規則庫	rule base
轨道	軌	rail
轨道长半径	軌道半長徑	semi-major axis of ellipse
轨道预报	軌道預報	orbit prediction
轨迹	軌道	track
贵重货	高值貨	valuable cargo
滚动式舱盖	滾動式艙蓋	rolling hatch cover
滚动轴承	滾動軸承	rolling bearing
滚卷式舱盖	滾動收放式艙蓋	roll stowing hatch cover
滚轮导缆器	滾子導索器	roller fairleader
滚球止推轴承	滾球止推軸承	ball thrust bearing
滚柱式扩管器	擴管器	roller tube expander
滚装	車輛駛上駛下船,車輛 運輸艦	roll on/roll off
滚装船	滾裝船	roll on/roll off ship
滚装货	滾裝貨	ro/ro cargo
锅炉安全阀	鍋爐安全閥	boiler safety valve
锅炉本体	鍋爐體	boiler body
锅炉舱	鍋爐艙,鍋爐間	boiler room
锅炉点火	鍋爐點火	boiler lighting up
锅炉点火泵	鍋爐點火泵	boiler ignition oil pump
锅炉点火设备	鍋爐點火設備	boiler firing equipment
锅炉二次鼓风机	鍋爐二次鼓風機	boiler secondary air blower
锅炉辅助蒸汽系统	鍋爐輔助蒸汽系統	boiler auxiliary steam system
锅炉附件	鍋爐裝具,鍋爐附件	boiler fittings
锅炉干汽管	鍋爐乾汽管	boiler dry pipe
锅炉给水泵	鍋爐給水泵	boiler feed pump
锅炉给水系统	鍋爐給水系統	boiler feed system
锅炉给水止回阀	鍋爐給水止回閥	boiler feed check valve
锅炉鼓风机	鍋爐鼓風機	boiler blower
锅炉排污阀	鍋爐放水閥	boiler blow down valve
锅炉牵条	鍋爐拉條,鍋爐牽條	boiler stay
锅炉牵条管	鍋爐拉條管	boiler stay tube
锅炉强制循环泵	鍋爐強力循環泵	boiler forced-circulating pump
锅炉燃油泵	鍋爐燃油泵	boiler fuel oil pump

祖国大陆名	台湾地区名	英 文 名
锅炉燃油系统	鍋爐燃油系統	boiler fuel oil system
锅炉受热面	鍋爐受熱面	boiler heating surface
锅炉水冷壁	鍋爐水管壁	boiler water wall
锅炉水位表	鍋爐水位計	boiler water gauge
锅炉水位调节器	鍋爐水位調整器	boiler water level regulator
锅炉外壳	鍋爐殼,鍋爐襯套	boiler clothing, boiler casing
锅炉烟箱	鍋爐煙道	boiler uptake
锅炉引风机	鍋爐誘導通風扇	boiler induced-draft fan
锅炉主蒸汽系统	鍋爐主蒸汽系統	boiler main steam system
锅炉自动控制系统	鍋爐自動控制系統	boiler automatic control system
国际安全管理规则	國際安全管理章程	International Safety Management Code, ISM Code
国际安全通信网	國際安全通信網	international safety NET
国际标准化组织	國際標準組織	International Organization for Standardization, ISO
国际标准化组织黏度分级	國際標準組織黏度分級	ISO viscosity classification
国际标准环境状态	國際標準組織環境狀況	ISO ambient reference condition
国际冰况巡视报告	國際冰況巡邏布告	international ice patrol bulletin
国际船舶吨位丈量公约	船舶噸位丈量國際公約	International Convention on Tonnage Measurement of Ships
国际低极轨道搜救卫星计划协议	國際衛星輔助搜救計劃協議	International COSPAs-SARSAT Programme Agreement
国际低极轨道搜救卫星系统计划协定	國際搜救衛星系統計劃協約	International COSPAs-SARSAT Programme Agreement
国际防止船舶造成污染公约	防止船舶污染國際公約	International Convention for the Prevention of Pollution from Ships
国际防止海洋油污染公约	防止海水油污染國際公約	International Convention for the Prevention of Pollution of the Sea by Oil
国际防止散装运输有毒液体物质污染证书	國際載運有毒液體物質防止污染證書	International Pollution Prevention For The Carriage of Noxious Liquid Substances in Bulk
国际防止生活污水污染证书	國際防止污水污染證書	International Sewage Pollution Prevention Certification, ISPP Certification
国际防止油污证书	國際防止油污證書	International Oil Pollution Prevention Certificate, IOPP Certificate
国际干预公海油污事故公约	油污染事故在公海行使干涉國際公約	International Convention Relating to Intervention on the High Seas in Case of

祖 国 大 陆 名	台 湾 地 区 名	英 文 名
		Oil Pollution Casualties
国际惯例	國際慣例	international custom and usage
国际海道测量组织	國際海道測量組織	International Hydrographic Organization
国际海港制度公约与规约	國際海港制度公約與規約	Convention and Stature on the International Regime of Maritime Ports
国际海上避碰公约	海上避碰國際公約	International Convention for Preventing Collisions at Sea
国际海上人命安全公约	海上人命安全國際公約	International Convention for Safety of Life at Sea, SOLAS
国际海上搜寻救助公约	國際海上搜索與救助公約	International Convention for Maritime Search and Rescue
国际海事委员会电子提单规则	國際海事委員會電子載貨證券規則	CMI Rules of Electronic Bills of Lading
国际海事委员会海运单统一规则	國際海事委員會海運單統一規則	CMI Uniform Rules for Sea Waybills
国际海事卫星	國際海事衛星	international maritime satellite
国际海事卫星船舶地球站	國際海事衛星船舶電台	international maritime satellite ship earth station, INMARSAT SES
国际海事卫星 A 船舶地球站	國際海事衛星 A 船舶電台	INMARSAT A ship earth station
国际海事卫星 B 船舶地球站	國際海事衛星 B 船舶電台	INMARSAT B ship earth station
国际海事卫星 C 船舶地球站	國際海事衛星 C 船舶電台	INMARSAT C ship earth station
国际海事卫星 M 船舶地球站	國際海事衛星 M 船舶電台	INMARSAT M ship earth station
国际海事卫星海岸地球站	國際海事衛星海岸電台	INMARSAT coast earth station, INMARSAT CES
国际海事卫星陆地地球站	國際海事衛星陸地電台	INMARSAT land earth station, INMARSAT LES
国际海事卫星网络协调站	國際海事衛星網路協調站	INMARSAT network coordination station
国际海事卫星系统	國際海事衛星系統	international maritime satellite system, INMARSAT
国际海事卫星移动号码	國際海事衛星行動碼	INMARSAT mobile number
国际海事卫星组织	國際海事衛星組織	International Maritime Satellite Organization, INMARSAT
国际海事卫星组织公约	國際海事衛星組織公約	Convention on the International Maritime

祖 国 大 陆 名	台 湾 地 区 名	英 文 名
		Satellite Organization
国际海事组织	國際海事組織	International Maritime Organization, IMO
国际海事组织公约	國際海事組織公約	Convention on the International Maritime Organization
国际海事组织类号	國際海事組織類號	International Maritime Organization Class, IMO class
国际海事组织危险类别	國際海事組織危險類別	IMO hazard class
国际海域	國際海域	international sea area
国际海员培训、发证和值班标准公约	航海人員訓練、發證及當值標準國際公約	International Convention on Standard of Training
国际航标协会	國際燈塔協會	International Association of Light house Authorities
国际航行警告业务	國際航行警告電傳業務	international NAVTEX service
国际合作 GPS 跟踪网	國際合作 GPS 網路	cooperative international GPS network
国际集装箱安全公约	安全貨櫃國際公約	International Convention for Safe Containers
国际救助公约	國際救助公約	International Convention on Salvage
国际 DSC 频率(=国际数字选择呼叫频率)		
国际气象组织	國際氣象組織	International Meteorological Organization
国际散装化学品规则	國際散裝化學品章程	International Bulk Chemical Code
国际时间局	國際時間局	Bureau International de'l Heure
国际数字选择呼叫频率,国际 DSC 频率	國際數位選擇呼叫頻率	international DSC frequencies
国际水道测量组织公约	國際海道測量組織公約	Convention on the International Hydrographic Organization
国际通岸接头	國際岸上接頭	international shore connection
国际偷渡公约	處理偷渡人國際公約	International Convention Relating to Stowaways
国际信号码	國際信號代碼碼組	international signal code
国际信号码组符号	國際信號代碼符號	international code symbol, INTERCO
国际信号旗	國際信號旗	international signal flag
国际信号旗"A"字硬质复制品	複製硬質國際代碼信號"A"旗	rigid replica of the International Code flag "A"
国际信息业务	國際資訊業務	international information service
国际移动卫星组织	國際海事衛星組織	International Mobil Satellite Organization
国际油污防备、响应和合作公约	油污防備、因應與合作國際公約	International Convention on Oil Pollution Preparedness, Response and Coopera-

祖国大陆名	台湾地区名	英 文 名
		tion
国际油污损害民事责任公约	油污损害民事責任國際公約	International Convention on Civil Liability for Oil Pollution Damage
国际载重线公约	國際載重線公約	International Convention on Load Lines, LL
国际直拨	國際直撥	international direct dialing
国家安全通信网	國家安全通信網	national safety NET
国家管辖海域	國家管轄海域	sea areas under national jurisdiction
国家询问	國家詢問	national enquiry
国内安全通信网	國家安全通信網業務	national safety NET service
国内奈伏泰斯业务	國內航行警告電傳業務	national NAVTEX service
国内数字选择呼叫频率	國內數位選擇呼叫頻率	national DSC frequencies
国内协调人	國家協調人	national coordinator
国旗	國旗,船籍旗	ensign
国有船舶豁免权	國有船舶豁免權	immunity of state-owned vessel
过驳	轉駁	lighterage
过电流	過量電流	over current
过电压	過電壓	over voltage
过度阶段	過渡階段	transition period
过度冷却	過冷	undercooling
过度磨损	過度磨損	overwear
过度应力	超應力	overstress
过渡导标	疊導標	transit leading mark
过渡工况	過渡工作情況	transient working condition
过河标	橫越標	crossing mark
过河点	渡河點	crossing river point
过境货,转口货	過路貨,轉口貨,接運貨	transit cargo
过境提单,转口提单	接運載貨證券	transit bill of lading
过境自由公约与规约	過境自由公約與規約	Convention and Stature on Freedom of Transit
过量空气常数	過量空氣係數	excess air coefficient
过期提单	過期載貨證券	stale bill of lading
过热器	過熱器	superheater
过滩吃水	淺灘低潮水深	bar draft
过[湍]滩	通過湍流	passing through the rapids
过稳船	高穩度船	stiff ship
过载能力	過載能力	overload capacity
过载试验	過負荷試驗,超載試驗	overload test

祖国大陆名	台湾地区名	英 文 名
过载脱扣	過載跳脫	overload trip

H

祖国大陆名	台湾地区名	英 文 名
[海]岸	海岸	coast
海岸带	海岸地帶	coastal zone
海岸带管理权	海岸帶管理權	right of management of coastal strip
海岸地球站	海岸衛星電台	coast earth station
海岸地球站识别码	海岸衛星電台識別碼	coast earth station identification
海岸电台	海岸電台	coast station
海岸电台表	海岸電台表	list of coast station
海岸电台群呼识别	海岸電台群呼識別	group coast station call identity
海岸电台识别	海岸電台識別	coast station identity
海岸雷达站	海岸雷達站	coast radar station
海岸无线电台	海岸無線電台	coast earth station
海岸效应	海岸效應	coastal effect
海冰	海冰	sea ice
海冰密集度	海冰密集度	sea ice concentration
海船	海船,海輪	sea-going vessel
海船所有人责任限制国际公约	海船所有人責任限制國際公約	International Convention Relating to the Limitation of the Liability of Owners of Seagoing Ships
海盗	海盜	pirate
海盗行为	海盜行爲	piracy
海道测量	海道測量	hydrographic survey
海道测量学	海道測量學	hydrography
海堤	海堤	sea wall, sea bank
海底电报	海底電報	cable gram
海底电缆	水底電纜	submarine cable
海底电缆传感器	電纜位置測定器	cable position sensor
海底管道	水底管線	submarine pipeline
海发光	海發光	luminescence of the sea
海风	海風	sea breeze
海关	海關	customs
海军导航卫星系统,子午仪系统	海軍衛星導航系統,子午儀系統	Navy Navigation Satellite System, NNSS, Transit System
海空两用灯标	海空兩用航行燈	marine and air navigation light

祖 国 大 陆 名	台 湾 地 区 名	英 文 名
海况	海面狀況	sea condition
海浪干扰抑制	抗海浪干擾	anti-clutter sea
海浪回波	海浪回波	sea echo
海浪预报	波浪預測	wave forecast
海里	浬,海里	nautical mile, n mile
海流(**洋流**)	洋流	ocean current
海流花	旋潮流圖	current rose
海锚	海錨	sea anchor
海绵	海綿	sponge
海面搜寻协调船	海面搜索協調船	surface search coordinator
海面温度	海面溫度	sea surface temperature
海难救助船	救難船	salvage and rescue ship
海难救助[打捞]	海上救助	marine salvage
海难救助合同	海難救助契約	salvage contract, salvage bond
海难救助作业	海難救助作業	salvage operation
海平面气压	海平面氣壓	sea-level pressure
A1 海区	A1 海域	sea area A1
A2 海区	A2 海域	sea area A2
A3 海区	A3 海域	sea area A3
A4 海区	A4 海域	sea area A4
海商法	海商法	maritime law, maritime code
海上安全监督	海上安全監督	marine safety supervision
海上安全信息	海事安全資訊	maritime safety information, MSI
海上保险	海上保險	marine insurance
海上泊位	海上泊位	sea berth
海上补给	海上整補	RAS(=replenishment-at-sea)
海上补油站	加油整補站	fueling-at-sea station
海上焚烧	海上焚化	incineration at sea
海上风险	海上危險	perils of the sea
海上航路标志	海上航標	sea mark
海上航行补给	海上整補	replenishment at sea
海上核材料运输民事责任公约	海上運載核子物質民事責任公約	Convention Relating to Civil Liability in the Field of Maritime Carriage of Nuclear Materials
海上护送	護航	escorting
海上货物运输法	海上貨物運輸法	Carriage of Goods by Sea, COGSA
海上急救	海上急救	first aid at sea
海上加油	海上加油	FAS

祖 国 大 陆 名	台 湾 地 区 名	英 文 名
海上加油站	海上加油整補傳送站	FAS delivery station
海上救助	海上救助	salvage at sea
海上救助协调中心	海上搜救協調中心	maritime rescue co-ordination center
海上救助中心	海上救助站	maritime rescue sub-center
海上旅客及其行李运输雅典公约	海上運送旅客及其行李雅典公約	Athens Convention Relating to the Carriage of Passengers and Their Luggages by Sea
海上旅游区	海上旅遊區	sea tourist area
海上平台	海上平台	offshore platform
海上气象数据	海上氣象數據	marine weather data
海上求生	海上求生	survival at sea
海上识别数字	水上識別碼	maritime identification digits, MID
海上事故	海上事故	sea accident
海上书信电报	海上書信電報	sea letter telegram, SLT
海上搜救	海上搜[索與]救[助]	marine search and rescue
海上搜救计划	海上搜救計劃	maritime SAR plan
海上速度	海流速率	sea speed
海上维修	海上維修	at sea maintenance
海上卫星无线电导航业务	水上無線電衛星導航業務	maritime radionavigation-satellite service
海上无线电导航业务	海上無線電助航業務	maritime radionavigation service
海上无线电话	海上無線電話	marine radiotelephone
海上无线电书信	海上無線電書信	radio maritime letter
海上询问	海上詢問	maritime enquiry
海上移动选择呼叫识别码	水上行動選擇呼叫識別碼	maritime mobile selective-call identify code
海上移动业务	海上行動業務	maritime mobile service
海上移动业务识别	水上行動業務識別	maritime mobile service identity
海上预报	海上預報	marine forecast
海上遇险信道	海上遇險頻道	maritime distress channel
海上蒸汽雾	海面蒸汽霧	sea smoke
海上资历	海勤資歷	sea service
海上自然保护区	海洋自然保護區	marine natural reserves
海上钻井架	海上鑽油台	drilling rigs at sea
海事报告	海事報告	marine accident report
海事调查	海事調查	maritime investigation
海事法庭	海事法庭	maritime court, marine court
海事分析	海事分析	marine accident analysis

祖国大陆名	台湾地区名	英 文 名
海事管辖	海事管轄權	maritime jurisdiction
海事和解	海事和解	maritime reconciliation
海事判例	海事判例	maritime case
海事请求	海事求償	maritime claim
海事诉讼	海事訴訟	maritime litigation
海事索赔责任限制公约	海事求償責任限制公約	Convention on Limitation of Liability for Maritime Claims
海事调解	海事調解	maritime mediation
海事援助	海事援助	maritime assistance
海事仲裁	海事仲裁	maritime arbitration
海水泵	海水泵	sea water pump
海水淡化装置	海水淡化裝置	sea water desalting plant
海水密度	海水密度	seawater density
海水染色标志	海水染色標誌	dye marker
海水水色	海水水色	seawater color
海水透明度	海水透明度	seawater transparency
海水温度	海水溫度	sea temperature
海水系统	海水系統	sea water service system
海水循环泵	海水循環泵	sea water circulating pump
海水盐度	海水鹽度	seawater salinity
海水蒸发器	海水蒸發器	sea water evaporator
海损担保函	海損擔保	average guarantee
海损分担保证书	共同海損承諾書	average bond
海损管制示意图	損害管制圖	damage control plan
海损理算人	海損清算人	average adjuster
海损理算书	海損理算書	average statement
海滩	海灘	beach
海图	海圖	nautical chart, chart
海图比例尺	海圖比例尺	chart scale
海图标题栏	海圖圖例	chart legend
海图基准面	海圖深度基準面	chart datum
海图夹编号	海圖夾編號	folio number
海图夹标签	海圖夾標籤	folio label
海图夹目录	海圖夾目錄	folio list
海图卡片	海圖卡	chart card
海图室	海圖室	chart room
海图桌	海圖桌	chart table
海图作业(=航迹绘算)		

祖 国 大 陆 名	台 湾 地 区 名	英 文 名
海图作业工具	海圖作業工具	chart work tools
海湾	海灣	gulf
海峡	海峽	strait
海峡船	海峽船	channel ship
海啸	海嘯	tsunami
海牙规则	海牙規則	Hague Rules
海洋磁力测量	海洋磁力測量	marine magnetic survey
海洋大气船	大洋測候船	ocean weather vessel
海洋调查	海洋調查	marine investigation
海洋调查船	海洋調查船	oceanographic research vessel
海洋定点船	海洋測候船	ocean station vessel
海洋腐蚀	海洋腐蝕	marine corrosion
海洋工程	海洋工程	oceaneering
海洋工程测量	海洋工程測量	marine engineering survey
海洋环境	海洋環境	marine environment
海洋环境保护	海洋環境保護	marine environmental protection
海洋环境调查	海洋環境調查	marine environment investigation
海洋监测	海洋監測	marine monitoring
海洋监测船	海洋監測船	ocean monitoring ship
海洋监视	海洋監視	marine surveillance
海洋科学技术	海洋科技	marine science and technology
海洋矿物资源	海洋礦產資源	mineral resources of the sea
海洋气象报告	海洋氣象報告	ocean weather report
海洋生态调查	海洋生態調查	marine ecological investigation
海洋生物资源	海洋生物資源	living resources of the sea
海洋水文	海洋水文	marine hydrology
海洋污染	海水污染	marine pollution
海洋污染物	海洋污染物	marine pollutant
海洋污染物报告	海水污染物報告	marine pollutants report
海洋渔业	海洋漁業	marine fishery
海洋重力测量	海洋重力測量	marine gravimetric survey
海洋主权	海洋主權	maritime sovereignty
海洋资源调查	海洋資源調查	marine resources investigation
海涌(台风或地震引起)	長浪,激湧	ground swell
海员	航海人員	seafarers
海员通常做法	海員常規	ordinary practice of seaman
海员证	船員證	seaman's book

祖 国 大 陆 名	台 湾 地 区 名	英 文 名
海运单	海運單	seaway bill
海运货	海運貨	sea-borne cargo
海运货物保险	海運貨物保險	maritime cargo insurance
氦氧潜水	氦氧潛水	helium-oxygen diving
含油舱底水	含油舭水	oily bilge water
含油混合物	含油混合物	oily mixture
寒潮	寒潮,寒流	cold wave
寒流	寒流	cold current
汉堡规则	漢堡規則	Hamburg Rules
焊剂	熔接劑	welding flux
焊接	熔接	weld
焊接型曲轴	焊接型曲柄軸	welded type crankshaft
焊条	焊條	welding wire
航摆角(**艏摇角**)	平擺角	yaw angle
航标表	燈標表	list of lights
航标船	浮標母船,浮標管理船	buoy tender
航标起重机	吊航標機	light buoy crane
航程	航程	distance run
航次	航次	voyage
航次保险	航次保險	voyage insurance
航次报告	航次報告	voyage report
航次结账单	航次帳單	trip account
航次期租合同	航次計時租船	time charter on trip basis, TCT
航次租船	航次傭船	trip charter, voyage charter
航道	主航道	fairway
航道标志	航道標誌	channel marker
航道标准尺度	航道標準尺度	standard dimension of channel
航道灯标	航道燈標	channel light
航道浮标	航道浮標	channel buoy
航道弯曲度	航道彎曲度	bend of channel
航海保证	航海勤務	nautical service
航海表	航海表	navigation table, nautical table
航海晨昏朦影	航海矇光	nautical twilight
航海法规	海事法規	maritime rules and regulations
航海顾问	航海諮詢	navigation consultant
航海过失	航海過失	nautical fault
航海健康申报书	航海健康申報書	maritime declaration of health
航海科学	航海科學	nautical science

祖国大陆名	台湾地区名	英 文 名
航海气象	航海氣象	nautical meteorology
航海日志	航海日誌,航泊日誌	log book
航海史	航海史	history of marine navigation, nautical history
航海天文历	航海曆	nautical almanac
航海通告	航行通告	notice to mariners
航海图书目录	航海圖書目錄	catalog of charts and publications
航海图书资料	航海圖書刊物	nautical charts and publications
航海心理学	航海心理學	marine psychology
航海性能	航海性能	seagoing qualities
航海学	航海學	marine navigation
航海医学	航海醫學	marine medicine
航海仪器	航海儀器	nautical instrument
航海专家系统	航海專家系統	marine navigation expert system
航弧,偏摆	船身迴擺(指錨泊急流中時)	sheer
航迹	航跡	track, TK
航迹冲程	航跡衝距	track reach
航迹灯	接近受補船艉燈	wake light
航迹分布	航跡分佈	track distribution
航迹绘算,海图作业	海圖作業	chart work
航迹积算仪	推算航跡儀	dead-reckoning tracer, DRT
航迹计算	航跡計算	track calculating
航迹推算	實際航跡	track made good
航空电台	航空電台	aircraft station
航空母舰	航空母艦	aircraft carrier
航路	航線,航路	route, passage
航路点	航路點	way point
航路设计图	航路圖	routing chart
航路指南	航行指南	sailing directions
航路指南补篇	航行指南增補篇	supplement of sailing directions
航速	速率,航速	speed
航速燃油消耗量条款	船速與耗油量條款	vessel's speed and fuel consumption clause
航速索赔	航速索賠	speed claim
航线间隔	航路間隔	track spacing
航线设计	航線設計	passage planning
航向	航向	course
航向记录器	航向記錄儀	course recording machine, course recorder

祖 国 大 陆 名	台 湾 地 区 名	英 文 名
航向稳定性	航向穩定性	course stability, stability of motion
航向线	航向線	course line, CL
航向向上	航向向上	course up
航向自动操舵仪	航向自動操舵裝置	course autopilot
航行安全通信	航行安全通信	navigation safety communication
航[行]标[志]	航路標誌	navigation mark
航行波	船行浪	ship wave
航行垂直补给	海上垂直補給	perpendicular replenishment at sea
航行灯照距测定	航行燈照距測定	determination of range of visibility for navigation light
航行风	航行風	navigation wind, ship wind
航行管理	航管	navigation management
航行横向补给	航行中橫向補給	abeam replenishment at sea
航行计划	航行計劃	navigational plan, sailing plan
航行计划报告	航行計劃報告	sailing plan report
航行警告	航行警告	navigational warning
航行警告[电传]系统，奈伏泰斯	航行警告電傳	navigational telex, NAVTEX
航行警告区	航行警告區	NAVAREA
航行警告区公告	航行區警告通告	NAVAREA warning bulletin
航行警告区警告	航行警告區之警告	NAVAREA warning
航行警告区业务	航行警告區業務	NAVAREA warning service
航行警告区域协调国	航行警告區域協調人	NAVAREA coordinator
航行警告信号	航行警告信號	navigational warning signal
航行期间	航程條款	term of voyage
航行权	航行權	right of navigation
航行试验	[海上]試航	sea trial
航行值班	航行當值	navigational watch
航行纵向补给	海上舺向補給	astern replenishment at sea
航修	航修	voyage repair
航用参考图	非航用海圖	non-navigational chart
航用海图	航海用海圖	navigational chart
航用行星	導航行星	navigational planets
航运法	航業法	shipping law
航运业务	航運業務	shipping business
行列编队	行列編隊	line formation
好能见度(7级)	能見度良好(7级)	visibility good
号灯	號燈	lights

祖 国 大 陆 名	台 湾 地 区 名	英 文 名
号灯垂直位置	號燈之垂直位置	vertical positioning of lights
号灯发光强度	號燈照明強度	intensity of lights
号灯间距	號燈間隔	spacing of lights
号灯能见距	號燈能見距	visibility of light
号灯水平位置	號燈水平位置	horizontal positioning of lights
号笛	汽笛,號笛	whistle
号笛音响度测定	音響信號聽距測定	determination of range of audibility of sound signal
号锣	鑼	gong
号型	號標,型材	shape
号钟	號鐘	bell
耗汽率	耗汽率	steam rate
耗热率	耗熱率	heat rate
合并舷灯	合併燈,聯合燈	combined lantern
合成波高	合成波高	height of wave and swell combined
合成油	合成油	synthetic oil
合格安全型设备	合格安全型設備	certified safe type apparatus
合格证书	符合證書	document of compliance
合理速遣	合理派遣	reasonable despatch
合力	合力	resultant
合同费率制	合約運費制	contract freight system
合作指数	協同指數	index of cooperation
合座舷灯	合併燈	sidelights combined in one lantern
河鼓二(天鹰 α)	河鼓二,牽牛[星]牛郎[星](天鷹 α)	Altair
河口	河口	river mouth, estuary
核测量系统	核子測量系統	nuclear measurement system
核动力船	核子動力船	nuclear [powered] ship
核动力船舶经营人责任公约	核子船舶營運人責任公約	Convention on the Liability of Operators on Nuclear Ships
核动力货船安全证书	核子貨船安全證書	nuclear cargo ship safety certificate
核动力客船安全证书	核子客船安全證書	nuclear passenger ship safety certificate
核动力驱动	核能驅動	nuclear driven
核动力推进装置	核能推進裝置	nuclear propulsion plant
核对数字	核對數位	check digit
核反应堆	核子反應器	nuclear reactor
核反应堆保护系统	核反應器保護系統	reactor protective system
核反应堆控制系统	核反應器控制系統	reactor control system

祖国大陆名	台湾地区名	英 文 名
核反应堆中毒	核子反應器中毒	nuclear reactor poisoning
核潜艇	核子潛艇	nuclear submarine
核燃料	核燃料	nuclear fuel
黑潮	黑潮	Kuroshio, Black stream
黑色金属	鐵金屬	ferrous metal
黑匣子	黑盒子	black box
痕迹	跡	trace
恒定风	恒定風	permanent wind
恒功率调速	恒功率調速	speed regulation by constant power
恒速特性曲线	等速特性曲線	characteristic curve at constant speed
恒位线	等方位線	line of equal bearing
恒温器	恒溫器	thermostat
恒温调节器	溫度調節器	thermostat regulator
恒向线	恒向線	rhumb line
恒向线方位	恒向線方位	rhumb line bearing, RLB
恒向线航线算法	恒向線航法	rhumb line sailing
恒星	星,恒星	star
恒星日	恒星日	sidereal day
恒星时	恒星時	sidereal time, ST
恒星视位置	恒星視位置	star apparent place
恒星图	星座圖	star chart, star atlas
恒星月	恒星月	sidereal month
[恒星]周年视差	週年視差	annual parallax
恒张力带缆绞车	自動張力繫船絞車	automatic constant tension mooring winch
恒张力拖缆机	自動等拉力拖攬機	automatic constant tension towing winch
恒转矩调速	恒轉矩調速	speed regulation by constant torque
桁	橫桁	yard
桁材深度	縱桁深度	girder depth
桁架桅	格式桅	lattice mast
桁拖网	桁拖網	beam trawl
横荡	橫移	sway
横队	橫隊	line abreast
横舵柄	舵軛	rudder yoke
横帆	橫帆,方帆	square sail
横帆船	橫帆船	square-rigged vessel
横风	橫風	cross wind
横风航驶	橫戧	reach
横隔板	隔膜	diaphragm

祖 国 大 陆 名	台 湾 地 区 名	英 文 名
横骨架式	横肋系統	transverse frame system
横桁闪光信号灯	横桁閃光信號燈	blinker yardarm
横江轮渡号型	横江渡輪號標	shape for crossing ferry
横距	迴轉横距	transfer
横缆	横纜	breast line
横浪	横浪	beam sea
横梁	甲板梁	deck beam
横流	横流,側流	cross current
横流标	横流標	cross-current mark
横流扫气	横驅氣	cross scavenging
横漂	横漂	crabbing
横剖面	横截面,横剖面	cross section
横剖面图	横剖面圖	transverse section plan
横强度	横向強度	transverse strength
横倾角	横傾角	angle of heel
横倾力矩	傾側力矩	heeling moment
横驶区	横越區	crossing area
横拖	横拖	girding
横稳性	横向穩度	transverse stability
横向	横切,横向(與龍骨成 直角之方向)	athwartships
横向补给装置	横向補給裝置	abeam replenishing rig
横向冲距	横向衝距	side reach
横向磁棒	横向磁棒	athwartships magnet
横向的	横向的	transverse
横向间隙	側向間隙	lateral clearance
横向振动	側向振動	lateral vibration
横摇	横摇	rolling
横摇周期	横摇週期	rolling period
横移率	横移率	rate of transverse motion
横越	横越	crossing ahead
红丹	紅丹漆	red lead paint
红色提单	附帶保險載貨證券	red bill of lading
红色信号	紅星信號	red star-signal
红套	紅套	shrink-on
洪峰	洪峰	flood peak
后八字	後八字方向	on the quarter
后部	船尾部	after body

祖 国 大 陆 名	台 湾 地 区 名	英 文 名
后方,向后,倒车	後方,向後,在……後,倒俥,後退	astern
后燃	後燃	after-burning
后燃期	後燃期	after burning period
后视图	後視圖	rear view
后桅	後桅	after mast
后续舰	後續船艦	follow-up ship
厚壁	厚壁	thick wall
厚壁轴承	厚鋼殼型,軸承	thick steel shell type bearing
厚度规	厚度規	thickness gauge
候潮港	淺灘港	bar port
候(蛇夫α)	候(蛇夫α)	Rasalhague
候选人,报考人	申請發證者	candidate
呼号	呼號	call sign, CS
呼叫	呼叫	calling
呼叫尝试	嘗試性呼叫	call attempt
呼叫点	呼叫點	calling-in-point, CIP
呼叫方	主叫用戶,發話方	calling party
呼叫各电台(无线电话用语)	呼叫各電台(無線電話用語)	CQ
呼吸阀	呼吸閥	breather valve
呼吸器	呼吸器	breathing apparatus
互光	變色燈	alternating light
互见中	互見中	in sight of one another
互见中的船舶	船舶互見	vessel in sight of one another
互调产物	互調産物	intermodulation products
互有责任碰撞条款	雙邊過失碰撞條款	both to blame collision clause
护航	護航	convoy
护送	船團護航	convoy escort
护卫舰	巡防艦	frigate
花水	急浪	rips
滑车	滑車	block
滑车组	滑車轆轤	tackle
滑动门	滑[拉式]門	sliding door
滑动配合	滑動配合	slide fit
滑动轴承箱	滑動軸承殼	sliding bearing housing
滑阀导程	滑閥導程	lead of slide valve
滑钩	滑鉤	pelican hook

祖国大陆名	台湾地区名	英 文 名
滑块	滑塊,偏龍骨,起重架之墊腳板	slipper, shoe
滑块制动器	塊狀軔,塊狀煞車	block brake
滑马(=移动式起重机)		
滑失	滑流	slip
滑巷	滑巷	lane slip
滑行艇	滑航艇	planing boat
滑油泵	[潤]滑油泵	lubricating oil pump
[滑油]承载特性	負載特性	load carrying properties
滑油间歇净化	潤滑油分批淨化	lubricating oil batch purification
滑油连续净化	潤滑油連續淨化	lubricating oil pass purification
滑油输送泵	[潤]滑油輸送泵	lubricating oil transfer pump
滑油消耗率	單位滑油消耗量	specific lubricating oil consumption
滑油注入管	[潤]滑油注入管	lubricating oil filling pipe
化学灯	指距用化學燈	chemical light
化学方程式	反應式(化學)	reaction formulas
化学腐蚀	化學腐蝕	chemical corrosion
化学灭火器	化學滅火器	chemical fire extinguisher
化学品船	化學品船	chemical cargo ship
化学物添加系统	化學品添加系統	chemical addition system
化学纤维绳	合成纖維索	synthetic fiber rope
划痕	刮痕,劃痕	scratch
划桨船	划槳船	row boat
划界渔业	區劃漁業	demarcated fishery
话传电报业务	話傳電報業務	voice messaging service
话传用户电报	話傳交換電報	phone telex, PHONETEX
还船	還船	redelivery of vessel
环管形燃烧室	環筒形燃燒器	can annular type combustor
环境温度	環境溫度	ambient temperature
环流理论	環流理論	circulation theory
环球航行	環航	circum-navigation
环索	回頭索,環索	endless rope
环形燃烧室	環形燃燒室	annular combustor
环形天线	環形天線	loop antenna
环形天线装调误差	環形天線對準誤差	loop alignment error
环照灯	環照燈	all-round light
环状星云	環狀星雲	annular nebula
缓冲器	緩衝器	buffer

祖国大陆名	台湾地区名	英 文 名
缓冲弹簧	緩衝彈簧	buffer spring
缓冲作用	緩衝作用	cushioning effect
缓流	憩流	slack stream
缓流航道	憩流航道	slack current channel
缓松	回舵,緩鬆	ease
幻日	幻日	anthelion
幻月	幻月	antiselena
换板	換板	changing plate
换能器	轉換器,轉發器	transducer
换能器充磁	轉變器充磁	magnetization of transducer
换能器指向性	轉換器指向性	transducer directivity
换气-压缩行程	驅氣-壓縮衝程	scavenging-compression stroke
换算图表	換算圖表	conversion chart
换向	逆轉	reversing
换向联锁	逆轉聯鎖	reversing interlock
换向起动程序	換向起動程序	reverse starting sequence
换向时间	換向時間	reversing time
换向伺服器	回動伺服電動機	reversing servomotor
换向装置	逆轉裝置	reversing arrangement
换证检验	換證檢驗	renewal survey
黄白交角	白道斜度	obliquity of the moon path
黄赤交角	黃道斜度	obliquity of the ecliptic
黄道	黃道	ecliptic
黄道带	黃道帶	zodiac
黄极	黃道天極	ecliptic pole
黄经	黃經	ecliptic longitude
黄纬	黃緯	ecliptic latitude
晃击	沖激,晃擊	sloshing
灰分	灰分含量	ash content
灰水	洗滌水	greywater
恢复	復原,回復	restoration
回程货	回程貨	return cargo
回答旗	回答旗,答應旗	answering pendant
回舵	回舵,鬆舵	ease her
回风	回風	return air
回归潮	回歸潮	tropic tide
回归习惯	回歸習性	homing behavior
回火[处理]	回火	tempering

祖 国 大 陆 名	台 湾 地 区 名	英 文 名
回流扫气	環狀驅氣	loop scavenging
回汽刹车	回汽煞俥	reverse steam brake
回热式汽轮机	回熱式蒸汽渦輪機	regenerative steam turbine
回热循环燃气轮机	回熱循環燃氣渦輪機	regenerative cycle gas turbine
回升(=［缆］回松)		
回声测冰仪	測冰儀	ice fathometer
回声测深	回音測深	echo sounding
回声测深仪	回聲測深儀	acoustic depth finder, echo sounder
回收索	回收索	recovery line
回头缆	滑索	slip wire, slip rope
回油阀式喷油泵	溢流閥式噴射泵	spill-valve injection pump
回油孔式喷油泵	布氏噴油泵	Bosch injection pump
回游路线	洄游路線	[fishing] migration route
回转不均匀	迴轉不規率度	cycle irregularity
回转吊杆绞车	［吊桿］迴旋絞車	slewing winch
回转流	旋轉流	rotary current
回转式起动空气分配器	旋轉式起動空氣分配器	rotary starting air distributor
回转式扫气阀	轉動驅氣閥	rotary scavenging valve
回转叶片式制冷压缩机	轉動滑葉冷凍壓縮機	rotary sliding-vane refrigerating compressor
汇流排	匯流排	busbar
会议电话	會議電話	conferencing call
会遇	會遇	encounter
会遇率	會遇率	encounter rate
昏影终	曙昏終了	end of evening twilight
混合编队	複列編隊	compound formation
混合潮	混合潮	mixed tide
混合导航系统	混合導航系統	hybrid navigation system
混合骨架式	混合肋骨系統	mixed frame system, combined frame system
混合航线算法	混合航法	composite sailing
混合货	混合貨物,混載貨	mixed cargo
混合流	混合潮流	mixed current
混合提单	合運載貨證券	omnibus bill of lading
混合调节	混合調節	mixing governing
混合循环	混合循環	mixed cycle
混流泵	混流泵	mixed flow pump
混杂不清货	混雜不清貨	commixture and unidentifiable cargo
混装舱间	共同裝載艙間	compartment loaded in combination

祖国大陆名	台湾地区名	英　文　名
混装船	混载船	combination carrier
豁免	豁免	exemption
豁免证书	豁免證書	exemption certificate
活动解拖钩	脫鉤	movable relieving hook
活动距标	可變距指標	variable range marker
活动梁	艙口活動梁	shifting beam
活动物货	牲口貨	livestock cargo
活塞	活塞	piston
活塞泵	活塞式泵	piston pump
活塞顶烧蚀	活塞頂燒蝕	piston crown ablation
活塞杆填料函	活塞桿填料函	piston rod stuffing box
活塞环搭口间隙	活塞環接口間隙	piston ring joint clearance, piston ring gap clearance
活塞环断裂	活塞環斷裂	piston ring breakage
活塞环磨损监测系统	活塞環磨損監測系統	piston ring wear monitoring system
活塞环黏着	活塞環膠著	piston ring sticking
活塞环平面间隙	活塞環軸向間隙	piston ring axial clearance
活塞冷却水泵	活塞冷卻水泵	piston cooling water pump
活塞平均速度	活塞平均速度	mean piston speed
活塞裙	活塞裙	piston skirt
活塞头	活塞頂	piston crown
活塞下部泵气功能	活塞下部泵效應	piston underside pumping effect
活塞销,轴头销	軸頭銷	gudgeon pin
活塞行程	活塞衝程	piston stroke
活塞运动装置失中	活塞連桿裝置欠對準	piston-connecting-rod arrangement mis-alignment
活塞组件	活塞組件	piston assembly
活套结	套馬扣	running bowline
火车轮渡	火車渡船	train ferry
火工矫形	火焰矯正	fairing by flame
火管锅炉	火管鍋爐	fire tube boiler
火箭	火箭	rocket
火箭降落伞信号	火箭式降落傘照明彈	rocket parachute flare
火警报警系统	火警報警系統	fire alarm system
火险	火險	fire insurance
火星	火星	Mars
火焰切割	火焰截割	flame cutting
伙食冷库	糧食冷凍庫	food stuff refrigerated storage

祖 国 大 陆 名	台 湾 地 区 名	英 文 名
货泵舱	貨泵室	cargo pump room
货舱	貨艙	cargo hold
货舱隔离	全艙隔離	separated by a complete compartment or hold from
货舱鉴定	貨艙檢查	inspection of hold
货舱空气干燥系统	貨艙空氣乾燥系統	cargo hold dehumidification system
货船	貨船	freighter, cargo ship
货船构造安全证书	貨船安全構造證書	cargo ship safety construction certificate
货船无线电安全证书	貨船安全無線電話證書	cargo ship safety radio certificate
货船无线电报安全证书	貨船安全無線電報證書	cargo ship safety radiotelegraphy certificate
货船无线电话安全证书	貨船安全無線電話證書	cargo ship safety radiotelephony certificate
货抵押贷款	船貨押貸	respondentia
货港未定租船合同	任務待定備船契約	open charter
货名	貨名	description of goods
货盘	托貨板	cargo pallet
货物伴生废弃物	貨物所生相關廢棄物	cargo-associated waste
货物残损单	貨損清單	damage cargo list
货物操作吨	貨物操作噸	tons of cargo handled
货物操作系数	貨物操縱係數	coefficient of cargo handling
货物查询单	貨物追查單	cargo tracer
货物记录簿	液貨紀錄簿	cargo record book
货物拒收险	貨物拒收險	rejection risks
货物冷藏装置检验	貨物冷凍裝置檢驗	survey of refrigerated cargo installation
货物平安险	單獨海損(不賠)	free from particular average, FPA
货物水渍险	水漬險	with average, WA
货物移位	貨物移位	cargo shifting
货物战争险	貨物戰爭險	cargo war risk
货物装卸设备	貨物裝卸設備	cargo gear
货物装载	貨物裝載	cargo-handling and stowage
货油泵	貨油泵	cargo oil pump
货油泵舱管系	貨油泵室管路	cargo oil pump room pipe line
货油舱	貨油艙	cargo oil tank
货油舱管系	貨油艙管路	cargo oil tank pipe line
货油舱气压指示器	貨油艙氣壓指示器	cargo oil tank gas pressure indicator
货油舱清洗装置	洗艙裝置	tank-cleaning plant
货油舱扫舱系统	貨油艙收艙系統	cargo oil tank stripping system
货油舱透气系统	貨油艙通氣系統	cargo oil tank venting system
货油舱洗舱设备	貨油艙清洗裝置	cargo oil tank cleaning installation

祖国大陆名	台湾地区名	英文名
货油舱油气驱除装置	貨油艙清除有害氣體裝置	cargo oil tank gas-freeing installation
货油阀(**液货阀**)	貨油閥	cargo oil valve
货油加热系统	貨油加熱系統	cargo oil heating system
货油控制室	貨油控制室	cargo oil control room
货油软管	貨油軟管	cargo oil hose
货油装卸系统	貨油裝卸系統	cargo oil pumping system
货油总管	貨油主管	main cargo oil line
货运	貨運	freight
货运代理人	轉運代理人	forwarding agent
货运合同	貨運契約	contract of affreightment
货运签约人	貨運簽約人	freight contractor
获得位置概率	獲得位置機率	location acquisition probability
获救财产价值	獲救財産價值	value of property salved

J

祖国大陆名	台湾地区名	英文名
机舱	輪機室,機艙	engine room
机舱污水井	機艙舭水	machinery space bilge
机舱应急舱底水阀	機艙應急舭水吸入閥	engine room emergency bilge suction valve
机舱照明系统	機艙照明系統	engine room lighting system
机舱自动化	機艙自動化	engine room automation
机动操纵	操縱	maneuver
机动船	動力船舶	power driven vessel
机动舰[船]	運轉船	maneuvering ship
机动救生艇	馬達救生艇	motor lifeboat
机帆船	機帆船	motor sailer
机工	機工	motor man
机架	框架	frame
机架定位	機架定位	positioning of engine frame
机炉舱(=机器处所)		
机旁控制	現場控制	local control
机器处所,机炉舱	機艙[空間],機器空間,機械室	machinery space
机体	引擎體	engine block
机械除锈	機械除銹	mechanical rust removal
机械传动	機械傳動	mechanical transmission

祖 国 大 陆 名	台 湾 地 区 名	英 文 名
机械滑车(**差动滑车**), 链条滑车	機械複滑車,鏈滑車	mechanical purchase, chain block
机械精加工	機械加工	machine finishing
机械喷射	無氣噴射	solid injection
机械气阀传动机构	機械致動閥機構	mechanically actuated valve mechanism
机械强度	機械強度	mechanical strength
机械式调速器	機械調速器	mechanical governor
机械式上支撑	機械式上支撐	mechanical top bracing
机械推进救生艇	機械推進救生艇	mechanically propelled lifeboat
机械[无气]喷射柴油 机	無氣噴射柴油機	solid-injection diesel
机械效率	機械效率	mechanical efficiency
机械杂质	機械雜質	mechanical impurities
机械增压	機械增壓	mechanical supercharging
机油老化	潤滑油老化	ageing of lubricating oil
机油稀释	潤滑油稀釋	lub-oil dilution
机组	單位	unit
机组有效功率	機組有效功率	unit effective power
机座	座板	bedplate
机座找平	定機座水平	leveling of engine bed
积差[率]	[累]積差率	accumulated rate
积分调节器	積分調整器	integral regulator
积分陀螺仪	積分迴轉儀	integrating gyroscope
积算船位	推算船位	dead reckoning position, DR
积碳	積碳	carbon deposit
积云	積雲	cumulus
积载图	貨物裝貨圖,裝載圖	stowage plan
积载因数	裝載因數	stowage factor
基本重复频率	基本重複頻率	basic repetition frequency
基本港	母港	basic port
基本恢复修理	修復	recovering repair
基本运费	基本運費	basic freight
基地电台	基[地]台	base station
基点	基點	base point, BP
基点风	四方位風	cardinal winds
基线	基線	baseline
基线误差	艏線誤差	lubber line error
基线延长线	基線延長(羅遠)	baseline extension

祖 国 大 陆 名	台 湾 地 区 名	英 文 名
基线延迟	基線遲延(羅遠)	baseline delay
基油	基油	base oil
基准舰	基準艦	datum ship
基准燃油低热值	基本燃料低熱值	fundamental fuel lower calorific value
基准纬度	基準緯度	standard parallel
激光测探仪	雷射測深儀	laser sounder
激励[振]频率	激振頻率	exciting frequency
1 级风	軟風(1 級)	light air
2 级风	輕風(2 級)	light breeze
3 级风	微風(3 級)	gentle breeze
4 级风	和風(4 級)	moderate breeze
5 级风	清勁風(5 級)	fresh breeze
6 级风	強風(6 級)	strong breeze
7 级风	疾風(7 級)	near gale
8 级风	大風(8 級)	gale
9 级风	烈風(9 級)	strong gale
10 级风	狂風(10 級)	storm
11 级风	暴風(11 級)	violent storm
12 级风	颶風(12 級)	hurricane
即期船	即期船	prompt ship
即期装船	即時裝載	prompt loading
极地冰	極冰	polar ice
极冠吸收	極冠吸收	polar cap absorption, PCA
极光	極光	aurora
极轨道卫星业务	繞極軌道衛星業務	polar orbiting satellite service
极好能见度(9 级)	能見度極佳(9 級)	visibility excellent
极化	極化	polarization
极化误差	極化誤差	polarization error
极距	極距	polar distance
极区航行	極區航行	polar navigation
极限的	極限的	ultimate
极限动倾角	最大動傾角	maximum angle of dynamic inclination
极限横倾力矩	最大傾側力矩	maximum heeling moment
极限开关	極限開關	proximity limit switch
极限强度	極限強度	ultimate strength
极限使用寿命	極限使用壽命	ultimate life
极限误差	極限誤差	limit error
极限重心高度	臨界重心高度	critical height of center of gravity

祖 国 大 陆 名	台 湾 地 区 名	英 文 名
极重要	極重要	vital
极坐标法	極坐標法	polar coordinate method
极坐标滑阀图	滑閥極坐標圖	polar slide valve diagram
急救	急救	medical first aid
急救医疗器具	急救醫藥用品	first-aid outfit
急流	急流	rapid stream
急闪光	快速閃光	quick flashing light
急运货	急運貨,急裝貨	distress cargo
棘轮装置	棘輪裝置	ratchet gear
集控室控制	主機控制室控制	engine control room control
集散式布风器	噴氣擴散器	air jet diffuser
集污舱	貯留艙	holding tank
集鱼灯	集魚燈	fishing lamp
集中操纵货油装卸系统	集中操縱貨油裝卸系統	centralized operation cargo oil pumping system
集中监测器	集中偵測器	centralized monitor
集中监视系统	中央監視系統	centralized monitoring system
集中式空气调节系统（**中央空调系统**）	中央空調系統	central air conditioning system
集装货	貨櫃裝載貨物	containerized cargo
集装箱	貨櫃	container
集装箱船	貨櫃船	container ship
集装箱导具	貨櫃導具	container guide fitting
集装箱吊架	貨櫃吊架	container lifting spreader
集装箱堆场	貨櫃場,貨櫃調度場	marshalling yard, container yard, CY
集装箱服务费	貨櫃服務費	container service charge
集装箱国家代号	貨櫃國碼	container country code
集装箱货运站	貨櫃集散站	container freight station, CFS
集装箱装箱单	貨櫃裝櫃圖	container load plan
集装箱[装卸]作业区	貨櫃終站基地	container terminal
辑私船	緝私船,緝私艇	revenue cutter
几何惯性导航系统	幾何慣性導航系統	geometric inertial navigation system
几何相似	幾何相似	geometrical similarity
脊(＝高压脊)		
脊线	脊線	ridge line
计程仪	計程儀	log
计程仪读数	計程儀讀數	log reading

祖 国 大 陆 名	台 湾 地 区 名	英 文 名
计程仪改正率	計程儀修正率	percentage of log correction
计程仪航程	測程儀航程	distance by log
计程仪航速	計程儀航速	speed by log
计费时间	計費時間	chargeable time
计划航迹向	預期航向	course of advance, CA
计划航速	前進速	speed of advance
计量泵	計量泵	metering pump
计算方位	計算方位	computed azimuth, calculated azimuth
计算风力力臂	計算風壓力臂	calculated wind pressure lever
计算风力力矩	計算風壓力矩	calculated wind pressure moment
计算高度	計算高度	calculated altitude
计算机辅助避碰	電腦輔助避碰	computer assisted collision avoidance
计算纬度	推算緯度	latitude by account
计算误差	計算誤差	error in calculation
计重货物	計重貨	weight cargo
记号(=标记)		
记名背书	記名背書	named endorsement
记名提单	記名載貨證券	straight bill of lading
技术规范	技術規範	technical specification
技术通报	技術通報	technical bulletin
季风	季節風,季風	monsoon
季节航路	季節性航路	seasonal route
季节期	季節期間	seasonal periods
季节性热带区域	季節性熱帶區域	seasonal tropical area
迹流	跡流	wake
继电器	繼電器	relay
继续	繼續操作	carry on
加标签	標籤	labeling
加负荷程序	加載方案	load-up program
加固(=加强)		
加强,加固	加強,補強	stiffening
加强肋	加強肋	reinforced rib
加热水倍率	加熱水率	heating water ratio
加热箱	加熱櫃	heating container
加热蒸汽	加熱蒸汽	heating steam
加湿器	給濕器	humidifier
加速计	加速計	accelerometer
加速燃气轮机	加力燃氣渦輪機	booster gas turbine

祖 国 大 陆 名	台 湾 地 区 名	英 文 名
加速性能	加速性能	acceleration performance
加速旋回	加速迴旋	acceleration turn
加压溶解气体	加壓溶解氣體	gases dissolved under pressure
加压式燃油系统	封密加壓燃油系統	closed and pressured fuel system
加油艇	加油艇	bowser boat
加油站	加燃油站	oil fuel filling station
加载性能	加載性能	loaded performance
岬角(岬)	岬角,岬	headland, cape
甲板	甲板	deck
甲板板	甲板板	deck plate
甲板边板	甲板緣板	deck stringer
甲板布置图	甲板位置圖,甲板佈置圖	deck plan
甲板部	艙面部門	deck department
甲板冲洗管系	甲板衝洗管路系統	deck washing piping system
甲板敷面	甲板被覆	deck covering
甲板货	艙面貨	deck cargo
甲板货油管系	甲板貨油管路	cargo oil deck pipe line
甲板机械	艙面機械	deck machinery
甲板进水角	甲板浸水角	angle of deck immersion
甲板列板	甲板列	deck strake
甲板漆	甲板漆	deck paint
甲板洒水系统	甲板灑水系統	deck sprinkler system
甲板室	甲板房艙	deck house
甲板水封	甲板水封	deck water seal
甲板水排泄管系	甲板水排洩管路系統	deck water piping system
甲板系索耳(=甲板羊角)		
甲板线	甲板線	deck line
甲板羊角,甲板系索耳	艙面繫索扣	deck cleat
甲板照明系统	甲板照明系統	deck lighting system
甲板纵骨	甲板縱材	deck longitudinal
甲板纵桁	甲板縱梁	deck girder
甲级分隔	"A"級區(防火)	A class division
假回波	假回波	false echo
假想出油量	假想油流出量	hypothetical outflow of oil
假信号	鬼信號	ghost signal
驾驶船舶过失	駕駛船舶過失	default in navigation of the ship

祖 国 大 陆 名	台 湾 地 区 名	英 文 名
驾驶和航行规则	操舵及航行规则	steering and sailing rules
驾驶台	駕駛台	bridge
驾驶台甲板	駕駛台甲板	bridge deck
驾驶台间通信	船橋間通信	bridge-to-bridge communication
驾驶台控制	指揮台操縱	bridge control
驾驶台遥控系统	駕駛台遙控系統	bridge remote control system
驾驶员	航行員	deck officer
驾助	助理船副	assistant officer
架板结	跳板結	plank stage hitch
架空电缆	架空電纜	overhead power cable
架空吊车	吊運車	trolley
尖端对接	錐尖相連	apexes together
尖端向上	錐尖向上	apex upwards
尖端向下	錐尖向下	apex downwards
尖头信号	跳波	blip
间接传动	間接傳動	indirect transmission
间接换装	間接換裝	indirect transshipment
间接回波	間接回波	indirect echo
间接冷却式空气冷却器	間接空氣冷卻器	indirect air cooler
监测	監測	monitoring
监测站	監視站	monitor station
监视屏	監測屏	monitoring panel
减额功率	降[低]額[定]馬力	derating
减负荷程序	減載方案	load-down program
减磨剂	抗磨劑	anti-friction composition
减速	減速	slacken speed, slow down
减速比	減速比,縮減比	reduction ratio
减速齿轮	減速齒輪	reduction gear
减温器	減溫器	attemperator
减压阀	減壓閥	pressure reducing valve
减摇泵	減搖泵	anti-roll pump
减摇控制设备	穩定器控制裝置	stabilizer control gear
减摇鳍装置	鰭板穩定器	fin stabilizer
减摇设备舱	穩定裝備室	stabilizer equipment room
减摇水舱	減搖水艙	anti-rolling tank
减载波发射	減載波發射	reduced carrier emission
减振联轴器	減振聯軸節	vibration-absorbing coupling
检查清单	檢查表	check-off list

祖 国 大 陆 名	台 湾 地 区 名	英 文 名
检查权	搜索權	right of search
检查日期	檢查日期	date of inspection
检修,拆卸检修	檢修,大修,翻修	overhaul
检验报告	檢驗報告	survey report
检疫	檢疫	quarantine
检疫船	檢疫船	quarantine vessel
检疫锚地	檢疫錨地	quarantine anchorage
检疫证书	檢疫完成	free pratique
简单循环	簡單循環	simple cycle
简式提单	簡式載貨證券	short form bill of lading
简易补给装置	暫時補給裝置	temporary replenishing rig
件杂货	零散雜貨	break bulk cargo
建造日期	建造日期	date of built
建造入级	建造入級	constructive classification
建造相应阶段	建造達類似階段	similar stage of construction
建造中船舶权利登记公约	建造中船舶權利登記公約	Convention Relating to Registration of Rights in Respect of Vessels under Construction
健康检查	健康檢查	medical examination
健身房	運動室	gymnasium
舰间间隔(船[舰]间横距)	船艦間橫距	beam distance between ships
舰间斜距	船艦間斜距	oblique distance between ships
舰间纵距	船艦間縱距	fore-and-aft distance between two ships
舰艇编队队形	艦艇編隊隊形	ship formation pattern
舰艇编队队形要素	船艦編隊隊形要素	elements of ship formation pattern
舰艇编队序列	艦艇編隊序列	order of ship formation
舰艇编队运动	艦艇編隊運動	ship formation movement
舰艇编队运动规则	船艦編隊運動規則	regulations for ship formation movement
舰艇编队转向	艦艇編隊轉向	ship formation course alteration
舰艇相遇圆	船艦相遇圓	ship's meeting circle
渐缩喷嘴	漸縮噴嘴	convergent nozzle
键槽	鍵槽	keyway
江心洲	江心洲	central island
将船撑开	使船橫著離開	breasting the ship apart
缆绳,系船索	叉索	bridle
桨	槳	oar
桨门	槳架	oarlock

祖 国 大 陆 名	台 湾 地 区 名	英 文 名
桨入水过深(出不了水面)	划空桨,桨入水過深	catch a crab
桨叶	輪葉	vane
降交点	降交點	descending node
降落伞信号	降落傘信號彈	parachute signal
降旗礼	低旗敬禮	dip
降凝剂	流動點下降劑	pour point depressant
"降速"按钮	"降速"按鈕	push button "down"
降雪	降雪	falling snow
降压起动	降壓起動	reduced-voltage starting
交变应力	交變應力	alternating stress
交叉轨迹角	交叉軌跡角	cross track angle
交叉曲线	交叉曲線	cross curve
交叉相遇局面	交叉相遇情況	crossing situation
交船	交船	delivery of vessel
交点月	交點月	draconitic month
交发日期	交發日期	filing date
交发时间	交發時間	filing time
交付	交貨	delivery
交会法	交會法	method of intersection
交会信号	信號漏失	spill-over signal
交流电	交流電,直線流	alternating current, rectilinear current
交流电力推进装置	交流電力推進裝置	A. C. electric propulsion plant
交流电站	交流電站	AC power station
交流三相三线制	交流三相三線制	AC three-phase three-wire system
交通	交通	communication
交通安全评估	交通安全評估	appraisal of traffic safety
交通管制区	交通管制區	traffic control zone
交通控制区	交通控制區	traffic control area
交通量	交通量	traffic volume
交通流	交通流量	traffic flow
交通密度	交通密度	traffic density
交通容量	交通容量	traffic capacity
交通艇	交通艇	traffic boat
角度传感器	角度感應器	angular position sensor
角阀	[折]角閥,肘閥	angle valve
角件	櫃角裝置	corner fitting

祖 国 大 陆 名	台 湾 地 区 名	英 文 名
角偏差	角偏差	angular misalignment
角宿一(室女 α)	角宿一(室女 α)	Spica
角形反射器	角形反射器,雷達波反射器	corner reflector
绞船索	拖索	warp
绞机	牽索絞機	hauling winch
绞缆滚筒	絞纜滾筒	gypsy
绞缆机	絞索絞機	gypsy winch
绞缆筒	捲索筒	warping head , warping end
绞缆移船	絞纜移船	warping the berth
1-1 绞辘	雙吊桿联合作業	double purchase
2-1 绞辘	2-1 辘轳	luff tackle
绞盘	絞盤,起錨機	capstan
铰接式装油塔	活節裝載塔	articulated loading tower
铰接塔系泊系统	活節塔繫泊系統	articulated tower mooring system
铰链连接	鉸鏈連接	link joint
铰链门	鉸鏈[式]門	hinged door
搅拌机	攪拌器	stirrer
叫号电话	叫號電話	station call
较强涨潮(一日中)	較強漲潮(一日中)	lesser flood
较弱落潮(一日中)	較弱落潮(一日中)	lesser ebb
校正龄期	校正齡期	age of correction
校直	鑿榫	tabling
校准	校準	calibration
校准杆	校準桿	calibrating lever
校准角(=看齐角)		
教学海图	教學海圖	instructional chart
阶梯形活塞	階段活塞,塔形活塞	stepped piston
阶跃操舵	階躍操舵	step steering
阶跃输入	分段輸入	step input
接地检查灯	接地檢查燈	ground detecting lamp
接收机	接收機	receiver
接收机灵敏度	接收機靈敏度	sensitivity of a receiver
接收机选择性	接收機選擇性	selectivity of a receiver
接收静止搜救卫星无效报警	接收無效之定置搜救衛星警報	GEOSAR invalid alert received
接收天线	接收天線	receiving antenna
接收站	接收電台,接收站	receiving station

祖国大陆名	台湾地区名	英文名
接受船	受補船	receiving ship
节	節	knot, kn.
节流	節流	throttle
节流阀	節流閥	throttle valve
节流调节	節流調速	throttle governing
节制闸灯	調整閘燈	regulating lock light
杰森条款	詹森條款(超額條款)	Jason clause
结冰	結冰	icing
结点任务控制中心	結點任務管制中心	nodal MCC
结构	結構,構造	structure
结关	結關,出港許可	clearance
结关单	出口結關證書,出航許可證	clearance certificate
结关清单	出口結關單,出港證書	bill of clearance
结胶	膠結	gumming
结[节]点	波節	node
捷水道	捷徑航路	short-cut route
截止阀	停止閥	stop valve
解开	解開	cast off
解扩	解擴散	de-spread
解码(=译码)		
解调	解調變	demodulation
解析式惯性导航系统	解析慣性航行系統	analytic inertial navigation system
解约	解約	canceling
解约日	解約日	canceling date
界面	界面,介面	interface
界限标	界限標	limit mark
金相检查	金相檢查	metallographic inspection
金星	金星	Venus
金属救生艇	金屬救生艇	metal lifeboat
金属片状粗滤器	金屬層片過濾器	metal-edge type strainer
金属蠕变	金屬潛變	creep of metal
襟翼舵	襟翼舵	flap-type rudder
紧固螺栓	緊固螺栓	fastening bolt
紧急	緊急	emergency, urgency
紧急报告	緊急信文	urgency message
紧急倒车	緊急倒俥	back emergency
紧急倒车冲程	緊急停俥距離	crash stopping distance

祖 国 大 陆 名	台 湾 地 区 名	英 文 名
紧急的	緊急的	urgent
紧急断缆工具	緊急斷纜工具	emergency breakaway tools
紧急关闭(=紧急停堆)		
紧急航行危险报告	緊急航行危險報告	urgent navigational danger report
紧急呼叫	緊急呼叫	emergency call, urgency call
紧急呼叫格式	緊急呼叫格式	urgency call format
紧急回转	緊急迴轉	emergency turn
紧急阶段	緊急階段	emergency phase
紧急警报	緊急警報	emergency warning
紧急气象危险报告	緊急氣象危險報告	urgent meteorological danger report
紧急强制用车	緊急運作	emergency overriding
紧急刹车	緊急軔,緊急煞俥	emergency brake
紧急示位发信机	應急示位發射機	emergency locator transmitter, ELT
紧急停车按钮	緊急停俥按鈕	emergency stop push button
紧急停堆,紧急关闭	緊急關閉	emergency shut-down
紧急通信	緊急通信	urgency communication
紧急通信程序	緊急通信程序	urgency communication procedure
紧急脱离	緊急脫離	breakaway emergency
紧急无线电示位标	應急指位無線電示標	emergency position-indicating radio beacon, EPIRB
紧急无线电示位标识别	應急指位元無線電示標識別	EPIRB identification
紧急危险(**紧迫危险**)	立即危險	immediate danger
紧急响应工作队	緊急應對工作隊	emergency response team
紧急信号	緊急信號	urgency signal
紧急优先等级	緊急優先順序	urgency priority
紧配合	緊配合	tight fit
紧迫局面	彼此接近	close quarters situation
紧转配合	緊轉配合	close running fit
紧追权	緊追權	right of hot pursuit
谨慎处理	必要的注意	due diligence
近岸流	近岸流	inshore current
近岸移动式钻井装置	可動式離岸鑽探平台	mobile offshore drilling unit, MODU
近地点	月近點	perigee
近点角	近點角	anomaly
近点月	近點月	anomalistic month
近海测量	海域測量	offshore survey
近海岛屿	外圍島嶼,離島	offshore islands

祖 国 大 陆 名	台 湾 地 区 名	英 文 名
近海航行	近海航行	offshore navigation
近海区	離島地區	offshore area
近海渔业	近海漁業	offshore fishery
近海钻井作业	近海鑽探作業	offshore drilling operation
近日点	近日點	perihelion
近日点前移	近日點前移	advance of perihelion
近艉(=[在]后)		
近星点前移	近星點前移	advance of periastron
近因	近因	proximate cause
近中天	近中天	ex-meridian
近中天高度改正	折合中天高度	reduction to the meridian
进倒车掉头	進退俥短迴轉掉頭	turning short round by ahead and astern engine
进口关单	進口[報]關單,進港證書	bill of entry
进口许可	輸入許可	import permit
进气阀	進氣閥	inlet valve
进气壳体	進氣殼體	air intake casing
进气提前角	進氣提前角	inlet advance crank angle
进气消音器	進氣消音器	intake silencer
进气滞后角	進氣滯後角	inlet lag crank angle
进气装置	進氣裝置	air inlet unit
进汽度	進汽度	degree of admission
进水角	泛水角	flooding angle
进水速度	泛水速度	speed of flooding
进坞	進塢	docking
进坞操纵	進塢操縱	docking maneuver
浸没式蒸发器	浸沒式蒸發器	flooded evaporator
浸水	水浸	flood
禁航区	禁航區	forbidden zone
禁渔期	禁漁期	[fishing] closed season
禁渔区	禁漁區	forbidden fishing zone
禁运	禁運	embargo
禁止船舶通航	禁止船舶通航	embargo on ship
禁止锚泊标	禁泊標誌	anchor prohibited mark
经差	經差	difference of longitude
经度改正量	經度修正值	longitude correction
经济工况	經濟工作狀況	economic working condition

祖 国 大 陆 名	台 湾 地 区 名	英 文 名
经济功率	經濟功率	economical power
经济航速(＝经济速度)		
经济器	節熱器	economizer
经济速度,经济航速	經濟速[率]	economical speed
经纬仪	經緯儀	altazimuth
晶体管	電晶體	transistor
精度	精度	precision accuracy
精对准	精校準	fine alignment
精密定位业务	精密定位業務	precise positioning service, PPS
精密星历	精密天文曆	precise ephemeris
精致货	精細貨	delicate cargo
警告	警告	warning
警告信号	警告信號	warning signal
警戒区	警戒區	precautionary area
净吨位	淨噸位	net tonnage, NT
净化系统	淨化系統	purification system
净化装置	淨化裝置	refining plant
净空高度	淨空高度	air draft, height clearance
净区	淨區	clear zone
净压头	有效落差	effective head
净正吸高	淨正吸入高	net positive suction height
净正吸入压头	淨正吸入水頭	net positive suction head
径流式涡轮	徑向流渦輪機	radial-flow turbine
径流式涡轮增压器	徑向流渦輪增壓器	radial flow turbocharger
径流式压缩机	徑向流壓縮機	radial-flow compressor
径向球轴承	徑向滾珠軸承	radial ball bearing
径向柱塞式液压马达	徑向活塞液力馬達	radial-piston hydraulic motor
静电放电	靜電放電	electrostatic discharge
静航向稳定性	靜航向穩定性	static course stability
静横倾角	靜橫傾角	static heeling angle
静默时间	靜默時間	silence period, SP
静配合	靜配合	stationary fit
静平衡	靜力平衡	static balance
静水力曲线图	靜水[性能]曲線圖	hydrostatic curves plan
静水弯矩	靜水彎曲力矩	still water bending moment
静水压力释放器	靜力釋放裝置	hydrostatic release unit
静索	固定索具,靜索	standing rigging
静态	靜態	static state

祖国大陆名	台湾地区名	英文名
静态定位	静態定位	static positioning
静稳性	静穩度	statical stability
静稳性曲线	静穩度曲線	curve of statical stability
静压调节器	静壓力調整器	static pressure regulator
静压头	静落差	static head
静叶片	固定葉片	stationary blade
静噪	消雜音(静音)	squelch
静止地球轨道	定置軌道,地球同步軌道	geostationary earth orbit
静止锋	滞留鋒	stationary front
静止气象卫星	定置氣象衛星	geostationary meteorological satellite
静止搜救卫星通信率	定置搜救衛星通訊比率	GEOSAR traffic ratio
静止搜救卫星系统	定置搜救衛星系統	geostationary SAR satellite system
静止卫星	定置衛星	geostationary satellite, stationary satellite
静止卫星报警通报时间	定置衛星警報通報時間	time of GEOSAR alert notification
静止卫星轨道	定置衛星軌道	geostationary satellite orbit
静止卫星通信容量	定置衛星通訊之容量	volume of GEOSAR traffic
静止卫星业务	定置衛星業務	geostationary satellite service
静止运行环境卫星	定置運作環境衛星	geostationary operational environmental satellite
救捞船	救撈船	salvage ship
救生部署表	救生部署表	boat station bill
救生担架	救生擔架	rescue litter
救生吊带	救生吊帶	rescue sling
救生吊蓝	救生吊籃	rescue basket
救生吊座	救生吊座	rescue seat
救生筏	救生筏	life raft
救生服	浸水衣	immersion suit
救生浮	救生浮具	life float
救生浮具浮力试验	救生浮具浮力試驗	buoyancy test for buoyant apparatus
救生浮索	救生浮索	buoyant lifeline
救生圈	救生圈	lifebuoy
救生圈试验	救生圈浮力試驗	buoyancy test for lifebuoy
救生设备	救生器具	life saving appliance
救生设备配备	救生設備配備	carriage of life saving appliances on board
救生手册	求生手冊	survival manual
救生索	扶手索,救生索	life line
救生艇	救生艇	lifeboat

祖 国 大 陆 名	台 湾 地 区 名	英 文 名
救生艇乘员定额	救生艇容载量	carrying capacity of lifeboat
救生艇筏	救生艇筏	survival craft
救生艇筏回收装置	救生艇筏回收装置	survival craft recovery arrangement
救生艇筏手提无线电设备	艇用輕便無線電設備	portable radio apparatus for survival craft
救生艇甲板	救生艇甲板	lifeboat deck
救生艇罗经	救生艇用羅經	lifeboat compass
救生艇试验	救生艇試驗	test for lifeboat
救生艇无线电报设备	救生艇無線電報裝置	radiotelegraph installation for lifeboat
救生网	救生網	rescue net
救生信号	救生信號	life saving signal
救生衣	救生衣	lifejacket
救生衣试验	救生衣浮力試驗	buoyancy test for life-jacket
救生属具	救生設備	life equipment
救助报酬	救助報酬	salvage remuneration
救助报酬请求	救助求償	claim for salvage
救助泵	救助泵	salvage pump
救助单位	救助單位	rescue unit, RU
救助分中心	救助分中心	rescue sub-center, RSC
救助人	施救者	salvor
救助艇	救難艇	rescue boat
救助义务	救助義務	obligation to render salvage service
居间障碍物	居間障礙物	intervening obstruction
居住舱,房舱	房艙,艙間	cabin
拘留权	扣押權	right of seizure
局部比例尺	局部比例尺	local scale
局部强度	局部強度	local strength
局部应力	局部應力	local stress
局域网络	局部區域網路	local area network
举证责任	舉證責任	onus of proof, burden of proof
矩阵信号	矩陣信號	matrix signal
巨浪(6级)	極大浪(6级)	very rough sea
距变率	距離變動率	rate of distance variation
距离差位置线	距離差位置線	position line by distance difference
距离定位	距離定位	fixing by distances
距离分辨力	距離分解[度]	range resolution
距离索	距離索	distance line
距离位置线	距離位置線	position line by distance

祖 国 大 陆 名	台 湾 地 区 名	英 文 名
卷(=折)		
卷层云	卷層雲	cirrostratus
卷帆索	捲帆索	brails
卷纲机	捲繩機	rope reel
卷积云	卷積雲	cirrocumulus
卷绳车	儲放捲盤	stowage reel
卷筒	捲筒(起重機)	drum
卷网机	捲網機	net drum
卷云	卷雲	cirrus
绝对辐射热星等	絕對熱星等	absolute bolometric magnitude
绝对计程仪	絕對測程儀	absolute log
绝对免赔额	自負額	deductible
绝对湿度	絕對濕度	absolute humidity
绝对视差	絕對視差	absolute parallax
绝对速度	絕對速度	absolute velocity
绝对温标	熱力學溫度標	Kelvin scale
绝对温度	絕對溫度	absolute temperature
绝对压力	絕對壓力	absolute pressure
绝对延迟	絕對遲延	absolute delay
绝热材料	絕熱材料	heat insulating material
绝热过程	絕熱過程	adiabatic process
绝缘材料	絕緣材料	insulating material
军船	軍艦	naval ship, warship
军事航海	軍事航海	military navigation
均功调节	均衡調節	equalizing regulation
均匀的	均匀的	uniform
均质货	匀質貨	homogeneous cargo
均质器	均質機	homogenizer
菌形锚	菌形錨	mushroom anchor
菌形通风筒	菌型通風筒	mushroom ventilator

K

祖 国 大 陆 名	台 湾 地 区 名	英 文 名
开杯试验	開杯法試驗	open cup test
开船旗	開船旗	blue peter
开底泥驳	開底泥駁	hooper barge
开底挖泥船	斗式挖泥船	hooper dredger

祖 国 大 陆 名	台 湾 地 区 名	英 文 名
开放网络	開放網絡	open network
开关	開關	switch
开罐器	開罐器	tin opener
开航权	航行權	right for sailing
开航日期	啓航日期	date of departure
开环系统	開環系統	open-loop system
开口滑车	活口滑車,開口滑車	snatch block
开口销	開口銷	split pin
开锚	離岸錨	offshore anchor
开式冷却水系统	開式冷卻水系統	open cooling water system
开式喷油器	開式燃油閥	open type fuel valve
开式燃烧室	開式燃燒室	open combustion chamber
开式循环燃气轮机	開口循環式燃氣輪機	open cycle gas turbine
开式液压系统	開式液壓系統	open type hydraulic system
开尾销	開口銷	cotter pin
看齐角，校准角	校準角	aligning angle
康索尔	康蘇(電子航海儀)	Consol
康索兰	康蘇蘭	Consolan
抗冰加强	抗冰加強	ice strengthening
抗拉强度	抗拉強度	tensile strength
抗磨剂	抗磨劑	anti-wear agent
抗乳化度	脫乳化數	demulsification number
抗弯强度	彎曲強度	bending strength
抗氧化剂	抗氧化劑	oxidation inhibitor
抗氧化抗腐蚀剂	抗氧化抗腐蝕劑	anti-oxidant anti-corrosion additive
靠把(=碰垫)		
靠泊	靠泊	alongside
靠泊表	靠泊表	parking meter
靠近加油装置	靠近加油裝置	close-in fueling rig
靠码头	靠[泊]碼頭	alongside berth wharf
科里奥利力	科氏力,自轉偏向力	Coriolis force
颗粒收集器	顆粒收集器	particulate trap
颗料排放物	排放顆粒	particulate emission
壳管式冷凝器	殼管式冷凝器	shell and tube condenser
壳管式热交换器	殼管式交換器	shell-and-tube heat exchanger
可编程只读存储器	可程式唯讀記憶體	programmable read only memory
可变的	變數	variable
可变换的集装箱船	可變換貨櫃船	convertible container ship

祖 国 大 陆 名	台 湾 地 区 名	英 文 名
可变排气阀关闭机构	可變排氣閥關閉機構	variable exhaust valve closing device
可变喷油正时机构	可變噴油定時機構	variable injection timing mechanism
可变压缩比	可變壓縮比	variable compression ratio
可拆链环	拆合環	detachable link
可倒转柴油机	可逆轉柴油機	reversible diesel engine
可航半圆	可航半圓	navigable semicircle
可航性	可航性	navigableness
可互换部件	可互換配件	interchangeable parts
可见的	可見的	visible
可见痕迹	可見痕迹	visible trace
可浸长度	可浸長度	floodable length
可靠性	可靠性	reliability
可控被动水舱式减摇装置	可控被動水艙穩定系統	controllable passive tank stabilization system
可控硅变流机组	閘流體換流器/組	thyristor converter set
可控硅励磁系统	閘流體勵磁系統	thyristor excited system
可控相复励磁系统	可控相之補償複激系統	controllable phase compensation compound excited system
可控自励恒压装置	可控自激等電壓設施	controllable self-excited constant voltage device
可能碰撞点	可能碰撞點	possible point of collision, PPC
可燃毒物元件	可燃有毒元素	burnable poison element
可听距离	聞距	range of audibility
可调螺距桨	可控距螺槳	controllable pitch propeller, CPP
可调螺距桨控制系统	可控距螺槳控制系統	controllable pitch propeller control system, CPP control system
可调叶片	可調葉片	adjustable vane
可卸螺旋桨叶	可卸螺槳片	loose propeller blade
可卸硬件	可卸艤品	loose hardware
可压缩性	可壓性	compressibility
可移式风机	可攜式風扇	portable fan
可折顶篷	可摺頂篷	foldable canopy
刻度板	刻度板	scale plate
刻度防锈饮水杯	有刻度之防銹飲器	rustproof graduated drinking vessel
客舱旅客	房艙乘客	cabin passenger
客船	客船	passenger ship
客船安全证书	客船安全證書	passenger ship safety certificate
客货船	客貨船	passenger-cargo ship

祖 国 大 陆 名	台 湾 地 区 名	英 文 名
客票	客票	passenger ticket
客厅	大餐廳,客廳	saloon
空白背书	空白背書	endorsement in blank
空白定位图	作業圖	plotting chart
空白提单	空白載貨證券	open bill of lading
空舱运费	空艙運費	deadfreight
空船排水量(**空载排水量**)	輕載排水量	light displacement
空隔舱(=围堰)		
空间段	太空段	space segment
空间段提供者	太空部分提供者	space segment provider
空间跟踪	太空追蹤	space tracking
空间无线电通信	太空無線電通信	space radiocommunication
空间系统	太空系統	space system
空间遥令	太空遙令	space telecommand
空间站	太空站	space station
空距	隙尺	ullage
空距尺	油面計	ullage scale
空泡腐蚀(=气蚀)		
空气分配器	空氣分配器	air distributor
空气浮力修正系数	空氣浮力修正係數	air floatation correction coefficient
空气加热器	空氣加熱器	air heater
空气冷却器	空氣冷卻器	air cooler
空气马达	氣動馬達	air motor
空气喷射式	空氣噴射式	air injection type
空气调节装置蒸发器(**空调蒸发器**)	空調器蒸發器	air conditioning evaporator
空气消耗率	單位耗氣量	specific air consumption
空气压缩机	空氣壓縮機	air compressor
空气预热器	空氣預熱器	air preheater
空气阻力	空氣阻力	air resistance
空燃比	空氣燃油比	air-fuel ratio
空投设备	空投設備	droppable equipment
空位	空艙	void space
空压机自动控制	空壓機自動控制	air compressor auto-control
空载	無載	no load
空载试验	空載試驗,無負載試驗	no-load test
空载状态	輕載船況	light condition

祖国大陆名	台湾地区名	英文名
空中探鱼	空中探測魚群,飛行探測魚群	aerial scouting
空转转速	惰速[率]	idle speed
孔雀(孔雀α)	孔雀十一(孔雀α)	Peacock
控制棒	控制桿	control rod
控制棒导管	操縱桿導管	control rod guide tube
控制棒驱动机构	操縱桿驅動機構	control rod drive mechanism
控制部位转换开关	控制站換向開關	control station change-over switch
控制点	控制點	control point
控制阀(=导阀)		
控制室操纵屏	控制室操縱板	control room maneuvering panel
控制塔	控制塔	control tower
控制台,仪表板	控制台,電子儀器座	console
控制涡流式[直流]扫气	控制漩流趨氣	controlled swirl scavenging
控制用空气压缩机	控制用空[氣]壓[縮]機	control air compressor
控制站	控制站	control station
口粮	口糧	food ration
扣船	扣留船舶	detention of ship
扣留	扣船	embargo
扣押船舶	扣押船舶	arrest of ship
跨索绞车	跨索絞機	spanwire winch
跨索系固眼板	跨索儲放眼板	spanwire storage padeye
跨运车	跨載機	straddle carrier
快速接头	快釋接頭	quick release coupling
快速稳定装置	快速固定設施	fast settling device
快速性	快速性	speedability
宽深比	寬深比	breadth depth ratio
宽限期	寬限期	period of grace
宽限日期	寬限日期	days of grace
狂浪(7级)	高浪	high sea
狂涛(8级)	浪甚高	very high sea
矿砂船	礦砂船	ore carrier
矿砂–石油船	礦砂與油兼用船	ore-oil carrier
矿物油	礦物油	mineral oil
框架误差	水平環誤差	gimballing error
框架箱	框架貨櫃	skeletal container

祖 国 大 陆 名	台 湾 地 区 名	英 文 名
亏舱	堆貨餘隙,貨載空隙	broken stowage, broken space
亏舱率	堆貨餘隙率	ratio of broken space
馈线链路	饋線鏈路	feeder link
捆包货	綑包貨	baled cargo
捆扎	捆紮	strapping
扩频信号	擴頻譜信號	spread spectrum signal
扩压器	擴散器	diffuser
扩展方形搜寻	擴展方形搜索	expanding square search
括纲	收縮網	purse line

L

祖 国 大 陆 名	台 湾 地 区 名	英 文 名
垃圾	垃圾	garbage
垃圾船	垃圾船	garbage boat
垃圾倾倒区	垃圾傾倒區	dumping ground
拉出	拉出錨鏈	rouse out
拉出绞车	拉出絞機	outhaul winch
拉到顶	滿懸	close up
拉缸	氣缸刮削	cylinder scraping
拉紧索具	拉緊索具	set up rigging
拉进	拉進錨鏈	rouse in
拉进绞车	拉進絞機	inhaul winch
拉开	曳開	haul away
拉条(=支索)		
来自信号,我是……	本台,是(無線電話用語)	Delta Echo, DE
兰氏平衡器(往复惯性力矩平衡器)	蘭氏均衡器	Lanchester balancer
拦索	攔索,吊貨控索	bull-rope
栏杆	欄桿	guard rail
篮子[号型]	籃形號標	basket [shape]
缆	纜,大索,大纜	hawser
揽货	攬貨	solicitation
[缆]回松,回升	緩緩放鬆	come up
缆绳周径	纜繩週長	circumference of rope
缆桩	雙繫纜椿	bollard, bitts
浪花	碎浪	breakers

祖 国 大 陆 名	台 湾 地 区 名	英 文 名
浪击落水	浪擊落水	washed overboard
浪级	波浪等級	wave scale
浪损	浪損	damage caused by waves
捞缆钩	撈纜機	cable grapnel
劳氏海事周报	勞氏海事週報	Lloyd's Weekly Casualty Report
老化	老化	aging
老人(船底α)	老人(船底α)	Canopus
[雷达避碰]试操纵	試操縱	trial maneuvering
雷达标绘	雷達測繪	radar plotting
雷达导航	雷達導航	radar navigation
雷达反射器	雷達反射器	radar reflector
雷达海图	雷達海圖	radar chart
雷达回波箱	雷達回波箱	radar echo-box
雷达模拟器	雷達模擬機	radar simulator
雷达识别	雷達識別	radar identification
雷达桅	雷達桅	radar mast
雷达信标	雷達信標	radar beacon, racon
雷达性能监视器	雷達性能監測器	radar performance monitor
雷达引航图	雷達導航圖	radar navigation chart
雷达应答器	雷達詢答器	radar transponder
雷达指向标	雷達航標	ramark
雷达最大作用距离	雷達最大效程	maximum radar range
雷达最小作用距离	雷達最小效程	minimum radar range
雷德蒸汽压力	瑞德蒸氣壓	Reid vapor pressure
雷光	雷光	thunderlight
雷诺数	雷諾數	Renold's number
雷氏黏度	雷氏黏度	Redwood viscosity
雷雨	雷雨	thunder storm
雷阵雨	雷陣雨	thunder-shower
肋板	底肋板	floor
肋骨	肋骨	frame
肋骨号数	肋骨編號	frame number
肋距	肋骨間距	frame space
肋片式蒸发器	鰭面蒸發器	finned-surface evaporator
A 类机舱(=A 类机器处所)		
A 类机器处所,A 类机舱	甲種機艙空間	machinery space of category A

祖 国 大 陆 名	台 湾 地 区 名	英 文 名
A类有毒液体物质	A 類有毒液體物質	category A noxious liquid substance
B类有毒液体物质	B 類有毒液體物質	category B noxious liquid substance
C类有毒液体物质	C 類有毒液體物質	category C noxious liquid substance
D类有毒液体物质	D 類有毒液體物質	category D noxious liquid substance
棱形系数	[縱向]稜塊係數,稜形係數	prismatic coefficient
冷藏船	冷凍船,冷藏船	refrigerator ship
冷藏货	冷藏貨	refrigeration cargo
冷藏货舱	冷凍貨艙	refrigerated cargo hold
冷藏货条款	冷凍貨條款	refrigerated cargo clause
冷藏间	冷凍室	refrigerated space, refrigerated room
冷藏箱	冷凍貨櫃	reefer container
冷吹运行	冷吹操作	cold blow-off operation
冷冻货	冷凍貨	frozen cargo
冷冻机油	冷凍油	refrigerator oil
冷冻间	冷凍室	refrigerated chamber
冷风机	冷氣機	air cooling machine
冷锋	冷鋒	cold front
冷高压	冷高壓	cold high
冷剂泵	冷媒泵	refrigerating medium pump
冷凝器真空度	冷凝真空	condenser vacuum
冷平流	冷平流	cold advection
冷气团	冷氣團	cold air mass
冷却法	冷卻法	cooling method
冷却水倍率	冷卻水率	cooling water ratio
冷却系统	冷卻系統	cooling system
冷水事故	冷水事故	cold-coolant accident
冷缩配合	收縮配合	shrinkage fit
冷态起动	冷溫起動	cold starting
冷停堆	冷關閉	cold shut-down
离岸(=出发)		
离岸价格	出口地船上交貨	free on board, FOB
离岸流	急浪流	rip current
离泊	離泊	clearing from alongside, unberthing
离风	離風	off the wind
离浮筒	離浮筒	clearing from buoy
离港出海	出海	stand out to sea
离合机构	離合機構	engaging and disengaging gear

祖 国 大 陆 名	台 湾 地 区 名	英 文 名
离码头	離碼頭	leaving wharf
离析	析離	segregation
离心泵	離心泵	centrifugal pump
[离心]分油机	油水離心分離器	centrifugal oil separator
离心惯性力矩	離心慣性力矩	centrifugal inertia moment
离心式压气机	離心壓縮機	centrifugal compressor
离心式制冷压缩机	離心冷凍壓縮機	centrifugal refrigerating compressor
里程表	浬程表	distance table
理货	理貨	tally
理货员	理貨員,記算員	tallyman
理论热循环	理論熱循環	theoretical heat cycle
理网机	理網機	net shifter
理想气体	理想氣體	perfect gas
理想循环	理想循環	ideal cycle
力矩平衡器	力矩調整器	moment compensator
力矩器	扭矩器	torquer
力锚	著力錨	riding anchor
历书	曆書,航海曆,天文曆	almanac
立标	標桿	beacon
立标塔	指標塔	beacon tower
立柜式空气调节器	自給式空調	self-contained air conditioner
立桨	舉槳	toss oars
立式的	立式	vertical
沥青	瀝青炭	pitch
沥青分	瀝青含量	asphaltenes content
沥青清漆	瀝青溶液	bituminous solution
例行复述	例行複述	routine repetition
例行呼叫	例行呼叫	routine calls
粒度	單位尺度	unit size
连带责任	連帶責任	joint and several liability
连动脱钩装置	連動脫鉤裝置	simultaneous disengaging unit
连杆	連桿	connecting-rod
连接补给	連接整補	connected replenishment
连接公海	與公海相通	connected with the high seas
连接件	連接裝具	connecting fitting
连接缆	連接線	connecting line
连接链环	連接鏈環	joining link, connecting link
连接卸扣	連接接環	connecting shackle, joining shackle

祖 国 大 陆 名	台 湾 地 区 名	英 文 名
连续日	連續自然日	running days, consecutive days
连续输出功率	額定連續常用出力	continuous service rating
连续值守	連續守值	continuous watch
联暗光	聯頓光	group occulting light
联邦无线电导航计划	聯邦無線電導航計劃	Federal Radionavigation Plan, FRP
联动补给装置	聯動補給裝置	housefall rig
联动脱钩装置	聯動脫鉤裝置	simultaneous disengaging gear
联杆操作补给装置	聯桿操作補給裝置	burton rig
联合船位	混合定位	combined fix, CF
联合国编号	聯合國編號	UN number
联合国船舶登记条件公约	船舶登記條件聯合國公約	United Nations Convention on Conditions for Registration of Ships
联合国国际货物多式联运公约	國際貨物多式聯運聯合國公約	United Nations Convention on International Multimodal Transport of Goods
联合国国际贸易运输港站经营人赔偿责任公约	國際貿易終端站營運人責任聯合國公約	United Nations Convention on the Liability of Operators of Terminals in International Trade
联合国海洋法公约	聯合國海洋法公約	United Nations Convention on the Law of the Sea
联合救助协调中心	聯合救助協調中心	associated rescue coordination center
联检	聯檢	joint inspection
联闪光	聯閃光	group flashing light
联运货	聯運貨	through transport cargo, through cargo
联运提单	聯運載貨證券	through bill of lading
练习生(**实习生**),学徒	實習生,學徒	apprentice
炼制品	精煉品	refined products
链缠锚	錨障	foul anchor
链(长度单位=1/10海里)	鏈(長度單位)	cab.
链传动	鏈傳動	chain drive
链斗装置	鏈斗裝置	bucket arrangement
链节	(錨鏈)節	shot
链轮	鏈輪	gypsy wheel, sprocket
链条滑车(=机械滑车)		
链条张紧机构	緊鏈器	chain tightener
链抓力	鏈抓著力	holding power of chain
良好船艺	良好船藝	good seamanship
良好能见度(8级)	能見度甚佳(8級)	visibility very good

祖 国 大 陆 名	台 湾 地 区 名	英 文 名
梁拱	弧高,拱高	camber
粮食库	糧食庫	provision room
两半角[转向]法	兩半角轉向法	method of altering course by two half-angles
两船间距	兩船間距	ship clearance
两级喷射系统	兩級噴射系統	two-step injection system
两脚插头	兩腳插塞	two-pin plug
量杆	量桿	metering rod
量隙规,塞尺	餘隙規	clearance gauge
瞭头	在艏瞭望	look-out on forecastle
瞭望	瞭望	look-out
列位	槽	slot
劣质燃料	劣質燃料	inferior fuel
猎户座	獵戶座	orion
猎雷舰	獵雷艦	mine hunter
猎潜艇	驅潛艇	submarine chaser
裂变产物	核子分裂産物,分裂産物	fission product
裂变能	分裂能	fission energy
裂变中子	分裂中子	fission neutron
裂冰	冰崩	calving
裂化燃料油	裂化燃料油	cracked fuel oil
邻角	鄰角	adjacent angle
邻图索引	鄰接海圖索引	index of adjoining chart
临检权	臨檢搜索權	right of visit and search
临界初稳性高度	臨界初定傾[中心]高度	critical initial metacentric height
临界倾覆力臂	臨界翻覆力臂	critical capsizing lever
临界倾覆力矩	臨界翻覆力矩	critical capsizing moment
临界实验	臨界試驗	criticality test
临界速度	臨界速度	critical speed
临界舷角	臨界相對方位	critical relative bearing
临界压力比	臨界壓力比	critical pressure ratio
临时检验	臨時檢查,臨時檢驗	occasional survey
临时通告	臨時通告	temporary notice
临时无灯浮标	臨時無燈浮標	temporary unlighted buoy
临时无线电信标业务	臨時無線電示標業務	temporary radio beacon service
临时修理	臨時修理	temporary repair

祖 国 大 陆 名	台 湾 地 区 名	英 文 名
临时许可	臨時放行	provisional release
灵敏部分	敏感元件	sensitive element
灵敏度	靈敏度	sensibility
凌波性	凌波性	seakindliness
菱形编队	菱形編隊	diamond formation
菱形号型	菱形號標	diamond [shape]
零	零	zero
零磁差线	無磁偏線	agonic line
零点误差	零點誤差	zero error
零功率实验	零功率實驗	zero-power experiment
零件明细表	零件明細表	parts list
零升力	無升力	zero lift
零升力角	零升力攻角	zero-lift angle
零时基点	零時基準點	zero time reference
零位调节	零位調整	zero adjustment
零下温度	零下溫度	subzero temperature
领班(=装卸长)		
领海	領海	territorial water, territorial sea
领海基线	領海基線	baseline of territorial sea
领事签证发票	領事簽證發票	consular invoice
另有相反的规定除外	除另有明文規定外	except otherwise herein provided
流冰	流冰	drift ice
流出物	流出物	effluent
流刺网	流網	drift net
流刺网振网机	流刺網振網機	drift net shaker
流动水分点	注動水分點	flow moisture point
流量表	流量計	flowmeter
流量控制阀	流量控制閥	flow-control valve
流量调节器	流量調整器	flow regulator
流网作业	流網捕魚	drift fishing
流线	流線	streamline
流线型舵	流線型舵	streamline rudder
流向	流向	direction of current
流压差(=漂角)		
留置权	留置權	right of lien, lien
硫分	硫含量	sulfur content
六分仪	六分儀	sextant
六分仪高度	六分儀高度	sextant altitude

祖 国 大 陆 名	台 湾 地 区 名	英 文 名
六分仪交会法	六分儀交會法	method of intersection by sextant
[六分仪]器差	儀器差,儀器誤差	instrument error
六分仪误差	六分儀誤差	sextant error
六分仪校正	六分儀校正	sextant adjustment
[六分仪]指标杆	[六分儀]指標桿	index arm [of sextant]
龙门桅	龍門柱	goal post
娄宿三(白羊 α)	婁宿三(白羊 α)	Hamal
娄宿一(白羊 β)	婁宿一(白羊 β)	Sheratan
漏气	漏气	blow-by
漏损和破损	漏損和破損	leakage and breakage
漏损检测	漏損檢查	examination of leakage and breakage
露出(＝出现)		
露点[温度]	露點	dew-point [temperature]
露天甲板	露天甲板	open deck
露天甲板空间	露天甲板空間	open deck space
炉膛	爐膛	furnace
卤化物灭火系统	海龍滅火系統	halon fire extinguishing system
陆标	岸標,岸上目標	landmark
陆标船位	地文定位	terrestrial fix, TF
陆标定位	岸標定位	fixing by landmark
陆地地球站	衛星陸上電台	land earth station, LES
陆地地球站时分多路复用信道	衛星陸上電台劃時多制頻道	LES TDM channel
陆地电台	陸上[電]台	land station
陆风	陸風	land breeze
陆基导航系统	陸基導航系統	ground-based navigational system
陆上移动卫星业务	陸上行動衛星業務	land mobile-satellite service
陆线费	陸線費	land-line charge
陆用发动机	陸用引擎	land engine
陆用式	陸用式	land type
滤尘器	濾塵器	dust filter
滤器自动清洗	過濾器自動清洗	automatic filter cleaning
滤水柜	濾水櫃	water filter tank
滤油设备	濾油設備	oil filtering equipment
旅客	旅客,乘客	passenger
旅客起居设备	旅客起居設備	passenger accommodation
旅客权利	旅客權利	right of passenger
旅客运输	旅客載運	carriage of passenger

祖 国 大 陆 名	台 湾 地 区 名	英 文 名
旅游船	遊覽船	tourist ship
铝粉漆	鋁漆	aluminum paint
铝基轴承合金	鋁基軸承合金	aluminum base bearing metals
绿闪光	綠閃光	green flash
绿星信号	綠星信號	green star-signal
乱水	亂水	broken water
伦敦引航公会	英國導航協會	Trinity House London
轮机长	輪機長	chief engineer
轮机管理	輪機管理	marine engineering management
轮机工程	輪機工程	marine engineering
轮机日志	輪機記事簿	engine room log book
轮机员	輪機員	engineering officer
轮廓灯	整補輪廓燈	contour light
轮型转子	葉輪轉子	blade wheel rotor
轮缘,齿圈	輪緣	rim
轮助	助理工程師,助理輪機員	assistant engineer
罗北	羅經北	compass north
罗方位	羅經方位	compass bearing, CB
罗航向	羅經航向	compass course, CC
罗经	羅經	compass
罗经点	羅經點	compass point
罗经花	圖上羅經	compass rose
罗经柜	羅經針箱	compass binnacle
罗经基点	四方(東、西、南、北),四向基點	cardinal point
罗经甲板	羅經甲板	compass deck
罗经盘	羅經盤	compass card
罗经盆	羅經碗	compass bowl
罗经校正浮标	校正羅經浮標	compass buoy
罗经液体	羅經液體	compass liquid
罗经座	羅經座	binnacle
罗兰	羅遠儀	Loran
罗兰 A	羅遠 A	Loran-A
罗兰 C	羅遠 C	Loran-C
罗兰表	羅遠表	Loran table
罗兰船位	羅遠定位	Loran fix
罗兰 C 告警	羅遠 C 警報	Loran-C alarm

祖 国 大 陆 名	台 湾 地 区 名	英 文 名
罗兰海图	羅遠海圖	Loran chart
罗兰 A 接收机	羅遠 A 接收機	Loran-A receiver
罗兰 C 接收机	羅遠 C 接收機	Loran-C receiver
罗兰天地波识别	羅遠天地波識別	identification of Loran ground and sky waves
罗兰位置线	羅遠位置線	Loran position line
逻辑阀	邏輯閥	logical valve
螺杆泵	螺泵	screw pump
螺杆式制冷压缩机	螺桿式冷凍壓縮機	screw type refrigerating compressor
螺距	螺距	pitch
螺距比	節圓直徑比	pitch ratio
螺距规	螺距規	thread pitch gauge
螺距角	螺距角,週節角	pitch angle
螺距角指示器	螺距角指示器	pitch angle indicator
螺纹	螺紋	thread
螺纹板牙	螺紋模	threading die
螺纹塞规	螺紋塞規	thread plug gauge
螺旋桨	螺[旋]槳	screw propeller
螺旋桨沉深	螺槳深沈	propeller submergence
螺旋桨横向力	螺槳橫向力	sidewise force of propeller
螺旋桨浸深比	螺槳浸深比	immersion ratio of propeller
螺旋桨浸深横向力	螺槳浸深橫向力	transverse force of propeller submergence
螺旋桨静平衡	螺槳靜平衡	propeller statical equilibrium
螺旋桨空转	空轉	racing
螺旋桨流	螺槳流	screw current
螺旋桨盘面积	螺槳盤面積	area of propeller disc
螺旋桨特性	螺槳特性	propeller characteristic
螺旋桨特性曲线	推進器特性曲線	characteristic curve of propeller
螺旋桨推力	螺槳推力	thrust of propeller
螺旋桨转矩	螺槳轉矩	torque of propeller
螺旋桨转速	螺槳轉速	revolution speed of propeller
螺旋试验	蝸旋試驗	spiral test
螺旋推进器船	螺槳推進船	screw propeller ship
螺旋线	蝸線	spiral
裸装货	裸貨	nude cargo
洛氏硬度	洛氏硬度	Rockwell hardness
落潮	落潮	falling tide
落潮持续时间	退潮持續期	duration of ebb

祖国大陆名	台湾地区名	英文名
落潮流	退潮流	ebb stream, ebb current
落潮流强度	退潮最大速率	ebb strength
落潮水道	退潮水道	ebb channel
落墩	坐墩	lying on the keel block
落旗致敬	落旗致敬	dip to

M

祖国大陆名	台湾地区名	英文名
马口铁,镀锡铁皮	鍍錫鐵片,馬口鐵皮	tin sheet
马力限制器	馬力限制器	horsepower limiter
码	電碼	code
C/A 码	C/A 碼	coarse/acquisition code, C/A code
P 码	P 碼	precision code, P code
码分隔制	碼分隔制	code division system
码头	碼頭	wharf
码头费(**港口费**)	碼頭費,船席費	berthage
码头护木	碼頭護木	camel
码头收货单	碼頭埠單	dock warrant
码相位	碼相位	code phase
埋缆机	埋纜機	cable burying machine
霾	霾	haze
脉冲式涡轮增压	脈衝渦輪增壓	pulse turbocharging
脉冲转换器增压	脈衝轉換器增器	pulse converter supercharging
脉 8 定位系统	脈八定位系統	pulse 8 positioning system
脉动磁场	脈動磁場	pulsating magnetic field
满舱货	整船貨	full cargo
满舱满载	完全滿載	full and down
满舵	滿舵	hard over
满帆通风	滿帆逆餞	full and bye
满月	滿月,望月	full moon
满载舱	滿載艙間	filled compartment
满载货	滿載貨	full and complete cargo
满载排水量	滿載排水量	full load displacement
满载水线	滿載水線	load water line
慢闪光	慢閃光	slow flash light
慢转阀	慢轉閥	slow turning valve
慢转起动程序	慢轉起動程序	slow turning starting sequence

祖 国 大 陆 名	台 湾 地 区 名	英 文 名
盲板法兰	管口蓋板	blank flange
盲发	盲發	blind sending
盲区	盲區	blind zone
猫爪结	貓爪結	cat's paw
毛毛雨	毛毛雨	drizzle
锚	錨	anchor
锚标	錨標	anchor buoy
锚冰	錨冰	anchor ice
锚泊	錨泊	anchoring
锚泊船	錨泊船	anchored vessel
锚泊偏荡	錨泊橫搖	yawing in anchoring
锚出水	錨出水	clear of water, anchor in sight
锚床	錨床	bill board
锚垂直	錨垂直	anchor up and down
锚灯	錨燈	anchor light
锚地	錨地	anchorage
锚地海图	泊地海圖	anchorage chart
锚端链节	[錨鏈]外短節	outboard shot
锚更	錨更	anchor watch
锚环	錨環	jew's harp
锚机舱	錨機室	windlass room
锚结	漁人扣	fisherman's bend
锚缆	錨索	anchor rope
锚缆标记	錨纜標誌	cable mark
锚离底	錨離地	anchor aweigh
锚链	錨鏈	anchor chain
锚链舱	錨鏈艙	chain locker
锚链钩	錨鉤	chain hook
锚链管	錨鏈管	naval pipe
锚链绞缠	纏鏈	fouling hawse
锚链孔盖	錨鏈孔蓋	buckler
锚链轮	嵌鏈輪	wildcat
锚链内端	錨鏈內端	bitter end
锚链收短	錨鏈收短	anchor at short stay
锚链筒	錨鏈筒	hawse pipe
锚啮入性	錨嵌入性	anchor penetration
锚球	錨球	anchor ball
锚设备	錨具	ground tackle

祖国大陆名	台湾地区名	英 文 名
锚设备检验	錨與錨設備檢驗	survey of anchor and chain gear
锚位	錨位	anchor position, AP
锚向船来	收錨	come home
锚穴	嵌錨穴,錨龕	anchor recess
锚爪	錨掌	fluke
锚抓力	錨抓著力	holding power of anchor
锚抓重比	錨抓著力與重量比	anchor holding power to weight ratio
卯酉圈	卯酉圈	prime vertical, PV
铆钉	鉚釘	rivet
贸易航线	貿易航線	trade route
梅雨	梅雨	plum rain, Meiyu
煤油发动机	煤油引擎	kerosene engine
每吨海里燃油消耗量	每噸海里燃油消耗量	fuel consumption per ton n mile
每分钟转数	每分鐘轉數	revolution per minute
每厘米吃水吨数	每公分吃水噸數	tons per centimeter immersion, TPC
每厘米纵倾力矩	每公分俯仰差力矩	moment to change trim per centimeter
每月检查	每月檢查	monthly inspection
每周检查	每週檢查	weekly inspection
美国船级社	美國船級協會	American Bureau of Shipping, AB
美国汽车工程师协会黏度分级	美國汽車工程師學會黏度分級	SAE Viscosity Classification
美国石油协会分级	美國石油學會分級	American Petroleum Institute Classification
门到门	戶到戶	door to door
门式起重机	門式起重機	transtainer
蒙气差,折光差	折光差,折射	refraction
密度	密度	density
密度温度系数	實密度–溫度修正係數	true density-temperature correction coefficient
密封甲板阀	封密甲板閥	hermetic deck valve
密封水系统	水封系統	seal water system
密封性	密封性	leakprofness
密封蒸汽系统	封閉蒸汽系統	sealing steam system
密码	密碼	cipher
密码电话学	密語電話	ciphony
密码器	密碼器	cipher device
密语	密語	secret language
免费运送	免費運送	carriage free
免予灭鼠证书	除鼠豁免證書	derating exemption certificate

祖 国 大 陆 名	台 湾 地 区 名	英 文 名
灭火器	滅火器	fire extinguisher
灭火装置	滅火裝備	fire extinguishing appliance
灭鼠证书	除鼠證明書	derating certificate
民船	民用船	civil ship
民船禁航	禁止民船出港	civil embargo
民用晨昏朦影	民用朦光	civil twilight
名牌	銘牌	name plate
明暗光	頓光	occulting light
明给注油器	顯給滑潤器	sight-feed lubricator
明礁	露礁	rock uncovered
明轮	明輪	paddle wheel
明轮推进器船	明輪船	paddle wheel vessel
明语	明語	plain language
鸣笛标	鳴笛標誌	whistle-requesting mark
模糊度解析	不明確訊息解析	ambiguity resolution
模糊效应	模糊效應	blurring effect
模件式脉冲转换增压	模組脈衝轉換增壓	modular pulse converter supercharging
模拟	模擬	simulation
模拟器,仿真器	模擬機設施	simulator
模拟试验	模擬試驗	simulation test
模–数转换器	模擬–數位轉換器	analog-to-digital converter
膜片	膜片	diaphragm
摩擦阻力	摩擦阻力	frictional resistance
磨合	適配運轉	running-in
磨料磨损	磨料磨損	abrasive wear
磨肉机	磨肉機(捏和機)	meat mill
磨损	磨耗	wear
磨损率	磨耗率	wear rate
末端环	尾環,鏈端環	end link
末端链节	船內端錨鏈節	inboard end chain
末端再加热空气调节系统	終端再加熱空調系統	terminal reheat air conditioning system
莫尔斯码	莫斯電碼	Morse code
莫尔斯码雾号	莫爾斯碼霧號	Morse code fog signal
墨卡托海图	麥卡托海圖	Mercator chart
墨卡托算法	麥氏航海	Mercator's sailing
墨卡托投影	麥卡托投影法	Mercator projection
木材船	木材[運載]船	lumber cargo ship

祖国大陆名	台湾地区名	英 文 名
木材条款	木材條款	timber clause
木材载重线	[装]载木[材]载重線	timber load line
木船	木船	wooden ship
木匠	木匠	carpenter
木排	木排	wood raft
木星	木星	Jupiter
目标录取	目標獲得	target acquisition
目测方位	目測方位	visual bearing
目的地码	目的地碼	destination code
目的港	目的港	port of destination
目的港船上交货	船上交货	ex ship
目的港码头交货	碼頭交货	ex pier, ex wharf, ex quay

N

祖国大陆名	台湾地区名	英 文 名
钠和钒含量	鈉釩含量	sodium and vanadium content
钠基润滑脂	鈉基滑脂	sodium grease
奈伏泰斯(=航行警告[电传]系统)		
奈伏泰斯电文编号	航行警告電傳信文編號	NAVTEX message numbering
奈伏泰斯紧急警告	航行警告電傳緊急警告	NAVTEX vital warnings
奈伏泰斯日常警告	航行警告電傳例行警告	NAVTEX routine warnings
奈伏泰斯优先电文	航行警告電傳優先信文	NAVTEX priority message
奈伏泰斯重要警告	航行警告電傳之重要警告	NAVTEX important warning
耐波性	淩海性	seakeeping quality
耐腐蚀钢	耐蝕鋼	corrosion-resisting steel
耐火	耐火	fire-tight
耐火材料	耐火材料	refractory material
耐火救生艇	防火救生艇	fire protected lifeboat
耐久性	耐久性	durability
耐磨装置	耐擦器	chafing gear
耐热材料	耐熱材料	heat resisting material
南方标	南方標	south mark
南河三(小犬 α)	南河三(小犬 α)	Procyon
南极	南極	south pole
南极光	南極光	aurora australis

祖 国 大 陆 名	台 湾 地 区 名	英 文 名
南极星座	南極星座	octans
南门二(半人马 α)	南門二(人馬 α)	Rigil Kent
南天极	南天極	south celestial pole
[难船]漂浮物	遇難船漂浮物	flotsam
挠度计	撓度計	deflectometer
挠性接头,辫子	豬尾式接頭	pigtail
挠性罗经	撓性電羅經	flexibility gyrocompass
挠性陀螺仪	撓性迴轉儀	flexibility gyroscope
挠性轴系	撓性軸系	flexible shafting
内波	内波,潛波	internal wave
内补偿法	内補償法	method of internal compensation
内部安全检查	内部安全稽核	internal safety audits
内底	内底	inner bottom
内底板	内底板	inner bottom plate
内底边板	舭緣板	margin plate
内底结构图	内底結構圖	inner bottom construction plan
内底纵骨	内底縱材	inner bottom longitudinal
内功率	内功率	internal power
内海	内海	internal sea, inner sea
内河船	内河船	river boat, inland vessel
内河分级航区	分級航區	graded region
内河航标	内陸水道助航標誌	inland waterway navigation aids
内河航道图	内水航道圖	chart of inland waterway
内河航行	内河航行	inland navigation
内河航行规则	内河航行規則	inland rules
内河航行基准面	内水航行海圖深度基準面	chart datum for inland navigation
内河引航	内水領航	inland waterway navigation and pilotage
内河引航图	内水引水圖	pilot chart of inland waterway
内进汽	内側進汽蒸汽機	inside admission
内径千分尺	内分厘卡,内測微計	inside micrometer
内陆水道	内陸水道	inland waterway
内陆站	内陸倉庫	inland depot
内燃机船	内燃機船	motor vessel, MV
内水	内水	internal waters
内线航道	内線航道	inside passage
内效率	内效率	internal efficiency
内斜轴系	漸縮軸系	converging shafting

祖国大陆名	台湾地区名	英 文 名
内旋	内向旋轉,内旋	inward turning
内置式平面轴承	内置平面軸承	inboard plain bearings
能耗制动	動力煞俥,動力制韌	dynamic braking
能见地平	視水平線	visible horizon
能见度	能見度	visibility
能见度不良(5级)	能見度不良(5級)	visibility poor
能见距	能見距	range of visibility
能量调节阀	能量調節閥	capacity adjusting valve
泥[浆]泵	泥泵	dredging pump
逆电流试验	逆流試驗	reverse current test
逆风	逆風	foul wind
逆风航驶	逆駛	by the wind
逆功率保护	逆功率保護	reverse power protection
逆功率试验	逆功率試驗	reverse power test
逆流船	逆流船	upstream vessel
逆螺旋试验	逆蝸旋試驗	reverse spiral test
逆时针旋转(=左向旋转)		
逆时针转动	逆時針旋轉	counter-clockwise rotation
逆水	逆流	up stream
年差	年磁差	magnetic annual change
年度检验	歲驗	annual survey
1974年SOLAS公约修正案	一九七四年海上人命安全國際公約修正案	Amendments to the 1974 SOLAS Convention
1974年国际海上人命安全公约	一九七四年海上人命安全國際公約	International Convention for the Safety of Life at Sea, 1974
1978年海员培训、发证和值班国际公约	一九七八年航海人員訓練、發證及當值標準國際公約	International Convention on Standard of Training, Certification and Watching-keeping for seafarers, 1978
黏度	黏度,黏性	viscosity
黏度分级	黏度分級	viscosity classification
黏度计	黏度計	viscosimeter, viscometer
黏度−温度图	黏度−溫度[曲線]圖	viscosity-temperature
黏度指数	黏度指數	viscosity index
黏度自动控制系统	黏度自動控制系統	viscosity automatic control system
捻缝	捻縫	calk, caulking
镍铬耐热合金	鎳鉻合金	nichrome
镍铬钛[耐热]合金	鎳鉻立克[耐熱合金]	nimonic alloy

祖 国 大 陆 名	台 湾 地 区 名	英 文 名
凝点	凝固點	freezing point, solidification point
凝汽式汽轮机	凝水式蒸汽渦輪機	condensing steam turbine
凝水泵	凝水泵	condensate pump
凝水系统	凝水系統	condensate system
凝水再循环管路	凝水再循環泵	condensate recirculating pipe line
牛眼环	牛眼環	bull eye ring
扭矩(=转矩)		
扭力	扭力	torsion
扭力计	扭力計,轉矩計	torsional meter
扭锁	扭轉鎖定器	twist lock
扭叶片	扭葉片	twisted blade
扭[振]共振	扭轉共振	torsional resonance
扭振减振器	扭轉振動減振器	torsional vibration damper
扭转	扭轉	torsion
扭转疲劳裂纹	扭轉疲勞裂痕	torsional fatigue cracks
扭转强度	抗扭強度	torsional strength
扭转振动	扭轉振動	torsional vibration
农用船	農用船	agricultural vessel
浓度	濃度	density
怒涛(9 级)	怒濤(9 级)	precipitous sea
暖锋	暖鋒	warm front
暖机	暖機	warming-up
暖机蒸汽系统	暖機蒸汽系統	warming-up steam system
暖流	暖流	warm current
暖平流	暖平流	warm advection
暖气团	暖氣團	warm air mass
挪威船级社	挪威驗船協會	Det Norske Veritas, NV

O

祖 国 大 陆 名	台 湾 地 区 名	英 文 名
呕吐袋	嘔吐袋	seasickness bag
耦合振动	偶合振動	coupled vibration

P

祖国大陆名	台湾地区名	英文名
耙吸装置	耙吸設施	drag and suction device
拍岸浪	潑岸浪, 拍岸浪	surf
拍击	顫抖	pounding
拍卖船舶	拍賣船舶	auction of ship
排出阀	排出閥	discharge valve
排出流	排出流	discharge current
排出流横向力	排流橫向力	transverse force of discharge current
排出压头	流出落差	discharge head
排筏(**木排筏**)	木排	log raft
排放标准	流出物標準	effluent standard
排放集管	排洩歧管	discharge manifold
排放浓度	流出物濃度	effluent concentration
排放条件	排洩條件	conditions of discharge
排放性能	排放性能	emission performance
排链	錨艙排列錨鏈	tiering
排泥机具	排泥設施	soil discharging facility
排气阀	排氣閥	exhaust valve
排气阀液压旋转系统	排氣閥液壓轉動系統	hydraulic exhaust valve rotation system
排气颗粒	排氣顆粒	exhaust particulate
排气壳体	排氣外殼	exhaust casing, exhaust hood
排气口	排氣口	exhaust port
排气温度	排氣溫度	exhaust temperature
排气行程	排氣衝程	exhaust stroke
排气有害成分	毒性排氣成分	poisonous exhaust composition
排气装置	排氣裝置	exhaust unit
排汽室	排氣室	exhaust chest
排汽系统	排汽系統	exhaust steam system
排水船	排水型船	displacement ship
排水法	排洩法	draining method
排水口	排水口	wash port
排水量	排水量	displacement
排水能力	排水能量	water discharge capacity
排泄	排洩	discharge
排盐泵	衝放泵	blow down pump

祖 国 大 陆 名	台 湾 地 区 名	英 文 名
排盐量	排鹽率	brine rate
排油阀卸载作用		delivery valve line retraction
排油监控装置	洩油偵控系統	oil discharge monitoring and control system
排油控制	排油管制	control of discharge of oil
盘车	盤俥	bar the engine
盘车联锁装置	盤俥裝置聯鎖裝置	turning gear interlocking device
盘面比	盤面比	disc ratio
盘线装置	盤繩機	line winder
旁瓣回波	旁波瓣回波	side-lobe echo
旁桁材	側縱梁	side girder
旁通阀	旁通閥	by-pass valve
旁通调节	旁路調節	by-pass governing
抛单锚	抛單錨	coming to a single anchor
抛锚	抛錨	let go anchor
抛锚掉头	抛錨短迴轉掉頭	turning short round with anchor
抛锚试验	錨泊試驗	anchoring test
抛起锚口令(**锚令**)	錨令	anchoring orders
抛弃	［海難］投棄,海難投棄 貨物	jettison
抛绳回收袋	射索回收袋	shot line return bag
抛绳设备	抛繩器	line throwing appliance
抛投	抛,投	cast
抛艉锚	抛後錨	anchor by the stern
泡沫灭火系统	泡沫滅火系統	foam fire extinguishing system
配餐间	配膳室	pantry
配电	配電	distribution
配电屏	饋電屏	feeder panel
配电系统	配電系統	distribution system
配合件	配合件	matching member
配汽机构	配汽設施	steam distribution device
配汽调整	配汽調整	steam distribution adjustment
配载图	載貨圖,貨物裝載圖	cargo plan
配置轴	序列軸	disposition axis
配装类	裝載種類	stowage category
喷淋式热交换器	噴霧式熱交換器	spray type heat exchanger
喷淋蒸发式冷凝器	噴灑蒸發冷凝器	spray evaporative condenser
喷气推进船	噴氣推進船	air jet ship
喷洒装置	噴灑裝置	sprinkler

祖 国 大 陆 名	台 湾 地 区 名	英 文 名
喷砂(除锈)	噴砂	sand blast
喷射泵	噴射泵	jet pump
喷水推进	噴射推進	jet propulsion
喷水推进船	噴水推進船	hydrojet boat, water jet vessel
喷油器	燃油噴射器	fuel injector
喷油器冷却泵	燃料噴射閥冷卻泵	fuel injection valve cooling pump
喷油器启阀压力	燃油閥開啓壓力	fuel valve opening pressure
喷油器试验台	噴油器試驗設備	injector testing equipment
喷油提前角	噴油提前角	fuel injection advance angle
喷油正时(**喷油定时**)	噴油定時	injection timing
喷油嘴滴漏	噴油嘴滴漏	nozzle dribbling
喷嘴	噴嘴	nozzle
喷嘴阀	噴嘴閥	nozzle valve
喷嘴环	噴嘴環	nozzle ring
喷嘴室	噴嘴室	nozzle chamber
喷嘴调节	噴嘴調節	nozzle governing
喷嘴组	噴嘴塊	nozzle block
砰击	波縈	slamming
膨胀比	膨脹比	expansion ratio
膨胀柜	膨脹櫃	expansion tank
膨胀–换气行程	膨脹–驅氣衝程	expansion-scavenging stroke
膨胀行程	膨脹衝程	expansion stroke
碰垫,靠把	碰墊	fender
碰角	碰撞角	collision angle
碰撞保险	碰撞保險	collision insurance
碰撞点	碰撞點	point of collision, PC
碰撞警报	碰撞警報	collision warning
碰撞速度	碰撞速度	collision speed
碰撞危险	碰撞危機	risk of collision
批注清单	批註清單	remark list
毗连区	鄰接區	contiguous zone
疲劳极限	持久限界	endurance limit
疲劳裂纹	疲勞裂紋	fatigue cracking
匹配	配合,相配	matching
片蚀	層蝕	layer corrosion
偏摆(=航弧)		
偏荡	縱横搖盪,[艍艎]平擺	yawing
偏航	偏航	off way, off course

祖 国 大 陆 名	台 湾 地 区 名	英 文 名
偏近点角	偏近點角	eccentric anomaly
偏缆灯	拖帶邊燈	towing side light
偏离风向	轉向下風	pay off
偏蚀	偏蝕,偏食	partial eclipse
偏心传动装置	偏心機構	eccentric gear
偏心显示	偏心顯示	off-centered display
偏移	偏位	offset
偏中值	偏中值	misalignment value
偏转仪	磁向偏差測算儀	deflector
漂浮式下水	自由浮離下水	float-free launching
漂浮物	漂浮物	floating object, floating substance
漂浮烟雾信号	浮煙信號	buoyant smoke signal
漂航	漂流	drifting
漂角,流压差	流偏角,偏流角	drift angle
漂流角,风流压差	偏流修正角	crab angle
漂流物	漂流物	derelict
漂流渔船	流網漁船	drift fishing boat
漂心距中距离	縱向浮面中心與舯距離	longitudinal distance of center of floatation from midship
撒缆	撒纜	heaving line
撒缆活结	撒纜索活結	heaving line slip knot
撒缆枪	射繩槍,撒纜槍,抛繩槍	line throwing gun
拼车货	拼車貨	less than truck load, LTL
拼合式	拼合式	split type
拼合轴燃气轮机	拼合軸燃氣渦輪機	split-shaft gas turbine
拼箱货	拼櫃貨	less than container load, LCL
频带	頻帶	frequency band
频道	頻道	channel
频率标准	頻率標準	frequency standard
频率表	頻率表	frequency list
频率分隔制	頻率分隔制	frequency division system
频率分集	頻率分集	frequency diversity
频率容限	頻率容許差度	frequency tolerance
频率指配	頻率指配	frequency assignment
频闪灯	閃光燈	strobe light
频移键控	移頻按鍵[制]	frequency shift keying, FSK
品质	品質	character
平板龙骨	平板龍骨	plate keel

祖国大陆名	台湾地区名	英 文 名
平舱	扒平,平艙	trimming
平舱费	平艙費	leveling charge
平插座	平式套座	flush socket
平吃水	縱平浮	even keel
平赤道	平赤道	mean equator
平衡舵	平衡舵	balance rudder
平衡阀	均壓閥,均衡閥	balanced valve
平衡活塞	均衡活塞	dummy piston
平衡陀螺仪	平衡迴轉機	balanced gyroscope
平衡重	配重,衡重	counterweigh
平黄道	平黄道	mean ecliptic
平甲板船	平甲板船	flush deck vessel
平桨!	平槳!	bank the oars!
平结	平結	reef knot
平近点角	平近點角	mean anomaly
平均包装容积	平均包裝容積	average bale capacity
平均潮面	中潮海平面	half-tide level
平均吃水	平均吃水	mean draft
平均低潮间隙	平均低潮間隙	mean low water interval, MLWI
平均高潮间隙	平均高潮間隙	mean high water interval, NHWI
平均海面	平均海平面	mean sea level, MSL
平均海面季节改正	平均海平面季節性變更	seasonal change in mean sea level
平均经度	平均經度	mean longitude
平均空档深度	平均空檔深度	average void depth
平均每吨容积	平均每噸容積	average per ton
平均纬度	平均緯度	mean latitude
平均误差	平均誤差	mean error
平均压力计	平均壓力計	mean pressure meter
平均有效压力	有效平均壓力	effective mean pressure
平均指示压力	指示平均有效壓力	indicated mean effective pressure
平离	平離	leaving bodily
平流(=憩流)		
平流层	平流層	stratosphere
平流区域	憩流區域	slack water area
平面传感器	平面感應器,平面探針	flat-surface sensor, flat-surface probe
平面图	平面圖	plane chart
平面位置显示器	平面位置指示器	plane position indicator, PPI
平时	平均時	mean time

祖 国 大 陆 名	台 湾 地 区 名	英 文 名
平台甲板	台甲板	platform deck
平台箱	平台貨櫃	platform container
平太阳	平均太陽,平太陽	mean sun
平纬航向	平緯航向	parallel course
平行航线搜寻	平行航線搜索	parallel track search
平行锚(=一点锚)		
平行舯体	平行舯段	parallel body
平旋推进器	擺線推進器	Voith Schneider Propeller, VSP
平扎	平纏	flat seizing
平整冰	平整冰	level ice
屏蔽式电热偶	遮蔽式電熱鍋	shielded thermocouple
泼水试验	澆水試驗	pouring water test
迫降	迫降	ditch
破冰船	破冰船	icebreaker
破冰[型]艏	破冰型艏	icebreaker stem, icebreaker bow
破布	破布	rags
破舱稳性	破損穩度,受損穩度	damaged stability, flooding stability
破断强度	裂斷強度	breaking strength, BS
剖视图	剖視圖	sectional view
扑鲸枪	捕鯨槍	harpoon gun
蒲福风级	蒲福風級	Beaufort [wind] scale
普遍呼叫	普通呼叫	general call to all station
普通操作员证书	普通值機員證書	general operator's certificate
普通船员	乙級船員	rating
普通货船	傳統式船	conventional ship

Q

祖 国 大 陆 名	台 湾 地 区 名	英 文 名
期满日,失效日	期滿日,失效日	expiry date
期票	期票	promissory note
期限条款	期限條款	duration clause
期修	定期檢修	regular repair
齐退	全體倒划	stern all
旗杆	旗桿	flag staff
旗柜	旗箱	flag chest
旗号通信	旗號通信	flag signaling
鳍轴	鰭軸	fin shaft

祖 国 大 陆 名	台 湾 地 区 名	英 文 名
启阀压力	噴射啓動壓力	injection start pressure
启线压力可调式喷油器	壓力控制式燃油閥	pressure-control fuel valve
启用试验	啓用試驗	commissioning test
起动	起動,開動	starting
起动故障报警	起動故障警報	start failure alarm
起动结束信号	起動結束信號	start finished signal
起动空气分配器	起動空氣分配器	starting air distributor
起动空气瓶	起動氣瓶	starting air reservoir
起动空气切断	起動空氣切斷	starting air cut off
起动空气系统	起動空氣系統	starting air system
起动空气总管	起動空氣歧管	starting air manifold
起动控制阀	起動控制閥	starting control valve
起动联锁	起動聯鎖	starting interlock
起动盲区	起動盲區	start-up blind-zone
起动器	起動機,起動器	starter
起动事故	起動事故	start-up accident
起动凸轮	起動凸輪	starting cam
起动性能	起動性能	starting performance
起动注油注水	起動注給	priming
起动转速	引擎起動轉速	starting engine speed
起动装置	起動裝置	starting device
起伏	起伏	heave
起航与加速工况管理	起動與加速操作型式管理	starting and accelerating operating mode management
起货机	吊貨機	cargo winch
起货设备	吊貨設備	cargo lifting equipment
起货设备吊重试验	貨物裝卸設備安全限試驗	proof test for ship cargo handling gear
起货设备定期检验	貨物裝卸設備定期檢驗	periodical survey of cargo gear
起居处所	起居艙空間	accommodation space
起居甲板	住艙甲板	accommodation deck
起居设备	起居設備	accommodation
起锚	起錨	weigh
起锚机	起錨機	anchor windlass, windlass
起锚设备(**锚设备**)	錨具	anchor gear
起锚完毕	起錨完畢	anchor is up
起锚系缆绞盘(**起锚绞盘**)	起錨絞盤,起錨機絞盤	anchor capstan

祖国大陆名	台湾地区名	英文名
起始搜寻点	起始搜索點	commence search point
起网	曳網	hauling net
起网机	起網機	net winch
起压	電壓建起	voltage build-up
起重船	起重船,水上起重機	floating crane
起重船打捞	起重船起吊	lifting by floating crane
起重机	起重機	crane
起重机起重臂(**起重机臂**)	起重機臂	crane boom
起重机伸距	起重機伸距	crane radius
气尘云	氣塵雲	gas-dust cloud
气垫船	氣墊船	air-cushion vehicle, hovercraft
气动放大器	氣力放大器	pneumatic amplifier
气动功率放大器	氣動功率放大器	pneumatic power amplifier
气动调节器	氣力調整器	pneumatic regulator
气阀间隙	閥餘隙	valve clearance
气阀升程	閥升程	valve lift
气阀升程图	閥升圖	valve lift diagram
气阀旋转机构	閥旋轉機構	valve rotating mechanism
气阀正时	閥動定時	valve timing
气缸常数	氣缸常數	cylinder constant
气缸窜气	氣缸漏氣	cylinder blow-by
气缸盖	缸蓋	cylinder cover
气缸工作容积	衝程容積	stroke volume
气缸[冷却]水套	缸套	cylinder jacket
气缸起动阀	氣缸起動閥	cylinder starting valve
气缸套	氣缸內襯套,[氣]缸[襯]套	cylinder liner
气缸体	氣缸體	cylinder block
气缸油	氣缸油	cylinder oil
气缸油输送泵	氣缸油輸送泵	cylinder oil transfer pump
气缸油注油量	氣缸油注量	cylinder oil dosage
气缸油注油率	單位氣缸耗油量	specific cylinder oil consumption
气缸直径	氣缸直徑,氣缸內徑	cylinder bore
气缸注油器	氣缸潤滑器	cylinder lubricator
气缸总容积	氣缸總容積	cylinder total volume
气焊	氣焊	gas welding
气候	氣候	climate

祖 国 大 陆 名	台 湾 地 区 名	英 文 名
气候航线	氣候航路	climate routing
气力离合器	氣力離合器	air clutch
气门	氣門	valve
气密	氣密	airtight
气密试验	氣密試驗	airtight test
气泡六分仪	氣泡六分儀	bubble sextant
气泡	舯膨出部	blister
气蚀,空泡腐蚀	孔蝕	cavitation erosion
气态燃料	氣體燃料	gaseous fuel
气体运输船	氣體載運船	gas carrier
气团	氣團	air mass
气味货	氣味貨	odorous cargo
气温	氣溫	air temperature
气温直减率	氣溫遞減率	temperature lapse rate
气象报告	氣象報告	meteorological bulletins
气象传真接收机	氣象傳真接收機	weather facsimile receiver
气象电文	氣象信文	meteorological messages
气象定线	天氣定航	weather routing
气象服务	氣象服務	meteorological service
气象辅助业务	氣象輔助業務	meteorological aids service
气象回波	氣象回波	meteorology echo
气象警告	氣象警告	meteorological warning, MET warning
气象雷达	氣象雷達	meteorological radar
气象能见度	氣象能見距	meteorological visibility
气象信息	氣象資訊	meteorological information
气象要素	氣象要素	meteorological element
气象要素图	氣象圖	meteorological chart
气象预报	氣象預報	meteorological forecast
气旋	旋風,氣旋	cyclone
气旋波	氣旋波	cyclonic wave
气旋风	氣旋風	cyclonic wind
气压(**大气压**)	大氣壓力	atmospheric pressure
气压表	氣壓表,氣壓計,晴雨計	barometer
气压计	氣壓計	barograph
气压趋势	氣壓趨勢	barometric tendency
气压系统	壓力系統	pressure system
气翼船	氣翼艇	aerofoil boat
气翼艇	氣翼艇	ram-wing craft

祖 国 大 陆 名	台 湾 地 区 名	英 文 名
气胀[救生]筏	充氣救生筏	inflatable liferaft
气胀救生衣	充氣救生衣	inflatable lifejacket
气胀设备	充氣設備	inflatable appliance
弃船	棄船	abandonment of ship
弃船信号	棄船信號	abandon ship signal
弃船演习	棄船演習	abandon ship drill
弃链器	釋纜器	cable releaser
汽车甲板	車輛甲板	wagon deck
汽车轮渡	車輛渡輪	car ferry, automobile ferry
汽车起重机	汽俥式起重機	auto crane
汽车运输船	車輛運輸船	pure car carrier, PCC
汽机鼓	蒸汽鼓,汽鼓	steam drum
汽轮机船	蒸汽渦輪機船	steam turbine ship
汽轮机单缸运行	蒸汽渦輪機單缸運轉	steam turbine single-cylinder operation
汽轮机级	蒸汽渦輪機級	steam turbine stage
汽轮机-燃气轮机联合装置	蒸汽與燃氣渦輪複合推進裝置	combined steam-gas turbine [propulsion] plant
汽轮机外特性	蒸汽渦輪機外特性	external characteristic of steam turbine
汽轮机油	輪機滑油	turbine oil
汽油	汽油	gasoline
汽油机	汽油機	petrol engine
憩流,平流	憩流,憩潮	slack water
千斤索	吊桿頂索	topping lift
千斤索绞车	吊桿頂索絞車	topping lift winch
牵索	內牽	inhaul [line]
牵条螺栓	牽條螺栓	stay bolt
牵引(=拖曳)		
铅锤	測錘,測深錘	lead
铅基白合金轴承	鉛基白合金軸承	lead base white metal bearing
铅基轴承合金	鉛基軸承合金	lead-base bearing alloy
签发日期	簽發日期	date of issue
前八字	前八字方向	on the bow
前导舰	前導船艦	leading ship
前帆	前帆	fore sail
前方	前方	ahead
前后吃水不当	前後吃水不當	out of trim
前进波	前進波	progressive wave
前倾[型]艏	斜艏	raked stem, raked bow

祖 国 大 陆 名	台 湾 地 区 名	英 文 名
前桅	前桅	fore mast
前向纠错	前向侦错	forward error correction
前向纠错方式	前向侦错方式	forward error correction mode，FEC
前行舰	正前方船	forward ship
前置放大器	前置放大器	pre-amplifier
潜水	潜水	diving
潜水泵	潜水泵	immersion pump，submersible pump
潜水病	潜水病	bends
潜水舱	潜水艙	submerged diving chamber
潜水导索	潜水導索	diver's descending line
潜水电话	潜水電話	diving telephone
潜水服	潜水衣	diving suit
潜水工作船	潜水工作船	diving boat
潜水供气系统	潜水供氣系統	gas distribution system for diver
潜水减压	潜水減壓	decompression of diving
潜水设备	潜水設備	diving equipment
潜水手语	潜水手語	diver's sign language
潜水头盔	潜水頭盔	diver's helmet
潜水靴	潜水靴	diver's boots
潜水员	潜水人，潜水員	diver
潜水钟	潜水鐘	diving bell
潜水装具	潜水器具	diving apparatus
潜艇	潜[水]艇	submarine
潜艇操纵	潜艇操縱	submarine handling
潜艇操纵强度	潜艇操縱強度	submarine's maneuvering strength
潜艇操纵性	潜艇操縱性	submarine's maneuverability
潜艇反操纵性	潜艇反操縱性	submarine's adverse maneuverability
潜艇航行状态	潜艇航行狀態	submarine's proceeding state
潜艇均衡	潜艇均衡	submarine trimming
潜艇母舰	潜艇母艦	submarine depot ship
潜艇平行上浮	潜艇平行上浮	submarine's trimmed surfacing
潜艇平行下潜	潜艇平行下潜	submarine's trimmed diving
潜艇起浮	潜艇上浮	submarine surfacing
潜艇水面航行状态	潜艇水面航行狀態	submarine's surface proceeding state
潜艇水下航行状态	潜艇水下航行狀態	submarine's proceeding state underwater
潜艇速浮	潜艇急浮	submarine quick surfacing
潜艇速潜	潜艇急潜	submarine quick diving
潜艇通气管航行状态	潜艇通氣管航行狀態	submarine's proceeding state with snorkel

祖 国 大 陆 名	台 湾 地 区 名	英 文 名
潜艇下潜	潛艇下潛	submarine diving
潜艇相对上浮	潛艇相對上浮	submarine's relative surfacing
潜艇相对下潜	潛艇相對下潛	submarine's relative diving
潜艇巡航状态	潛艇巡航狀態	submarine's cruising state
潜望六分仪	潛望六分儀	periscope sextant
潜越	潛越	passing underneath
潜在缺陷	潛在缺陷	latent defect
潜在渔业资源	潛在的漁業資源	potential fisheries resources
浅浪登陆	淺浪登陸	landing through surf
浅水潮	淺水潮	shallow water tide
浅水效应	淺水效應	shallow water effect
浅水与窄航道航行工况 管理	淺窄水道航行操作模式 管理	shallow and narrow channel navigation op- erating mode management
欠频	欠頻	under-frequency
欠稳船	低穩度船	tender ship
欠压	欠壓	under-voltage
欠压试验	欠壓試驗	under-voltage test
欠折射	次折射	sub-refraction
强度	強度	intensity, strength
强化群呼	強化群呼	enhanced group call, EGC
强力滑油系统	強力潤滑油系統	forced lubricating oil system
强酸值	強酸值	strong acid number, SAN
强胸横梁	抗拍梁	panting beam
强涌	強湧	high swell
强制打捞	強制打撈沉船	compulsory removal of wreck
强制循环锅炉	強制循環鍋爐	forced circulation boiler
强制引航	強制引水	mandatory pilotage, compulsory pilotage
强制振动	強制振動	forced vibration
墙(=壁)		
抢风	迎風	close haul
抢滩	搶灘	beaching
敲缸	柴油爆震	diesel knock
桥规	橋形規,橋形軸規,軸規	bridge gauge
桥规值	橋形規值	bridge gauge value
桥涵标	橋通路標	bridge opening mark
桥式联结器	橋式聯結器	bridge fitting
桥柱灯	橋柱燈	bridge pier light
切变线	剪切線	shear line

祖 国 大 陆 名	台 湾 地 区 名	英 文 名
切除残损物	切除殘骸	cutting away wreck
切断	切斷	make dead
切割	切割	cutting
切换到使用重油位置	切換使用重油	switching over to heavy oil
切缆机	切纜機	cable cutter
切片机	切片機	slicing machine
切向分量	切線方向分量	tangential component
切向应力	切線應力	tangential stress
侵蚀	浸蝕	erosion
青铜条	青銅條	bronze strip
氢氧气焊	氫氧熔接	oxy-hydrogen welding
氢氧切割	氫氧截割	oxy-hydrogen cutting
轻合金	輕合金	light alloy
轻合金船	輕合金船	light alloy ship
轻滑配合	輕滑配合	easy slide fit
轻浪(3 级)	微波	slight sea
轻泡货	輕笨貨(泡貨)	light cargo, bulky cargo
轻雾	中霧	moderate fog
轻型	輕型	low-duty
轻压配合	輕壓配合	light press fit
轻质炼制品	輕質精煉油	light refined products
轻质燃料油	輕燃料油	light fuel oil
轻转配合	輕轉配合	easy running fit
倾差仪	傾差儀	heeling error instrument, heeling adjustor
倾倒污染	傾倒污染	damping pollution
倾点	流動	pour point
倾废	廢[棄]物傾倒	dumping of wastes
倾覆	傾覆	overturn, capsize
倾斜	傾斜	tilting
倾斜试验	傾側試驗	inclining test
倾斜仪	傾斜儀	clinometer
倾斜自差	傾側自差	heeling error, heeling deviation
清舱设备	清艙設施	tank cleaning facilities
清除水雷船	掃雷船	vessel engaged in mineclearance operation
清除水雷作业	清除水雷作業	mineclearance operation
清洁船体螺旋桨特性曲线	潔淨船體螺槳特性曲線	clean hull propeller curve
清洁货	潔淨貨	clean cargo

祖 国 大 陆 名	台 湾 地 区 名	英 文 名
清洁提单	無批註載貨證券,清潔提單	clean bill of lading
清洁压载泵	清潔壓艙水泵	clean ballast pump
清洁压载舱	清潔壓艙水艙	clean ballast tank, CBT
清洁压载舱操作手册	清潔壓艙水艙操作手冊	clean ballast tank operation manual
清洁压载水	清潔壓艙水	clean ballast
清解锚链	解錨鏈	clearing hawse
清净分散剂	清潔分散劑	detergent/dispersant additive
晴天工作日	晴天工作日	weather working day, WWD
请求权	請求權,求償權	right of claim
秋分大潮	秋分大潮	autumnal equinoctial spring tide
秋分点	秋分	autumnal equinox
球鼻[型]艏	球形艏	bulbous bow
球阀	球閥	globe valve
球面滚柱轴承	球面滾子軸承	spherical roller bearing
球体[号型]	球形號標	ball [shape]
球窝节	球接頭	ball and socket joint
球形浮标	球形浮標	spherical buoy
球形艉	球形艉	bulb stern
区号	區號	zone letter
区间货	區間貨	local cargo
区配电板	分段配電板	section board
区时	區域時,區時	zone time, ZT
区域	區域	area
区域覆盖	區域覆蓋	local-mode coverage
区域码	地區碼	area code
区域协调人	區域協調人	area co-ordinator
区域性作业模式	區域性作業模式	local mode of operation
区域再加热空气调节系统	區域再熱空調系統	zone reheat air conditioning system
曲柄半径	曲柄推程	crank throw
曲柄臂	曲柄臂	crank web, crank arm
曲柄臂间距(开档)	曲臂間距	crank spread
曲柄连杆比	曲柄半徑與連桿長比	crank radius-connecting rod length ratio
曲柄销	曲柄軸銷	crank pin
曲径式密封	曲折填函蓋	labyrinth gland
曲折机动	曲折操縱	zigzag maneuvre
曲轴	曲柄軸	crankshaft

祖 国 大 陆 名	台 湾 地 区 名	英 文 名
曲轴红套滑移	曲柄軸短縮滑移	crankshaft shrinkage slip-off
曲轴疲劳断裂	曲柄軸疲勞破壞	crankshaft fatigue fracture
曲轴平衡重	曲柄軸衡重	crankshaft counterweight
曲轴箱	曲柄軸箱	crankcase
曲轴箱爆炸	曲柄軸箱爆炸	crankcase explosion
曲轴箱防爆门	曲柄軸箱防爆門	crankcase explosion relief door
曲轴箱透气管路	曲[柄]軸箱通氣管	crankcase vent pipe
曲轴转角	曲柄角	crank angle
驱气	消除油氣,清除有害氣體	gas-freeing
驱逐舰	驅逐艦	destroyer
屈服	降伏	yield
屈服极限	降伏極限	yield limit
趋势分析	趨向分析	trend analysis
取样器	取樣管,取樣器	sampling probe
去污系统	除污系統	decontamination system
全部缆绳松掉	各纜鬆開	cast of all lines
全部容量	總括數	all told
全封闭救生艇	全圍蔽救生艇	totally enclosed lifeboat
全封闭式制冷压缩机	全封閉冷凍壓縮機	hermetically sealed refrigerating compressor unit
全浮动式活塞销	全浮動式活塞銷	full floating gudgeon pin
全浮动式轴承套筒	全浮動式套筒	fully-floating sleeve
全集装箱船	全貨櫃船	full container ship
全降区标志	全降區標誌	full landing area mark
全潜船	全潛船	underwater ship
全球导航卫星系统	全球導航衛星系統	global navigation satellite system, GLONASS
全球定位系统	全球定位系統	global positioning system, GPS
全球覆盖	全球覆蓋	global-mode coverage
全球覆盖方式	全球覆蓋模式	global coverage mode
全球海上遇险安全系统	全球海上遇險及安全系統	global maritime distress and safety system, GMDSS
全球海上遇险安全系统区域	全球海上遇險及安全系統區域	GMDSS area
全球航行警告业务	全球航行警告業務	world wide navigational warning service, WWNWS
全日潮	週日單潮,週日潮	diurnal tide

祖 国 大 陆 名	台 湾 地 区 名	英 文 名
全松	鬆脫	by the run
全速后退	全速後退	back full
全损	全損	total loss
全体	全部,全體,總計	total
全向导航	全向導航	omnirange navigation
全向导缆器	全向導索器	omnidirectional fairleader
全向推进器	全向推進器	all direction propeller, Z propeller
全向无线电测距	萬向無線電測距	omnidirectional radio range
全向无线电信标	萬向無線電示標	omnidirection radio beacon
全载波发射	全載波發射	full carrier emission
全组合曲轴	全組合曲軸	full built-up crankshaft
缺员	缺額	short handed
确定证据	確定證據	conclusive evidence
确认遇险报警收妥	確認收妥遇險警報	acknowledgement of distress alert
确信位置	經確定之位置	resolved position
群波	波群	group of waves
群岛水域	群島海域	archipelago sea area
群岛通过权	群島間通過權	right of passage between archipelagoes
群呼广播业务	群呼廣播業務	group call broadcast service
群速	(波)群速度	group velocity

R

祖 国 大 陆 名	台 湾 地 区 名	英 文 名
燃点	點火點	ignition point
燃耗	燒盡,燒光,燒完	burn-up
燃料包壳	燃料包蓋	fuel cladding
燃料舱	燃料艙,煤艙	bunker
燃料烧毁(烧尽)	燒壞,燒毀	burn-out
燃料油	重油	bunker oil
燃料元件	燃料元素	fuel element
燃气发生器	氣體發生器	gas generator
燃气轮机船	燃氣渦輪機船	gas turbine ship
燃烧不完全	不完全燃燒	incomplete combustion
燃烧器	燃燒器	burner
燃烧室	燃燒室	combustion chamber
M 燃烧室	M 燃燒室	M combustion chamber
燃烧室外壳	燃燒室外殼	combustor outer casing

祖 国 大 陆 名	台 湾 地 区 名	英 文 名
燃烧效率	燃燒效率	combustion efficiency
燃烧自动控制	自動燃燒控制	automatic combustion control
燃油	燃油	fuel oil
燃油驳运系统	燃油轉駁系統	fuel oil transport system
燃油加热器	燃油加熱器	fuel oil heater
燃油净化系统	燃油淨化系統	fuel oil purifying system
燃油均化器	燃油均化器	fuel oil homogenizer
燃油黏–温图	燃油黏度溫度圖	fuel oil viscosity-temperature diagram
燃油输送泵	燃油輸送泵	fuel oil transfer pump
燃油消耗量	燃料消耗量	fuel consumption
燃油泄放系统	燃油排洩系統	fuel oil drain system
燃油预热图	燃油預熱圖	fuel oil preheating chart
燃油注入管	燃油注入管	fuel oil filling pipe
燃油装置	燃油裝備組	oil fuel unit
让路船	避讓船,讓路船舶	give-way vessel
绕航	變更航程	deviation
绕航变更报告	偏航報告,變更航程報告	deviation report
绕一道(圈)	繞轉	round turn
绕住	繞住	catch a turn
热带淡水载重线	熱帶淡水載重線	tropical fresh water load line
热带低压	熱帶低壓	tropical depression
热带风暴	熱帶風暴	tropical storm
热带辐合带	熱帶輻合地帶	intertropical convergence zone, ITCZ
热带气旋	熱帶氣旋	tropical cyclone
热带区带	熱帶地帶	tropical zone
热带扰动	熱帶擾動	tropical disturbance
热带载重线	熱帶載重線	tropical load line
热电动势	熱電動勢	thermo electromotive force
热电流	熱電流	thermocurrent
热电偶	熱電偶	thermocouple
热电式空气调节器	熱電式空調器,半導體空調	thermal electric type air conditioner, semi conductor air conditioner
热机	熱機	heat engine
热力膨胀阀	熱力膨脹閥	thermostatic expansion valve
热力学温标	熱力學溫度標	thermodynamic temperature scale
热裂	熱裂煉,熱分解	thermal cracking
热敏电阻	熱阻[半導]體	thermistor

祖 国 大 陆 名	台 湾 地 区 名	英 文 名
热疲劳	熱疲勞	heat fatigue, thermal fatigue
热疲劳裂纹	熱疲勞裂痕	heat fatigue cracking
热平衡	熱量平衡表	heat balance
热屏蔽	熱遮蔽	thermal shielding
热气融霜	熱氣除霜	hot gas defrost
热容量	熱容量	thermal capacity
热蠕变	熱潛變	thermal creep
热湿比	熱濕比	heat-humidity ratio
热水供暖系统	熱水加熱系統	hot water heating system
热水柜	熱水箱,熱水櫃	hot water tank
热水井	熱阱,熱井	hot well
热水循环泵	熱水循環泵	hot water circulating pump
热态起动	熱起動	hot starting
热停堆	熱關閉	hot shut-down
热稳定性	熱穩定性	thermostability
热盐水融霜	熱鹽水除霜	hot brine defrost
热值	發熱量	calorific value
人工操纵的	人工操縱者	manned
人工岛	人工島	artificial island
人工地平	人工水平儀	artificial horizon
人工港	人工港	artificial harbor
人工航槽	濬深水道	dredged channel
人工用户电报业务	人工電報交換業務	manual telex service
人工鱼礁	人工魚礁	fish shelter
人工智能语言	人工智慧語言	artificial intelligent language
人–机通信系统	人–機通信系統	man-machine communication system
人孔	人孔	manhole
人命救助	(船)生命救助費	life salvage
人身伤亡赔款限额	人身傷亡責任限制	limit of liability for personal injury
人员定位标	人員定位示標	personnel locator beacon, PLB
人员落水	人員落水	man overboard
人造卫星	人造衛星	artificial satellite
人造纤维绳	人造纖維繩索	synthetic rope
人造纤维渔网	人造纖維漁網	synthetic fishing net
人造行星	人造行星	artificial planet
人字队	V 形隊	v-shaped formation
人字桅,双脚桅	雙腳桅	bipod mast
认星	識星	star identification

祖 国 大 陆 名	台 湾 地 区 名	英 文 名
任务控制中心	任務管制中心	mission control center, MCC
任务控制中心服务区	任務管制中心服務區	MCC service area
日本船级社	日本海事協會	Japanese Maritime Corporation, NK
日标	日間助航標誌,畫標	day mark
日常例行维修	例行保養	routine maintenance
日常优先等级	日常優先順序	routine priority
日潮不等	週日差	diurnal inequality
日出	日出	sun rise
日光信号镜	日光信號鏡	daylight signaling mirror
日光仪	量日儀	heliometer
日晷投影	日晷投影	gnomonic projection
[日]环食	[日]環食	annular eclipse
日界线	日界線	calendar line, date line
日没	日沒	sun set
日冕	日冕	corona
日内瓦公海公约	日内瓦公海公約	Geneva Convention on the High Seas
日内瓦领海和毗连区公约	日内瓦領海與鄰接區公約	Geneva Convention on Territorial Sea and Contiguous Zone
日蚀	日食	solar eclipse
日用柜	日用櫃	daily tank, service tank
容积	容積	volume
容积泵	排量式泵	positive displacement pump
容积吨	呎碼噸	measurement ton
容积货物	呎碼貨,容積貨	measurement cargo
容积效率,充气系数	容積效率,體積效率	volumetric efficiency
容量	容量	capacity
容许磨损	容許磨損	permissible wear
容许误差	容許誤差	admissible error, tolerance error
熔点	熔點	melting point
融霜储液器	除霜接受器	defrost receiver
柔性支持板	柔性撐板	flexible stay plate
柔性转子	柔性轉子	flexible rotor
肉库	肉庫	butchery
肉眼检查	肉眼檢查	macroscopic test
乳化油	乳化油	emulsifying oil
入级	船級	classification
入级证书	船級證書	classification certificate
入网	上網	log in

祖 国 大 陆 名	台 湾 地 区 名	英 文 名
软波导	撓性導波管	flexible waveguide
软管	軟管	hose
软管固定架	油管固定柵架	hose tie rack
软管喷嘴	軟管噴嘴	hose nozzle
软铁球	軟鐵球	soft-iron sphere
闰秒	閏秒	leap second
闰年	閏年	leap year
闰日	閏日	leap day
闰月	閏月	leap month
润滑系统	潤滑系統	lubrication system
润滑脂	滑脂	lubricating grease
弱的	弱的	weak
弱链环	弱鏈	weak link
弱酸值	弱酸值	weak acid number
弱弹簧示功图	弱彈簧示功圖	weak spring diagram
弱涌	弱湧	weak swell

S

祖 国 大 陆 名	台 湾 地 区 名	英 文 名
洒水系统	噴水系統	sprinkler system
洒水装置	噴水裝置	sprinkler installation
萨巴蒂循环	定壓定容混式循環(柴油機)	Sabbath cycle
塞尺,量隙规	厚度規	feeler
赛氏黏度	色博黏度	Saybolt viscosity
赛艇	競賽艇	racing boat
三[层]金属轴承	三[層]金屬軸承	tri-metal bearing
三差	三差	triple difference
三岛型船	三島型船	three island vessel
三副	三副	third officer, third mate
三杆分度器	三臂定位器	three-arm protractor, station pointer
三缸星形发动机	三缸星形引擎	y-engine
三拐曲轴	三拐曲軸	three-thrown crank shaft
三管轮	三管輪	fourth engineer
三角插塞	三腳插塞	three-pin plug
三角皮带	三角皮帶	v-beet
三角形三(南三角 α)	三角形三(南三角 α)	Atria

祖 国 大 陆 名	台 湾 地 区 名	英 文 名
三角洲	三角洲	delta
三脚桅	三腳桅	tripod mast
三绕组变压器	三卷線變壓器	three-winding transformer
三通阀	三通閥	three-way valve
三通接头	三通接頭	t-junction
三位四通换向阀	三位四通方向控制閥	three-position four way directional control valve
三相点	三相點	triple point
三相反向变流机	三相變流機	three-phase inverter
三相四线制	三相四線制	three-phase four-wire system
三相异步电动机	三相異步電動機	three-phase asynchronous motor
三元流动	三維流動	three-dimensional flow
三胀式蒸汽机	三段膨脹蒸汽機	triple expansion steam engine
三字点	中間點	intermediate point
三字母信号码	三字母信號碼	three letter signal code
散波	發散波	divergent wave
散货	散裝貨	bulk cargo
散货船	散裝貨船	bulk-cargo ship, bulk carrier
散货捆包	大包捆	bundling of bulk
散开纵队命令	疏開編隊	column open order
散装谷物捆包	散裝穀類捆包	bundle of bulk grain
散装货条款	散裝貨條款	bulk cargo clause
散装容积	散裝容積	grain capacity, bulk capacity
散装时危险物质	散裝時危險物質	materials hazardous in bulk
扫舱泵	殘油泵,收艙泵	stripping pump
扫海	掃艙地腳	sweeping
扫雷队形	掃雷隊形	mine-sweeping formation
扫雷航海勤务	掃雷航海勤務	mine-sweeping navigation service
扫气	驅氣	purge
扫气口	驅氣孔	scavenging air port
扫气箱	驅氣歧管	scavenging air manifold
扫气箱着火	驅氣箱著火	scavenging box fire
扫气效率	驅氣效率	scavenging efficiency
扫气压力燃油限制器	驅氣空氣壓力燃油限制器	scavenging air pressure fuel limiter
沙暴	沙暴	sandstorms
沙脊	沙畦	sand ridge
砂轮	盤輪	disc wheel

祖 国 大 陆 名	台 湾 地 区 名	英 文 名
山区河流		mountain river
山字钩	雙鉤	double hook
闪点	閃點	flash point
闪发室	急驟蒸發室	flash chamber
闪发蒸发	急驟蒸發	flash evaporation
闪光灯	閃光燈	flashing light
闪光复位	閃爍復位	flicker reset
扇形	扇形	sector
扇形搜寻	扇形搜索	sector search
扇形艉	扇形艉	fantail stern
商船	商船	merchant ship
商船旗	商船旗	merchant ship flag
商船搜救手册	商船搜救手冊	merchant ship search and rescue manual
上部的	上部	upper
上层建筑	船艛建築,上層建築	superstructure
上层建筑甲板	船艛甲板	superstructure deck
上风岸	上風岸	weather shore
上风船	上風船	vessel to windward
上风舵	上風舵	down helm
上风满舵	上風滿舵	hard up
上风锚	上風錨	weather anchor
上风舷	上風舷	weather side
上甲板	上甲板	upper deck
上升风	上升風,飆,上坡風	anabatic
上升流	湧升流	upwelling
上弦	上弦	first quarter
上限越界	上限越界	off-normal upper
上行(**开往**)	開往…	bound to
上行船	上行船	up-bound vessel
上行行程	上行衝程	upward stroke
上游	上游	upper reach
上止点	上死點	top dead center, TDC
上中天	上中天	upper meridian passage, upper transit
上中天潮	順潮	direct tide
烧毁热负荷(烧尽热负荷)	燒盡熱通量	burn-out heat flux
烧球式柴油机	半柴油機	semidiesel
稍松(=打住)		

祖 国 大 陆 名	台 湾 地 区 名	英 文 名
设备交接单	設備接收單	equipment receipt
设计纬度	設計緯度	designed latitude
设立国际油污损害赔偿 　基金国际公约	設立油污損害國際賠償 　基金國際公約	International Convention on the Establishment of an International Fund for Compensation for Oil Pollution Damage
射电六分仪	無線電六分儀	radio sextant
X射线	X射線	X-ray
X射线检查	X射線檢查	X-ray examination
X射线照相探伤	放射線檢查	radiographic inspection
申报(=报关)		
伸缩接头	伸縮接頭	expansion joint
伸缩式减摇鳍装置	伸縮式鰭板穩定器	retractable fin stabilizer
伸缩套管	套筒伸縮管	telescope pipe
深舱	深艙	deep tank
深层流	深水流	deep current
深度记录器	測深計	depth recorder
深度千分尺	深度測微計	micrometer depth gauge
深度指示器	深度指示錶	depth indicator
深度自动操舵仪	深度自動操舵裝置	depth autopilot
深海分区	深海分區	benthic division
深潜救生艇	深潛救生艇	deep submersible rescue vehicle
深潜器	深潛器	deep diving submersible, bathyscaphe
深水航路	深水航路	deep water way
深水抛锚	深海抛錨	deep sea anchoring
深水拖网	深水拖網	deep water trawl
参宿七(猎户β)	參宿七(獵戶β)	Rigel
参宿四(猎户α)	參宿四(獵戶α)	Betelgeuse
甚低频通信	特低頻通信	VLF communication
甚高频紧急无线电示位 　标	特高頻應急指位無線電 　示標	VHF emergency position-indicating radio-obeacon
甚高频通信	特高頻通信	VHF communication
甚高频无线电测向仪	特高頻無線電測向儀	very high frequency radio direction finder, VHF RDF
甚高频无线电话设备	特高頻無線電話裝置	VHF radiotelephone installation
甚高频无线电设备	特高頻無線電裝置	VHF radio installation
渗漏试验	漏洩試驗	leakage test
渗碳	滲碳	cementation
渗透率	浸水率,浸透性	permeability

祖 国 大 陆 名	台 湾 地 区 名	英 文 名
升船机	船舶升降機	ship elevator, ship lift
升降口	升降口	companion way
升降式舱盖	吊式艙蓋	lift hatch cover
升降式平台	升降式鑽油台	jack-up rigs
升降梯	升降梯	companion ladder
升降系统	頂舉系統	jacking system
升交点	昇交點	ascending node
升力	升力	lift force
升压泵	增壓泵,加力泵	booster pump
升压伺服器	增壓伺服馬達	booster servomotor
生活废弃物	生活廢棄物	domestic waste
生活污水	污水,穢水	sewage
生活污水标准排放接头	穢水標準排洩接頭	sewage standard discharge connection
生活污水处理装置	穢水處理裝置	sewage treatment unit
生活污水柜	穢水櫃	sewage tank
生活污水排泄系统	穢水管路系統	sewage piping system
生活用水系统	生活用水系統	domestic water system
生命支持系统	生命支援系統	life support system
生物资源	生物資源	living resources
生效日	生效日	effective date
声道	聲道	sound channel
声号通信	音響通信	sound signaling
声级计	音強度計	sound-level meter
声力电话	聲力電話	sound powered telephone
声呐	聲納	sonar
声呐罩涂料	聲納罩塗料	sonar dome coating
声速误差	音速誤差	sound velocity error
声速校准	音速校準	sound velocity calibration
声响信号	音響信號	sound signal
牲畜运输船	牲口運輸船	livestock carrier
绳	索,纜,繩	rope
绳车	捲盤	reel
绳的弯曲部	繩[索]套	bight
绳结	繩結	bends and hitches
绳扣	回頭索	bight
绳索	繩索	cordage
绳梯	繩梯	rope ladder
绳头插接	反插接	backsplice

祖国大陆名	台湾地区名	英文名
绳头结	倒纽结	crown knot
绳头卸扣	钢丝索扣	bulldog grip
绳锥结	椎套结	marline spike hitch
盛货帆布袋	载货帆布袋	canvas cargo bag
盛行风	盛行风	prevailing wind
剩余动稳性	剩余动稳度	residual dynamical stability
剩余功率	剩余动力	surplus power
剩余自差	剩余自差	remaining deviation
失火警报	火警警报	fire alarm
失火自动报警系统	自动火警警报系统	automatic fire alarm system
失火自动报警与探测系统	自动火警警报与探火系统	automatic fire alarm and fire detection system
失控船	操纵失灵船	vessel not under command
失水事故	冷却水损失事故	loss of coolant accident
失速	失速	speed loss
失速角	失速角	stalling angle
失吸现象	失吸现象	suction loss
失效日(=期满日)		
施放烟幕机动	施放烟幕操纵	smoke screen laying maneuvre
湿底润滑	湿油槽润滑	wet sump lubrication
湿度	湿度	humidity
湿度表	湿度表	hygrometer
湿度计	湿度计	hygrograph
湿喷砂除锈	湿喷砂	wet sand blasting
湿气	湿	wet
湿式缸套	湿式缸衬[套]	wet cylinder liner
十六烷值	十六烷值	cetane number
十字架二(南十字α)	十字架二(南十字α)	Acrux
十字缆桩	十字形系桩	cross bitt
十字头	十字头	crosshead
十字头滑块	十字头履	crosshead slipper, crosshead shoe
十字头式柴油机	十字头型柴油机	crosshead type diesel engine
石棉	石棉	asbestos
石棉绳	石棉绳	asbestos cord
石英压力传感器	石英压力测感器	quartz pressure sensor
时标	定时记号	timing mark
时差	时差	equation of time, ET
时分复用	割时多工制	time division multiplexing, TDM

祖 国 大 陆 名	台 湾 地 区 名	英 文 名
时号	對時信號,報時信號	time signal
时基	時基	time base
时基[信号]发生器	時基[信號]發生器	time-base generator
GPS 时间	GPS 時間	GPS time
时间差	時差	time difference, TD
时间分隔	時間分割	time division
时间分隔制	時分制	time division system
时间校正	時間校正	time calibration
时角	時角	hour angle, HA
时区号	時區標號	zone description, ZD
时区图	時區圖	time zone chart
时圈	時圈	hour circle
时限继电器	緩動繼電器	time-lag relay
时效	時限	time limitation
识别	識別	identify, ID
识别器	識別器	identifier
识别数据	識別數據	identification data
识别信号	識別信號	identity signal
实际承运人	實際運送人	actual carrier
实际航迹向	實際航跡	actual track
实际航速	對地速度	speed over ground
实际全损	實際全損	actual total loss
实际循环	實際循環	actual cycle
实践	實務	practice
实时方式	實時模式	realtime mode
实习船	訓練船	training ship
实验室化验报告	實驗室分析報告	laboratory analysis report
实作训练	實作訓練	practice training
蚀耗极限	蝕耗極限	corroded limit
食物垃圾	食物廢棄物	food wastes
矢量	向量,矢量	vector
矢量显示	向量顯示	vector display
使船停住	拋錨停航	bring to
使用一种燃油船舶	使用一種燃油船	one fuel ship
始发港	始航港	port of origin, port of sailing
始航向	起程航向	initial course
驶出	駛出	put out
驶帆船	揚帆行駛之船舶	vessel under sail

祖 国 大 陆 名	台 湾 地 区 名	英 文 名
驶风	小艇驶帆	boat sailing
驶离	駛離	put away
驶向海岸	駛向海岸	stand into land
世界气象组织	世界氣象組織	world meteorological organization
世界时	世界時	universal time, GMT
世界卫生组织	世界衛生組織	World Health Organization, WHO
世界无线电行政大会	世界行政無線電會議	World administrative radio conference
示功阀	指示閥	indicator valve
p-ϕ 示功图	p-ϕ 示功圖	p-ϕ indicated diagram
p-V 示功图	p-V 示功圖	p-V indicated diagram
示位标	示位標	position indicating mark
势能(=位能)		
事故	事故	incident
事故报告	損傷報告	casualty report, CASREP
事故树分析	故障樹分析	fault tree analysis
事故污染	事故污染,意外污染	accidental pollution
事故修理	損害修理	damage repair
事故资料	事故數據	incident data
事实推定过失	事實推定過失	factual presumption of fault
事务长	勤務長	chief steward
事务员	事務員	purser
视差	視差	parallax
视差角	天體角	parallactic angle
视差三角形	視差三角形	parallactic triangle
视差移位	視差位移	parallactic displacement
视赤道坐标	視赤道坐標	apparent equatorial coordinates
视赤纬	視赤緯	apparent declination
视出没	視出沒	apparent rise and set
视地平	視海平線	apparent horizon
视顶距	視天頂距	apparent zenith distance
视风	視風向,似風	apparent wind
视风速(=船行风速)		
视高度	視高度,校正高度	apparent altitude, rectified altitude
视功率	視[在]功率	apparent power
视觉通信	視覺通信	visual signaling
视觉信号	視覺信號	visual signal
视力识别	視力識別	visual identification

祖 国 大 陆 名	台 湾 地 区 名	英 文 名
视密度	觀測密度	observed density
视情维修	視情況維修	on-condition maintenance
视太阳	視太陽,真太陽	apparent sun
视[太阳]时	視時,視太陽時	apparent [solar] time
视位置	視位	apparent position
视线角	視[線]角	angle of sight
视周期	視週期	apparent period
试车	試俥	engine trial
试舵	操舵裝置試驗	test the steering gear
试航条件	試航條件	sea trial condition
试验报告	試驗報告	test report
试验负荷	安全載重,安全負荷	proof load,PL
试样(=样品)		
室宿一(飞马α)	室宿一(飛馬α)	Markab
适航吃水差	適航俯仰差	seaworthy trim
适航性	適航性	seaworthiness
适货	適於運貨之(船舶)	cargo worthiness
适任	稱職	performance of competence
适拖	適拖	tow worthiness
适淹礁	平水礁	rock awash
适用航速	適用船速	operating ship speed
适用纬度	適用緯度	operating latitude
适于当值	適於當值	fitness for duty
适运水分限	適運水分限	transportable moisture limit
释放	釋放	release
收报局	收報局	office of destination,O/D
收报人	收信人	addressee
收报人名址	地址	address
收报台	收信台	station of destination
收发开关	收發開關	T-R switch
收好锚	收錨固定	secure the anchor
收回	救回	retrieval
收货待运提单	候裝載貨證券	received for shipment bill of lading
收货单	收貨單	mate's receipt
收货人	受貨人,收貨人,受貨單位	consignee
收桨	收槳	boat the oars
收紧(=归位)		

祖国大陆名	台湾地区名	英 文 名
收拉绞辘	收繁辘轳	bowsing tackle
收锚	收錨	housing anchor
收受设施	收受設施	reception facilities
收妥	收到	acknowledge
收鱼船	收魚船	fish buying boat
手操舵装置	手操舵裝置,笨舵	hand steering gear
手持火焰信号	手把火焰信號	hand flare
手电筒	手電筒	electric torch
手钓	手釣	hand line
手动调整	人工調整	manual setting
手动膨胀阀	手動膨脹閥	hand expansion valve
手斧	手斧	hatchet
手纲	手網	sweep line
手孔	手孔	hand hole
手旗通信	手旗通信	semaphore signaling, signaling by hand flags
手起动	手起動	hand starting
手摇泵	手搖泵	manual pump
守护船	待命船	stand-by ship
守位浮标	示位置浮標	watch buoy
首航(=处女航)		
首要条款	首要條款	paramount clause
艏	艏	bow
艏标志	艏標誌	heading marker
艏波	艏波	bow wave
艏部	艏部	fore body
艏垂线	艏垂標	fore perpendicular
艏导缆孔	艏導索椿,分水艏板	bow chock
艏倒缆	前倒纜	fore spring
艏灯	艏燈	head light
艏舵	艏舵	bow rudder
艏钩篙	掌篙手	bow hook
艏横缆	頭腰纜	bow breast
艏尖舱	艏尖艙	fore peak tank
艏尖舱壁,防撞舱壁	艏艙壁	forepeak bulkhead
艏缆	艏纜	head line
艏离	艏先離	leaving bow first
艏楼	艏樓	forecastle

祖 国 大 陆 名	台 湾 地 区 名	英 文 名
艏楼甲板	艏樓甲板	forecastle deck
艏锚	艏錨	bow anchor, bower
艏门跳板	艏大門跳板	bow ramp
艏碰垫	艏碰墊	bow pudding
艏旗杆	艏旗桿	jack staff
艏上浪	覆浪	green water
艏艉导标	艏艉標	head and stern mark
艏艉锚泊	艏艉碇泊	mooring head and stern
艏艉线	艏艉線	fore-and-aft line
艏舷		bow
艏舷浪	艏側浪	bow sea
艏向	艏向	heading, Hdg
艏向后倒缆	艏向後倒纜	forward back spring leading aft
艏向上	艏向上	head up
艏斜桅	艏斜桅	bowsprit
艏斜桅支索	艏斜桅拉索	bobstay
艏摇	平攏	yaw
艏柱	艏柱,艏材	stem
受风舷	上風舷	windward side
受话人付费电话	受話人付費電話	collect call
受理点	接收點	receiving point
受载期	到港期限	laydays
售后服务	售後服務	after-sales service
枢心	樞心	pivoting point
疏水系统	排洩系統	draining system
疏浚	濬深	dredging
输出阀	輸出閥	delivery valve
输出轴	輸出軸	output axis
输入轴	輸入軸	input axis
输鱼槽	輸魚槽	fish channel
熟练	熟練	proficiency
熟悉培训	熟悉訓練	familiarization training
鼠患检查	鼠患檢查	inspection of rat evidence
鼠笼式电动机	鼠籠式電動機	squirrel-cage motor
数据报告	數據報告	data report
数据电话机	數據電話	dataphone
数据分配计划	數據配送計劃	Data Distribution Plan
数据分组交换网	數據分組交換網	packet switching data network

祖 国 大 陆 名	台 湾 地 区 名	英 文 名
数据复原单元	數據復原單元	data recovery unit
数据记录器	數據記錄器	data logger
数据库使用率	數據庫使用率	database availability ratio
数据库有效率	數據庫有效比率	database effectiveness ratio
数据通信	資訊,數據通信	data communication
数据线路终端	數據線路終端設備	data circuit terminating equipment
数据终端设备	數據終端設備	data terminal equipment
数–模转换器	數位–模擬轉變器	digital-to-analog converter
数字拼读法	數位拼音	figure of mark pronunciation
数字旗	數字旗	numeral flag
数字无线系统	數位無線電系統	digital radio system
数字显示装置	數位顯示裝置	digital display unit
数字选择呼叫	數位選擇呼叫	digital selective calling, DSC
数字选择呼叫设备	數位選擇呼叫裝置	digital selective calling installation
数字选择呼叫系统	數位選擇呼叫系統	digital selective calling system
数字有线系统	數位有線系統	digital line system
衰落	衰落,衰落現象	fading
甩竿钓	甩竿釣(蚊鉤釣)	fly fishing
双半结	雙半套結	two half hitches
双绑	帶雙[纜]	double up
双绑系牢	各纜打雙	double up and secure
双边带	雙邊帶	double side-band, DSB
双层船壳	雙層船殼	double hull
双层的	雙層	two-ply
双层底	[二]重底	double bottom
双层底舱	[二]重底艙	double bottom tank
双差	雙差	double difference
双船围网	雙船圍網	double boat purse seine
双吊联合作业	雙吊桿作業	union crane service
双方责任	雙方責任	dual responsibility
双方责任碰撞	雙邊過失碰撞	both to blame collision
双风管空气调节系统	雙風管空調系統	dual-duct air conditioning system
双杆作业	雙桿固定合吊裝置	union purchase system
双工	雙工	duplex
双工操作	雙工作業	duplex operation
双管设备	雙管設備	double-hose rigs
双航道运河	雙向道運河	two lane canal
双横队	雙橫隊	double line abreast

祖 国 大 陆 名	台 湾 地 区 名	英 文 名
双机单轴式	雙機單軸系統	twin-engine single-shaft system
双机系统	雙機系統	dual system
双机组	雙機組	two-unit
双节点振动	雙節點振動	two-noded vibration
双卷筒绞车	雙筒絞車	double drum winch
双脚桅(=人字桅)		
双壳船	雙殼船	double-skin ship
双联滤器	複式過濾器	duplex strainer
双联曲柄	二拐曲柄	two-throw crank
双流程过热器	雙通道過熱器	two-pass superheater
双曲线导航系统	雙曲線導航系統	hyperbolic navigation system
双曲线位置线	雙曲線位置線	hyperbolic position line
双燃料柴油机	雙燃料柴油機	dual-fuel diesel engine
双刃的	雙刃的	two-edged
双扇形系统	雙扇形系統	two sector system
双速电动机	雙速電動機	two-speed motor
双索吊送传递法	雙索吊送傳遞法	burton method of transfer
双态罗经	雙態羅經	double-state compass
双套结	兜腰稱人結,腰結	bowline on the bight
双套设备	雙套設備	duplication of equipment
双体船	雙體或三體船	catamaran
双筒望远镜	雙筒遠望鏡	binoculars
双头螺栓	螺栓	stud
双头螺纹	雙頭螺紋	two-start screw
双凸轮换向	雙凸輪換向	double cam reversing
双推进器	雙螺槳	twin propellers
双桅帆船	雙桅帆船	schooner
双位	雙位	two-position
双位式调节器	雙位開關調整器	on-off two position regulator
双线制	雙線制	two-wire system
双向航路	雙向航路	two-way route
双向甚高频无线电话设备	雙向特高頻無線電話	two-way VHF radiotelephone apparatus
双向先合后断触点	雙向先合後斷觸點	two-way make-before break contact
双向信标	雙向示標	two course beacon
双向止回阀	雙向止回閥	double check valve, double non-return valve
双胀式蒸汽机	複膨脹式蒸汽機	compound expansion steam engine

祖 国 大 陆 名	台 湾 地 区 名	英 文 名
双支承舵	雙支承舵	double bearing rudder
双职高级船员	雙專長甲級船員	dual purpose officer
双轴系	雙軸系	twin shafting
双转子摆式罗经		twin gyro pendulous gyrocompass
双字母信号码	雙字母信號碼	two letter signal code
双纵队	雙縱隊	double column
双组空气抽逐器	雙組空氣抽射器	two-element air ejector
双作用汽缸	雙動汽缸	double-acting cylinder
双作用式发动机	雙動[動力]機	double-acting engine
水	水	water
水舱式减摇装置	減搖水艙穩定系統	anti-rolling tank stabilization system
水舱涂料	水艙塗料	water tank coating
水产养殖	水産養殖	aquaculture
水产资源	水産資源	fishery resource
水尺检量	測量水尺	draught survey
水处理	水處理	water treatment
水道	水道,航道	channel
水动力	流體動力	hydrodynamic force
水动力力矩	流體動力力矩	moment of hydrodynamic force
水动力系数	流體動力係數	hydrodynamic force coefficient
水分	含水量	water content
水垢沉淀物	水垢沈積	scale deposits
水鼓	水鼓	water drum
水管锅炉	水管鍋爐	water tube boiler
水果船	青果船	fruit carrier
水火成形	水火成形	flame and water forming
水击	水鎚	water hammer
水雷危险区	水雷危險區	areas dangerous due to mines
水力效率	液力效率	hydraulic efficiency
水量调节阀	水量調整閥	water regulating valve
水龙带,消防软管	水龍帶	fire hose
水密	水密	watertight
水密舱壁	水密艙壁	bulkhead resistant to water, watertight bulkhead
水密舱室	水密艙區	watertight compartment
[水密]分舱	艙區劃分	subdivision
水密门	水密門	watertight door
水密完整性	完整水密	watertight integrity

祖 国 大 陆 名	台 湾 地 区 名	英 文 名
水密型	水密型	watertight type
水面饱和水汽压	水面平衡蒸汽壓	water surface equilibrium vapor pressure
水面航行	水面航行	surface navigation
水灭火系统	噴水滅火系統	water fire extinguishing system
水平波束宽度	水平波束寬度	horizontal beam width
水平舵	横舵	horizontal rudder
水平工作范围	水平工作範圍	horizontal working range
水平光弧	水平弧區	horizontal sector
水平夹角位置线	水平角位置線	position line by horizontal angle
水平角定位	水平角定位	fixing by horizontal angle
水平轴阻尼法	水平轴阻尼法	damped method of horizontal axis
水汽压	水汽壓	water vapor pressure
水溶性酸(强酸)	水溶性酸(強酸)	water-soluble acids
水润滑	水潤滑	water lubricating
水上飞机	水上飛機	seaplane
水上通信	水上通信	marine communication
水上运输	水上運輸	marine transportation
水上助航标志系统	浮標系統	buoyage system
水上作业	海上作業	operation at sea
水深测量	水深測量	bathymetric survey
水深信号标	水深信號標誌	depth signal mark
水声对讲机	水聲對講機	hydrophone intercommunicator
水手长	水手長	boatswain, bosun
水手刀	水手刀	jack-knife
水损	海[水漬]損	sea damage
水委一(波江 α)	水委一(屬銀河星座)	Achernar
水位	水位	water level
水位表	水位計	water-level indicator
水下爆破切割	水下爆切	underwater explosive cutting
水下测量作业	水下測量作業	submarine survey work
水下储油罐	水下儲油櫃	underwater oil storage tank
水下倒车	水下倒俥	submerged running astern
水下航行	水下航行	underwater navigation
水下机器人	水下機器人	underwater robot
水下阶段减压法	水下階段減壓分	underwater stage decompression
水下居住舱	水下起居艙	underwater habitat
水下抛锚	水下抛錨	submerged anchor dropping
水下起锚	水下起錨	submerged anchor weighing

祖国大陆名	台湾地区名	英 文 名
水下声标	水下聲標	underwater sound projector
水下系泊装置	水下繫泊設施	underwater mooring device
水下悬浮	水下懸浮	underwater hovering
水下旋回	水下迴旋	underwater turning
水下音响信号	水中音響信號	submarine sound signal
水下游览船	水下觀光船	underwater sightseeing boat
水下游览艇	觀光潛艇	tourist submersible
水下障碍物	水下障礙物	sunken danger
水下作业船	水下作業船	underwater operation ship
水线	水線	waterline
水线面积	水線面積	area of water plane
水线面系数	水線面[積]係數	waterplane coefficient
水线漆	水線漆	boot-topping paint
水星	水星	Mercury
水循环系统	水循環系統	water circulation system
水压计程仪	水壓計程儀	pitometer log
水翼艇	水翼船	hydrofoil craft
水运经济学	海運經濟學	shipping economics
水准测量	水平測量	leveling survey
水准点	基準標誌	bench mark
水准仪	液位計	level gauge
顺风	順風	favorable wind, fair wind
顺风航驶	順風航駛	running free
顺风航行	乘風航行	scudding
顺风换抢	轉向迎風行駛	wearing
顺浪	順浪	following sea
顺浪航行	順浪航行	running with the sea
顺流	順流	favorable current
顺流船	順流船	downstream vessel
顺流掉头	藉流短迴轉掉頭	turning short round with the aid of current
顺时针旋转	順時針旋轉	clockwise rotation
顺水	下游	down stream
顺序单频编码	順序單頻編碼	sequential single frequency code
顺序阀	順序閥	sequence valve
顺应式结构或系统	順應式鑽油台	compliant structures or systems
瞬发临界事故	瞬發臨界事故	prompt critical accident
瞬时过载	瞬時過載	momentary overload
瞬态	瞬時狀態	instantaneous state

祖 国 大 陆 名	台 湾 地 区 名	英 文 名
朔望月	朔望月	synodical month
司太立合金	史斗鉻鈷,銘鉻鎢合金	Stellite
私用助航标	私用助航標	private aid to navigation
四冲程柴油机	四衝程柴油機	four stroke diesel engine
四点方位	四點方位	four point bearing
四点方位法	四點方位法	bow and beam bearing
伺服,随动	伺服	servo
伺服电动机,伺服马达	伺服電動機	servo-motor
伺服控制单元	伺服控制系統	servo control unit
伺服马达(=伺服电动机)		
松出	鬆出	pay out
松掉	鬆掉	cast loose
松紧螺旋扣	伸縮螺絲,緊索螺絲	rigging screw
松开	放鬆,鬆開	ease off
松配合	鬆配合	loose fit
送网管	輸網管	net carrying pipe
搜查证	搜查令狀	search warrant
搜救程序	搜救程序	search and rescue procedure
搜救重发器接收天线	搜救重發器接收天線	SARR receive antenna
搜救单元	搜救單位	search and rescue unit
搜救雷达应答器	搜救雷達詢答器	search and rescue radar transponder
搜救联络点	搜救聯絡點	SAR point of contact
搜救区	搜救區	search and rescue region, SRR
搜救任务协调员	搜救任務協調人	search and rescue mission coordinator, SMC
搜救卫星辅助跟踪	搜救衛星輔助追蹤	search and rescue satellite aided tracking
搜救卫星系统	搜救衛星系統	search and rescue satellite system
搜救协调通信	搜救協調通信	search and rescue coordinating communication, SAR coordinating communications
搜救协调中心	搜救協調中心	Rescue Coordination Center, RCC
搜救业务	搜救業務	search and rescue service, SAR service
搜救中继器	搜救重發器,搜救中繼器	search and rescue repeater
搜索机动	搜索策略	search maneuvre
搜索线	搜索線	sweep
搜寻半径	搜索半徑	search radius

祖 国 大 陆 名	台 湾 地 区 名	英 文 名
搜寻方式	搜索方式	search pattern
搜寻航线	搜索航跡	search track
搜寻基点	搜索基點	search datum
搜寻遇险船舶空间系统	搜索遇險船舶太空系統	space system for search of distress vessels
速闭阀	快閉閥	quick closing valve
速闭装置	快速關閉設施	quick action closing device
速长比	速長比	speed length ratio
速度	速度	velocity
速度级	速度級(輪機)	velocity stage
速度降	速度降	speed drop
速度降旋钮	降速鈕	speed drop knob
速度控制器	速度控制器	speed control assembly
速度设定值	速度設定值	speed setting value
速度误差	速度誤差	speed error
速度误差表	速度誤差表	speed error table
速度误差校正器	速度誤差校正器	speed error corrector
速遣	派遣	despatch
速脱扣	速脫扣	quick release buckle
塑料垃圾袋	塑膠垃圾袋	plastic garbage bag
塑料艇	塑膠艇	plastic boat
随边	殿緣(螺槳)	trailing edge
随船入坞货	船塢貨	dock cargo
随动(=伺服)		
随动控制	追蹤控制,追隨控制	follow-up control
随动速度	追蹤速度	follow-up speed
随动系统	追蹤系統	follow-up system
随动系统灵敏度	隨動系統靈敏度	sensitivity of follow-up system
随机取样法	隨機取樣法	method of random sampling
随机误差	偶發誤差	random error
岁差	歲差	precession
碎冰	碎冰	brash ice
碎冰机	碎冰機	ice crusher
碎冰山	碎冰山	berg-bit
碎浪海面	三角波近海	short sea
碎肉机	碎肉機	meat chopper
损管器材	損[害]管[制]設備	damage control equipment
损害赔偿	損害賠償	compensation for damage
缩帆	縮摺帆葉	reef

祖 国 大 陆 名	台 湾 地 区 名	英 文 名
缩结	縮短結	sheep shank
索股	索股	strand
索具	索具	rigging
索赔代理让书	代位求償書	letter of subrogation
索头环	鋼索接眼	rope socket
索星	尋星	star finding
索星卡	辨星儀,尋星盤	star identifier, star finder
索引图	索引圖,目錄示圖	index chart
锁定开关	鎖定開關	key lock switch
锁轴试验	鎖軸試驗	shaft locked test

T

祖 国 大 陆 名	台 湾 地 区 名	英 文 名
台风	颱風	typhoon
台风警报	颱風警報	typhoon warning, TW
台风路径	颱風路徑	typhoon track
台风眼	颱風眼	typhoon eye
台架试验	試驗台試驗	testing-bed test
台卡	迪凱(電子航儀)	Decca
台卡船位	迪凱船位	Decca fix
台卡导航仪	迪凱導航儀	Decca navigator
台卡海图	迪凱海圖	Decca chart
台卡活页资料	迪凱活頁數據	Decca data sheet
台卡计	相位計	decometer
台卡链	迪凱鏈	Decca chain
台卡位置线	迪凱位置線	Decca position line
台链		chain
太平斧	太平斧	axe
太平洋区	太平洋區域	Pacific Ocean Region, POR
太阳方位表	太陽方位表	sun's azimuth table
太阳黑点	太陽黑點	solar spot
太阳罗经,天体罗经	日規	sun compass
太阳日	太陽日	solar day
太阳质子效应	太陽質子效應	solar proton event
太阳周年视运动	太陽週年[視]運動	solar annual [apparent] motion
太阴全日潮	太陰日週潮	lunar diurnal tide
弹[挠]性联轴器	撓性聯結器	flexible coupling

祖 国 大 陆 名	台 湾 地 区 名	英 文 名
弹射座机构	彈座機構	ejection seat mechanism
弹性极限	彈性限界	elastic limit
弹性联轴器	彈性聯軸節	elastic coupling
弹性流体动力润滑	彈性流體動力潤滑	elasto-hydrodynamic lubrication
弹性模数	彈性係數	elastic modulus
弹性滞后	彈性遲滯	elastic hysteresis
探测	探查	detection
探测系统	探火系統	detecting system
探伤	探傷	crack detection
探照灯	探照燈	search light
碳弧焊	碳[極電]弧熔焊	carbon arc welding
碳环式密封	碳精填料,碳精迫緊	carbon ring gland
糖浆船	糖蜜船	molasses tanker
逃生	逃出	escape
逃生方法	逃生方法	means of escape
逃生通道	安全通道,逃生通道	escape trunk
陶瓷绝缘	陶瓷絕緣	ceramic insulation
陶器	陶器	crockery
套管式冷凝器	二管冷凝器	double-pipe condenser
特别工作灯(=专用灯)		
特别检验	特別檢驗	special survey
特别培训	特別訓練	special training
特别提款权	特別提款權	special drawing right, SDR
特别业务费	特別費用	special charge, SC
特大的	特大	king-size
特大高度	特高高度	very high altitude
特定航次	指定航程	voyage specified
特高频通信	超高頻通信	UHF communication
特急操纵	緊急操縱	crash maneuvering
特殊重复频率	特殊重複頻率	specific repetition frequency
特殊强制办法	特殊強制方法	special mandatory method
特殊情况	特殊狀況	special circumstances
特殊区域	特別海域	special area
特性数	示性數	characteristic number
特种处所	特種空間	special category space
特种货	特種貨	special cargo
特种业务	特種業務	special service
特种业务客船协定	特殊貿易客船協約	Special Trade Passenger Ships Agreement

祖 国 大 陆 名	台 湾 地 区 名	英 文 名
特种业务旅客	特殊貿易客船	special trade passenger
特种证书	特種證書	special certificate
藤壶	海生介	barnacle
梯道	梯道	stairway
梯度风	梯度風	gradient wind
梯队	梯隊	echelon formation
梯级结	梯級結	ratline hitch
梯形牌	梯形板	trapezoidal board
提单	提單,載貨證券	bill of lading, BL
提单背书	載貨證券之背書	endorsement of bill of lading
提单持有人	載貨證券持有人	holder of bill of lading
提单转让	轉讓載貨證券	transfer of bill of lading
提货单	提貨通知單,提貨單	delivery order
体积温度系数	容積穩定修正係數	volume-temperature correction coefficient
体积系数	容積換算係數	volume conversion coefficient
天波	天波	sky wave
[天波]分裂	脫裂	splitting
天波改正量	天波修正量	sky wave correction
天波延迟	天波延遲	sky wave delay
天波延迟曲线	天波延遲曲線	sky wave delay curves
天赤道	天球赤道	celestial equator
天船三(英仙α)	天船三(英仙α)	Mirfak
天窗	天窗	skylight
天底	天底點	nadir
天顶	天頂	zenith
天顶距	天頂距	zenith distance
天顶投影	天頂投影法	zenith projection
天极	天極	celestial pole
天津四(天鹅α)	天津四(天鵝α)	Deneb
天空状况	天空狀況	sky condition
天狼(大犬α)	天狼(大犬α)	Sirius
天幕	天遮	awning
天气	天氣	weather
天气报告	氣象報告	weather report
天气符号	天氣符號	weather symbol
天气公报	天氣公報	weather bulletin
天气过程	天氣過程	synoptic process
天气图	天氣圖	synoptic chart

祖国大陆名	台湾地区名	英 文 名
天气现象	天氣現象	weather phenomena
天气形势	綜觀[天氣]大勢	synoptic situation
天气预报	天氣預報	weather forecast
天桥	連橋	connecting bridge
天球	天球	celestial sphere
天囷一(鲸鱼 α)	天囷一(鯨魚 α)	Menkar
天枢(大熊 α)	天樞,北斗一(大熊 α)	Dubhe
天体	天體(星)	celestial body
天体出没	天體出沒	rise and set of celestial body
[天体]方位角	方位角	azimuth
天体高度	天體高度	celestial altitude
天体罗经(=太阳罗经)		
天体视运动	天體視運動	celestial body apparent motion
天体与太阳角距	天體與太陽之角距	elongation
天文潮	天文潮汐	astronomical tide
天文晨昏朦影	天文曦光	astronomical twilight
天文船位	天文定位	astronomical fix, AF
天文定位	天體定位	celestial fixing
天文定向	天文定向	astronomical orientation
天文观测	測天	celestial observation
天文航海	天文航海	celestial navigation
天文经度	天文經度	astronomical longitude
天文三角形	天文三角形	astronomical triangle
天文纬度	天文緯度	astronomical latitude
天文钟	天文鐘,船鐘	chronometer
[天文钟]日差	天文鐘日差率	chronometer rate, daily rate
天文钟误差	天文鐘誤差	chronometer error, CE
天文子午圈	天文子午圈	astronomical meridian
天文子午线	天文子午線	astronomical meridian
天文坐标	天文坐標	astronomical coordinate
天线调谐	天線調諧	antenna tuning
天线开关	天線開關	antenna switch
天线效应	天線效應	antenna effect
天线罩	天線外罩	radome
天象纪要	天文現象	phenomena
天象图,星图	天象圖	sky diagram
天璇(大熊 β)	天璇,北斗二(大熊 β)	Merak
天轴	天軸	celestial axis

祖国大陆名	台湾地区名	英文名
添注漏斗	灌斗,舱口灌斗	feeder
填舱货	補充貨載	berth cargo
填角焊	填角熔焊	fillet welding
填空货	填空貨	short stowage cargo
填隙货	填隙貨	filler cargo
调节棒	調整桿	regulating rod
调节级	調節級	governing stage
调距桨传动	可控距螺槳傳動	controllable pitch propeller transmission
调速阀	調速閥	speed regulating valve
调速特性	調速特性	speed regulating characteristic
调相,相位调制	相位調變	phase modulation
调谐	調諧	tune
跳板	跳板	gang board
跳板台	跳板著陸架	brow landing
跳周	跳週	cycle skipping
铁梨木轴承	鐵梨木軸承	lignum vitae bearing
铁模	縱移	swage
停泊费,滞留费	船舶滯留費,碇泊費,入港稅	groundage
停泊值班	碇泊值班	harbor watch
停潮	平潮	water stand
停车冲程	慣性停俥距離	inertial stopping distance
停车	停轉	shutdown
停船试验	停船試驗	stopping test
停船性能	停船性能	stopping ability
停堆深度		shut-down depth
停桨!	到了(操艇口令)	way enough
停拉	停曳	avast hauling
停用	停播,停用,停電	outage
停油位置	停供位置	cut-out position
停租	停租	off-hire
艇	小艇	boat
艇长	司艇	coxswain
艇底塞	艇底塞	boat plug
艇筏乘员定额	艇筏乘載量	carrying capacity of craft
艇筏配员	救生艇筏人員配額	manning of lifecraft
艇滑架	艇滑道	boat skate
艇碰垫	小碰墊	dolphin

祖 国 大 陆 名	台 湾 地 区 名	英 文 名
艇首缆	艇首索	bow painter, painter
艇天幕	小艇天遮	canopy
艇系紧带	小艇扣带	boat gripe
艇下水装置	小艇下水設施	boat launching appliance
艇罩	艇罩	boat cover
通报表	通報表	traffic list
通播发射台	通播發射台	collective broadcast sending station, CBSS
通播接收台	通播接收台	collective broadcast receiving station, CBRS
通道	通道	alley way
通风口	通氣孔	vent
通风帽	通風帽	ventilating cowl
通风筒	通風筒,通風器	ventilator
通风与空调机间	通風與空氣調節機室	ventilation and air-conditioning room
通海阀	海水閥	sea valve
通海阀箱	海底門	sea chest
通海接头	通海裝置	sea connection
通航分道	航行巷道	traffic lane
通航桥孔	可航橋孔	navigable bridge-opening
通航水域	適航水域	navigable waters
通话	通話	call
通话完毕(无线电话用语)	通話完畢(無線電通話用語)	over
通气管	通氣管	air pipe
通气箱(**储气箱**)	空氣儲蓄器	air container
通商航海条约	通商航海條約	treaty of commerce and navigation
通信	通信	communication
通信记录	通信紀錄簿	communication log
通信闪光灯	通信閃光燈	flashing light for signaling
通信卫星	通信衛星	communication satellite
通信协议	通信協定	communication protocol
通信询问	通信詢問	traffic enquiry
通信业务量	業務通報量	traffic
通行权	通行權	right of passage
通行信号标	通航標誌	traffic mark
通用泵	通用泵,常用泵	general service pump
通用的,万用的	通用的,萬用的	universal
通用证书	通用證書	general certificate

祖国大陆名	台湾地区名	英文名
通知方	通知方	notify party
同步	起伏一致	in step
同步发电机	同步發電機	synchronous generator
同步感应电动机	同步感應電動機	synchronous induction motor
同步器	同步器,協調器	synchronizer
同步器旋钮	同步旋鈕	synchronizer knob
同步卫星	同步衛星	synchronous satellite
同步指示灯	同步指示燈	synchro light
同步指示器	同步儀	synchroscope
同步阻抗	同步阻抗	synchronous impedance
同文电报	同文電文	common text message
同一的	同一的	uniform
同一责任制	同一責任制	uniform liability system
同轴发电机	共軸發電機	integral shaft generator
铜基轴承合金	銅基軸承合金	copper base bearing metals
铜铅轴承	銅鉛軸承	copper-lead bearing
统舱	統艙	steerage
统长甲板	連續甲板	continuous deck
统一船舶碰撞或其他航行事故中刑事管辖权某些规定的国际公约	關於碰撞或其他航行事件之統一刑事管轄國際公約	International Convention for the Unification of Certain Rules Relating to Penal Jurisdiction in Matters of Collision or Other Incidents of Navigation
统一船舶碰撞某些法律规定的国际公约	船舶碰撞法律統一規定國際公約	International Convention for the Unification of Certain Rules of Law with Respect to Collision between Vessels
统一对水上飞机的海难援助和救助及由水上飞机施救的某些规定的国际公约	救助與撈救海上航空器及由航空器施救之統一規定國際公約	International Convention for the Unification of Certain Rules Relating to Assistance and Salvage of Aircraft or by Aircraft at Sea
统一国有船舶豁免某些规定的国际公约	國有船舶豁免權統一規定國際公約	International Convention for the Unification of Certain Rules Concerning the Immunity of State-owned Ships
统一海船扣押某些规定的国际公约	統一海船假扣押規定國際公約	International Convention for the Unification of Certain Rules Relating to the Arrest of Seagoing Ships
统一海难援助和救助某些法律规定公约	海上救助及撈救統一規定公約	Convention for Unification of Certain Rules of Law Relating to Assistance and Salvage at Sea

祖 国 大 陆 名	台 湾 地 区 名	英 文 名
桶	桶(燃油之單位)	barrel
桶形浮标	桶狀浮標	barrel buoy
筒	筒	trunk
筒钩	筒鉤	can hook
筒形活塞式柴油机	筒狀活塞型柴油機	trunk piston type diesel engine
偷渡	偷渡者	stowaway
头尾对换	首尾顛倒換索	end for end
投网	投網	cast net
透气管	通風管	vent pipe
凸轮	凸輪	cam
凸轮控制器	凸輪控制器	cam controller
凸轮轴	凸輪軸	camshaft
突堤	海堤	mole
突码头	碼頭,突堤	jetty
突然倾斜	[船]突傾側	lurch
图号	圖號	chart number
涂层	塗層	coating, paint
涂料	塗料	coating, paint
涂煤油试验	煤油試驗	kerosene test
涂漆	塗刷	painting
土星	土星	Saturn
湍流,紊流	擾動,擾流	turbulence
推船	推船	pusher, pushboat
推定全损	推定全損	constructive total loss
推荐航线	推薦航路	recommended route
推进器	螺槳,推進器	propeller
推进式	推進式	push type
推进特性	推進特性	propulsion characteristic
推进装置	推進設施	propulsion device
推理机	推理機	inference engine
推力	推力	thrust
推力环	推力軸環	thrust collar
推力块	推力履片	thrust shoe
推力轴	推力軸	thrust shaft
推力轴承	推力軸承	thrust bearing
推算船位	估計船位	estimated position, EP
推算航程	實際距離,終結距離	distance made good
推算航迹向	估計航向	estimated course

祖 国 大 陆 名	台 湾 地 区 名	英 文 名
推算航速	實在速率, 終結速率	speed made good
推算经度	估計經度	estimated longitude
推算始点	出航點, 出發點	departure point
推算纬度	估計緯度	estimated latitude
退关	退貨	shut out
退关货	退關貨	shut-out cargo
退化	退化	degenerate
退火	退火	annealing
退桨!	反划!	back water!
托架	托架	cradle
托列莫利诺斯国际渔船 安全公约	托里莫列路斯漁船安全 國際公約	Torremolinos International Convention for Safety of Fishing Vessels
托盘	托板	pellet
托盘货	托板貨	palletized cargo
托运人	託運人	shipper
拖材结	曳索套結	timber and half hitch
拖船	拖船	tug, towing vessel
拖带长度	拖帶長度	length of tow
拖带灯	拖航號燈	towing light
拖带责任	拖帶責任	liability of towage
拖动电动机	拖動馬達	drive motor
拖钩	拖纜鉤	towing hook
拖航合同	拖船合約	towage contract
拖缆	拖纜	towing line
拖缆承架	拖索承梁	towing beam
拖缆弓架	拖纜拱架	towing arch
拖缆机	拖纜絞機	towing winch
拖缆限位器	拖纜限位器	stop posts for towline
拖缆桩	拖纜柱, 拖纜椿	towing bitt
拖轮费	拖船費	tug hire
拖锚, 走锚	拖錨	dragging anchor
拖锚航行	拖錨航行	dredging
拖锚滑行	拖錨溜行	clubbing
拖网	拖網	trawl
拖网浮标	拖網浮標	trawl buoy
拖网渔船	拖網漁船	trawler
拖网作业	拖網作業	trawling, trawl fishing
拖曳, 牵引	拖曳, 牵引	towing

祖国大陆名	台湾地区名	英 文 名
拖曳船队	拖曳船隊	towing train
拖曳设备	拖曳設備	towing gear
拖曳作业工况管理	拖曳作業方式管理	towing operating mode management
拖桩	拖纜樁	towing post
脱钩链段	鵜嘴形滑脱鉤鏈段	senhouse slip shot
脱扣	跳脱	trip
脱扣线圈	跳脱線圈	tripping coil
脱扣装置	跳脱設施	trip device
脱浅	浮飄	refloat
脱网	下網	log out
脱轴试验	脱軸試驗	shaft disengaged test
陀罗北	電羅經北	gyrocompass north
陀罗差	電羅經誤差	gyrocompass error
陀罗方位	電羅經方位	gyrocompass bearing, GB
陀罗航向	電羅經航向	gyrocompass course, GC
陀螺磁罗经	電磁羅經	gyro-magnetic compass
陀螺六分仪	電動水平六分儀	gyro sextant
陀螺罗经	電羅經	gyrocompass
陀螺罗经室	電羅經室	gyro compass room
陀螺漂移	迴轉儀漂移	gyro drift
陀螺球	迴轉球	gyro sphere
陀螺式减摇装置	迴轉穩定器	gyro[scopic] stabilizer
陀螺仪	迴轉機,迴轉儀	gyroscope, gyro
椭球体	椭球體	ellipsoid
椭圆极化	椭圓橜化	elliptical polarization
椭圆[型]艉	椭圓艉	elliptical stern

W

祖国大陆名	台湾地区名	英 文 名
挖泥船	挖泥船,挖泥機	dredger
挖泥船自控程序	挖泥船自控程序	automated control system of dredging process
挖泥工具	挖泥設施	dredging facility
外板展开图	外板展開圖	shell expansion plan
外补偿法	外補償法	method of outer compensation
外观检查	外觀檢查	observation check
外海渔业	外海漁業	off-sea fishery

祖 国 大 陆 名	台 湾 地 区 名	英 文 名
外航路	外航路	offshore tracks
外进汽	外邊進氣	outside admission
外径千分尺	外分厘卡,外測微計	outside micrometer
外牵索	外牽索	outhaul line
外燃机	外燃機	external combustion engine
外围设备	週邊設備	peripheral equipment
外舷	外檔	outboard
外斜轴系		diverging shafting
外旋	往外轉	outward turning
外置式圆柱滚子轴承	外置滾子軸承	outboard roller bearings
外置轴承	船外軸承	outboard bearing
弯曲疲劳破裂	彎曲疲勞破裂	bending fatigue cracks
完全燃烧	完全燃燒	complete combustion
完全运行能力	完全運作能力	full operation capability
完全运行状态	完全運作狀況	full operational status
完善性监测台	整合監視器	integrity monitor
完整货	混載貨	completing cargo
完整稳性	完整穩度	intact stability
万能装卸机	堆高機	fork lift truck
万向导缆器	萬向導纜器	universal fairleader
万向接头	萬向接頭	universal joint
万用的(=通用的)		
王良四(仙后 α)	王良四(仙后 α)	Schedar
王良一(仙后 β)	王良一(仙后 β)	Caph
网板	網板	otter board
网板架	網板架	trawl gallows
网船	網船	net boat
网档间距	網間距	distance between twin trawl
网络控制中心	網路控制中心	network control center
网络协调站	網路協調電台	network coordination station, NCS
网络协调站到陆地地球 站信令信道	網路協調電台與陸上衛 星電台之信號頻道	NCS/LES signaling channel
网络协调站到网络协调 站信令信道	網路協調電台間之信號 頻道	NCS/NCS signaling channel
网络协调站共用时分多 路复用信道	網路協調電台共用劃時 多工制頻道	NCS common TDM channel
网位仪	網位儀	net monitor
网衣	網片	netting

祖 国 大 陆 名	台 湾 地 区 名	英 文 名
网状责任制	網狀責任制	network liability system
往复泵	往復泵	reciprocating pump
往复式发动机	往復機	reciprocating engine
往复式制冷压缩机	往復冷凍壓縮機	reciprocating refrigeration compressor
往复式转舵机构	往復式操舵裝置	reciprocating type steering gear
望远镜方位仪	望遠鏡照準儀	telescopic alidade
危机处理和人的行为	危機處理及行爲管理	crisis management and human behaviors
危险半圆	危險半圈	dangerous semicircle
危险标志	危險物品標誌	dangerous mark
危险电文	危險消息	danger message
危险海岸	危險海岸	dangerous coast
危险货物报告	危險貨物報告	dangerous goods report
危险货物锚地	危險貨物錨地	dangerous cargo anchorage
危险货[物],危险品	危險貨物	dangerous cargo
危险津贴	危險加給	danger money
危险品(=危险货[物])		
危险品清单	危險貨物清單	dangerous cargo list
危险区域	危險區域	hazardous areas
危险水域	航行危險水域	foul water
危险天气通报	危險天氣通報	hazardous weather message
危险[物]	危險	danger
危险象限	危險象限	dangerous quadrant
威廉逊旋回法	威廉生掉頭法	Williamson turn
微波测距系统	微波測距系統	microwave ranging system
微处理器	微處理機	micro processor
微分调节器	微分調整器	derivative regulator
微机	微電腦	microcomputer
微机控制系统	微電腦控制系統	microcomputer control system
微机控制主机遥控系统	微電腦遙控主機系統	microcomputer remote control system for main engine
微浪(1 级)	静海,浪静(1 级)	calm(rippled) sea
微量	微量	trace
围井	圍壁	trunk
围裙救生圈	圍裙救生圈	breech buoy
围网	圍網,巾著網	purse seine
围网渔船	圍網漁船	purse seiner
围网作业	旋網漁業	surrounding fishing
围堰,空隔舱	堰艙,圍堰	cofferdam

祖 国 大 陆 名	台 湾 地 区 名	英 文 名
围油栏	攔油索	oil boom, oil fence
违反	違反	violation
违禁物品(**违禁品**)	違禁品	prohibited articles
违章排放	違規排洩	discharge in violation of regulations
桅	桅	mast
桅灯	桅頂燈	masthead light
桅顶横杆	桅頂橫桿	crosstree
桅冠灯	桅冠燈	truck light
桅套	靴,桅跟帆套	boot
维护,保养	維護,保養	maintenance
维护保养废弃物	維護保養所生廢棄物	maintenance waste
维斯比规则	威斯比規則	Visby Rules
维修前平均使用时间	平均修復間隔時間	mean-time-to-repair
伪距	偽距	pseudo range
伪卫星	偽訊號	pseudolite
伪装商船	偽裝商船	decoy ship, q-ship
尾流	航跡流	trailing wake
尾门	艉門	stern door
尾随行驶	尾隨行駛	following at a distance
尾拖网	尾拖網	stern trawl
纬差	緯差	difference of latitude
纬度改正量	緯度修正值	latitude correction
纬度渐长率	緯度漸長比	meridional parts, MP
纬度渐长率差	緯度漸長比數差	difference of meridianal parts, DMP
纬[度]圈(=纬线)		
纬度误差	緯度誤差	latitude error
纬度误差校正器	緯度誤差校正器	latitude error corrector
纬度效应	緯度效應	latitude effect
纬度因数	緯度因數	latitude factor
纬线,纬[度]圈	緯度平行圈	parallel of latitude
委付	委付	abandonment
艉	船尾	stern
艉部接近法(**艉接近法**)	艉接近法	astern approaching method
艉垂线	後垂標	aft perpendicular
艉灯	艉燈	sternlight
艉风	艉風	wind aft
艉横缆	艉橫纜	stern breast
艉机型船	艉機艙船	stern engined ship

祖 国 大 陆 名	台 湾 地 区 名	英 文 名
艉迹	艉跡	back track
艉尖舱	艉[尖]艙	aft peak tank
艉尖舱壁	艉艙壁	after peak bulkhead
艉缆	艉纜	stern line
艉离	艉先離	leaving stern first
艉龙骨墩(=[吊杆]跟 部滑车)		
艉楼	艉艛	poop
艉楼甲板	艉艛甲板	poop deck
艉锚	艉錨	poop anchor, stern anchor
艉门跳板	艉門跳板	stern ramp
艉膨出部	艉膨出部	bossing
艉旗杆	艉旗桿	ensign staff
艉倾	艉俯,艉坐	trim by stern, by the stern
艉升高甲板船	高艉主甲板船	raised quarter-deck vessel
艉舷	艉部	quarter
艉舷浪	艉侧浪	quartering sea
艉向后倒缆	艉向後倒纜	stern aft spring
艉向前倒缆	艉向前倒纜	stern forward spring
艉斜跳板	艉斜跳板	quarter ramp
艉轴	艉軸	tail shaft, stern shaft
艉轴承	艉軸承	stern bearing
艉轴管	艉軸套	stern tube
艉轴管滑油	艉軸管潤滑油	stern tube lubricating oil
艉轴管密封装置	艉軸管密封裝置	stern tube sealing
艉轴管填料函	艉軸管填料函	stern tube stuffing box
艉轴管轴封泵	艉軸管軸封油泵	stern tube sealing oil pump
艉轴架	艉軸架	shaft bracket
艉柱	艉柱	stern post
鲔钓船	鮪釣船	tuna long liner
鲔钓母船	鮪釣母船	tuna mother ship
卫生泵	衛生泵	sanitary pump
卫生水系统	衛生系統	sanitary system
卫生水压力柜	壓力衛生水櫃	sanitary pressure tank
卫星标准频率和时间信 号业务	衛星標準頻時信號業務	standard frequency and time signal satel- lite service
卫星船位	衛星定位	satellite fix
卫星导航系统	衛星導航系統	satellite navigation system

祖 国 大 陆 名	台 湾 地 区 名	英 文 名
卫星导航仪	衛星導航儀	satellite navigator
卫星电文	衛星信文	satellite message
卫星多普勒定位	衛星都卜勒定位	satellite Doppler positioning
卫星覆盖区	衛星覆蓋區	satellite coverage
卫星广播业务	衛星廣播業務	broadcasting-satellite service
卫星轨道	衛星軌道	satellite orbit
卫星海上移动业务	衛星水上行動業務,水上行動衛星通信業務	maritime mobile satellite service
卫星紧急无线电示位标	衛星應急指位無線電示標	satellite emergency position-indicating radiobeacon
[卫星]历书	[衛星]曆書	[satellite] almanac
卫星链路	衛星鏈路	satellite link
卫星气象业务	衛星氣象業務	meteorological-satellite service
卫星摄动轨道	衛星攪動軌道	satellite disturbed orbit
卫星搜救跟踪系统	衛星輔助搜救追蹤系統	search and rescue satellite aided tracking
卫星通信	衛星通信	satellite communication
卫星网络	衛星網路	satellite network
卫星无线电测定业务	衛星無線電測定業務	radio determination-satellite service
卫星系统	衛星系統	satellite system
[卫星]星历	衛星曆表	[satellite] ephemeris
卫星星座	衛星星座	satellite constellation
卫星移动业务	衛星行動業務	mobile-satellite service
卫星云图	衛星雲圖	satellite cloud picture
卫星转发海上安全信息	衛星轉發之海上安全資訊	maritime safety information via satellite
未包装货	未包裝貨	non-packed cargo
未定位警报	未能定位之警報	unlocated alert
未精测的等高线	近似等高線	approximate contour line
未精测的等深线	近似等深線	approximate depth contour
未列名货	未列名貨	N. O. S. cargo
未入级的	未入級	unclassified
位变率	方位變動率	rate of bearing variation
位能,势能	位能,勢能	potential energy
位势高度	地勢高度	geopotential height
位置	部位,位置	location
位置稳定性	位置穩定性	positional stability
位置线	位置線	line of position, LOP
位置线标准差	位置線標準誤差	position line standard error

祖国大陆名	台湾地区名	英文名
位置线梯度	位置線梯度	gradient of position line
位置信号码	位置信號碼	position signal code
温带气旋	溫帶氣旋	extratropical cyclone
温度补偿	溫度補償	temperature compensation
温度调节器	溫度調節器	thermoregulator
温度继电器	溫度開關	temperature switch
温控系统	控溫系統	temperature controlling system
温暖	溫暖	warm
文本	原文,本文	text
紊流(=湍流)		
稳定的	穩定的	steady
稳定回路	穩定迴路	stabilized loop
稳定时间	安定時間	settling time
稳定索	穩定索	steadying lines
稳定位置	調定位置	settling position
稳定性	穩度	stability
稳索	牽索	guy
稳索眼板	牽索眼板	guy eye
稳态	穩定狀態	steady [state]
稳心	定傾中心	metacenter
稳心半径	定傾半徑	metacentric radius
稳心高度	定傾[中心]高度	metacentric height
稳性衡准数	穩度基準數	stability criterion numeral
稳性力臂	穩度力臂	stability lever
稳性力矩	穩度力矩	stability moment
稳性消失角	穩度消失角	vanishing angle of stability
涡流室式柴油机	渦流室柴油機	swirl-chamber diesel engine
涡流阻力	興渦阻力	eddy making resistance
涡轮发电机	渦輪發電機	turbogenerator
涡轮机	渦輪機	turbine
涡轮可变几何形状喷嘴装置	渦輪可變幾何形狀噴嘴裝置	variable geometric turbine nozzle device
涡轮驱动	渦輪驅動	turbo-
涡轮增压器	渦輪增壓器	turbocharger, turboblowen
卧具储藏室	寢具儲存室	linen locker
握索结	扶手索結	manrope knot
握住	挽住	hold
污底	積垢	fouling

祖 国 大 陆 名	台 湾 地 区 名	英 文 名
污底螺旋桨特性曲线	污船體螺槳特性曲線	fouled hull propeller curve
污底阻力	積垢熱阻	fouling resistance
污秽货	污穢貨	dirty cargo
污泥,油泥	油泥	sludge
污染类别	污染類別	pollution category
污染物	污染物	pollutants
污染源	污染源	cause of pollution
污染指数	污染指數	contamination index
污水	艙底污水,舭水	bilge water
污水泵	污水泵	sewage pump
污水处理设备	污水處理設備	sewage treatment plant
污水道	通水小孔	limber
污水道盖板	通水道蓋板	limber board
污水柜	舭櫃,舭水艙	bilge tank
污水井	污水井	bilge well
污水排吸装置	舭水抽排裝置	bilge pumping arrangement
污水系统	舭水系统	bilge system
污水自动排除装置	舭水自動排除設施	bilge automatic discharging device
污油泵	污水泵,排渣泵	sludge pump
污油舱	污油櫃	dirty oil tank
污油水	污油水	slop
污油水舱	污油[水]櫃	slop tank
无冰区	無冰區	ice free
无潮点	無潮點	amphidromic point
无潮区	無潮區	amphidromic region
无灯标志	無燈標誌	unlit mark
无方向性信标	無方向性示標	non-directional beacon, NDB
无杆锚	無桿錨,山字錨	stockless anchor
无功负荷	無功負載	wattless load
无功功率	無功功率	wattless power, reactive power
无功功率自动分配装置	無功功率自動分配裝置	automatic distributor of reactive power
无害通过权	無害通過權	right of innocent passage
无级的	無段	stepless
无级调速	無段調速	stepless speed regulation
无缆系结	無纜索連結	non-line connection
无浪	無風,浪靜	calm
无铺位旅客	艙面旅客	deck passenger
无气喷射	無空氣噴射	airless injection

祖国大陆名	台湾地区名	英 文 名
无人管理的	無人當值	unmanned
无人机舱	無人化機艙	unmanned machinery space
无人值班的	無人當值	unattended
无人[值班]机舱,无人机舱	無人當值機艙,無人化機艙	unattended machinery space
无刷交流发电机	無刷交流發電機	brushless AC generator
无线电报	無線電報	radiotelegram
无线电报设备	無線電報裝置	radiotelegraph installation
无线电报学	無線電報術	Wireless telegraphy
无线电报员	無線電人員	radio officer
无线电报自动报警[器]	無線電報自動警報[器]	radiotelegraph auto-alarm
无线电操作员	無線電操作人員	radio operator
无线电测定	無線電測定術	radio determination
无线电测定电台	無線電測定電台	radio determination station
无线电测定业务	無線電測定業務	radio determination service
无线电测向	無線電測向	radio direction finding
无线电测向电台	無線電探向電台	radio direction-finding station
无线电测向仪	無線電測向儀	radio direction finder
无线电测向仪自差	無線電測向儀自差	radio direction finder deviation
无线电大圆方位	無線電大圓方位	radio great circle bearing
无线电导航	無線電導航	radio navigation
无线电导航陆地电台	陸上無線電導航電台	radio navigation land station
无线电导航图网	無線電導航網路圖	radio navigational lattice
无线电导航业务	無線電導航業務	radio navigation service
无线电导航移动电台	行動無線電導航電台	radio navigation mobile station
无线电定位陆地电台	陸上無線電定位電台	radiolocation land station
无线电定位[术]	無線電定位[術]	radiolocation
无线电定位业务	無線電探向電台	radio location service
无线电定位移动电台	行動無線電定位電台	radiolocation mobile station
无线电方位	無線電方位	radio bearing
无线电方位位置线	無線電方位位置線	radio bearing position line
无线电规则	無線電規則	radio regulation
无线电航行警告	無線電航行警告	radio navigational warning
无线电话报警信号发生器	無線電話警報信號産生器	radiotelephone alarm signal generator
无线电话呼叫	無線電話呼叫	radiotelephone call
无线电话设备	無線電話裝置	radiotelephone installation
无线电话学	無線電話術	radio telephony

祖 国 大 陆 名	台 湾 地 区 名	英 文 名
无线电话业务	無線電話業務	radiotelephone service
无线电话遇险频率	無線電話遇險頻率	radiotelephone distress frequency
无线电话员	無線電話務員	radiotelephone officer
无线电接力系统	無線電中繼系統	radio relay system
无线电经纬仪	無線電經緯儀	radio theodolite
无线电免检电报	無線電檢疫信文	radio pratique message
无线电气象业务	無線電氣象業務	radio weather service
无线电人员	無線電人員	radio personnel
无线电时号	無線電對時信號	radio time signal
无线电室	無線電室	radio room
无线电台	無線電台	radio station
无线电台日志	無線電日誌	radio log
无线电通信	無線電通信	radiocommunication
无线电通信业务	無線電通信業務	radiocommunication service
无线电舷角	無線電相對方位	relative bearing of radio
无线电信标	無線電指示標	radio beacon
无线电信标电台	無線電示標電台,無線電示台	radiobeacon station
无线电信号表	無線電信號表	list of radio signals
无线电业务	無線電業務	radio service
无线电用户电报	無線電電報交換	telex over radio, TOR
无线电用户电报电文	無線電傳電報	radiotelexogram
无线电用户电报呼叫	無線電傳呼叫	radio telex call
无线电用户电报书信	無線電交換電報書信	radio telex letter
无线电用户电报业务	無線電電報交換業務	radio telex service
无线电真方位	無線電真方位	radio true bearing, RTB
无线电值守	無線電當值	Radio Watch
无线线路	無線電鏈	radio link
无效果–无报酬	無效無償	no cure-no pay
无叶式扩压器	無葉擴散器	vaneless diffuser
无用发射	無用發射	unwanted emission
无用能	無用能	unavailable energy
无油润滑空气压缩机	無油潤滑空氣壓縮機	oil free air compressor
无阻尼的	無阻尼	undamped
五车二(御夫 α)	五車二(御夫 α)	Capella
午圈	上子午線	upper branch of meridian
坞内检验	入塢檢驗	docking survey
坞修	進塢檢修	dock repair

祖 国 大 陆 名	台 湾 地 区 名	英 文 名
物标能见地平距离	物標水平視距	distance to the horizon from object
物理起动	物理起動	physical start-up
物料单	物料單	stores list
误差理论	誤差理論	theory of errors
误发遇险警报	假遇險警報	false distress alerts
误卸	誤卸	mislanded
雾	霧	fog
雾泊(**雾中抛描**)	霧中抛錨	anchoring in fog
雾笛	霧笛	fog whistle
雾号	霧號	fog signal
雾化	霧化	atomization
雾化器	霧化器	atomizer
雾角	霧角	fog horn
雾警报	霧警報	fog warning
雾雷	霧雷	fog siren
雾锣	霧鑼	fog gong
雾中航行	霧航	navigating in fog

X

祖 国 大 陆 名	台 湾 地 区 名	英 文 名
西大距(行星)	西大距(行星)	greatest western elongation
西方标	西方標	west mark
吸井	吸引井	suction well
吸泥泵	吸泥泵	air lift mud pump
吸气行程	吸入衝程	suction stroke
吸入流	吸流	suction current
吸入压头	吸入高度	suction head
吸收材料	吸收材	absorbent material
吸收剂	吸收劑	absorption agent
吸收制冷	吸收冷凍	absorption refrigeration
吸扬式挖泥船	泵吸挖泥船	pump dredger
吸油口加热盘管	貨油吸入加熱盤管	cargo oil suction heating coil
吸鱼泵	吸魚泵	fish pump
牺牲阳极保护	耗蝕性陽極防護	sacrificial anode protection
舾装数	船具規號, 屬具數	equipment number
锡基白合金衬层	錫基白金襯層	tin-base white-metal linings
习惯装卸速度	依慣例快速處理	customary quick despatch, CQD

祖 国 大 陆 名	台 湾 地 区 名	英 文 名
洗舱泵	洗艙泵,巴特華斯泵（洗艙用）	Butterworth pump, tank cleaning pump
洗舱机	洗艙機	tank washing machine
洗舱口	洗艙開口	tank washing opening
洗舱水	洗艙水	tank washing water
洗涤器	清洗器,擦洗器,洗氣器,清除器	scrubber
洗碗水	洗盤水	dishwater
洗衣间	洗衣間	laundry
喜玛拉雅条款	喜瑪拉雅條款	Himalaya clause
系泊	舷緣殼板	berthing
系泊工况管理	繫泊作業管理	mooring operating mode management
系泊绞车	繫船絞車	mooring winch
系泊绞盘	繫泊絞盤	mooring capstan
系泊锚	繫留錨,碇泊錨	mooring anchor
系泊试验	繫泊試俥	dock trial, mooring trial
系船浮[筒]	繫泊浮筒,繫船浮筒	mooring buoy
系船索(=缆绳)		
系船桩	繫船樁	dolphin
系浮标的投海货物	附有浮標之投海貨物	lagan
系浮筒	繫浮筒	securing to buoy
系缆	繫泊纜索,繫纜	mooring line
系缆活结	繫纜活结	slip racking
系缆桩	繫船樁	mooring bitt
系艇杆	繫艇桿	boat boom
系索耳(=羊角)		
系统	系統,制度	system
系统操作测试	系統操作測試	system operability test
系统操作图	系統流程圖	system flow chart
系统故障	系統故障	system fail
系统观察	系統觀測	systematic observation
系统误差	系統誤差	systematic error
系统响应	系統回應	system response
系统信息	系統資訊	system information
系住	繫住	bend on
系桩拉力	繫纜樁拖力	bollard pull
细牙螺纹	細牙螺絲,密螺紋	fine thread
隙缝波导天线	槽嵌波導天線	slotted waveguide antenna

祖 国 大 陆 名	台 湾 地 区 名	英 文 名
狭长流冰区	狭長流冰	stream ice
狭水道	狭窄水道	narrow channel
狭水道操纵	狭水道中操縱	maneuvering in narrow channel
狭水道航行	狭水道航行	navigating in narrow channel, channel navigation
下半旗	下半旗	flag at halfmast
下二层甲板	下方中甲板	lower tween deck
下风岸	下風岸	lee shore
下风船	下風船	vessel to leeward
下风满舵	下風滿舵	hard down
下风舷	下風側,下風舷	lee side
下纲	下綱	foot line
下甲板	下甲板	lower deck
下降风	頹風	katabatic
下潜平台	下潛平台	submersible platform
下水	下水	launching
下水日期	下水日期	date of launching
下水设备	下水設備	launching appliance
下水站	下水站	launching station
下水装置	下水裝置	launching arrangement
下弦	下弦	last quarter
下限越界	下限越界	off-normal lower
下行(来自)	由…來	bound from
下行船	下行船	down-bound vessel
下行行程	下行衝程	downstroke
下游	下游	lower reach
下止点	下死點	bottom dead center, BDC
下中天	[天體]下中天	lower transit, lower culmination
夏季区带	夏期地帶	summer zone
夏季用润滑油	夏季用潤滑油	summer oil
夏季载重线	夏期載重線	summer load line
夏令时	夏令時	summer time
夏至点	夏至	summer solstice
纤维绳	纖維[繩]索	fiber rope
闲置船	停航	lay up
舷边角钢	舷緣板角鐵	stringer angle
舷侧	舷側	broadside
舷侧板	側邊板	side plate

祖国大陆名	台湾地区名	英 文 名
舷侧拖网船	舷側拖網船	beam trawler
舷侧纵桁	側縱材	side stringer
舷窗	舷窗	side scuttle, scuttle
舷窗内盖	舷窗内蓋	deadlight
舷灯	舷燈	sidelight
舷灯遮板	舷燈遮光板	screen of sidelight
舷顶列板	舷側厚板列	sheer strake
舷角,相对方位	相對方位	relative bearing
舷门	舷門	side port
舷门跳板	舷門跳板	side ramp
舷内	船内	inboard
舷旁排出口	舷外排出口	overboard discharge outlet
舷墙	舷牆	bulwark
舷墙排水口	舷牆排水口	bulwark freeing port
舷墙系索器	舷牆繫索器	bulwark gripper
舷梯	舷梯	accommodation ladder
舷梯绞车	舷梯絞機	accommodation ladder winch
舷梯强度试验	舷梯安全限試驗	proof test for accommodation ladder
舷外吊杆	舷外吊桿	outboard boom
舷外排出阀	舷外排洩閥	overboard discharge valve
舷外排水孔	舷外排水孔	overboard scupper
舷外作业	舷外作業	outboard work
显示方式	顯示方式	display mode
CRT 显示器	陰極射線管顯示器	cathode-ray tube display
险恶地	障礙地區	foul ground
现场纪录	現場紀錄	record on spot
现场校正	現場矯直	straightened up in place
现场试验	現場試驗	field test
现场通信	現場通信	on-scene communication
现场通信管制	現場通信管制	control of on scene communication
现场指挥	現場指揮	on-scene commander, OSC
现有船	現存船舶,現成船	existing ship
限界线	邊際線	margin line
限量危险品	限量危險品	dangerous goods in limited quantity
限流起动器	限流起動器	current-limiting starter
限速	限制速度	limiting speed
限速器	限速器	speed-limiting governor
限用操作员证书	限用值機員證書	restricted operator's certificate, ROC

祖 国 大 陆 名	台 湾 地 区 名	英 文 名
限于吃水船	吃水受限船	vessel constrained by her draught
限制	限制	restriction
限制降落区标志	限制降落區域標誌	restricted landing area mark
限制特性	限制特性	limited characteristic
限制纬度	限制緯度	limiting latitude
相对定位	相對定位	relative positioning
相位方位(＝舷角)		
相对风速	相對風速	relative speed of wind
相对计程仪	相對計程儀	relative log
相对密度	相對密度	relative density
相对免赔额	起賠限額	franchise
相对湿度	相對濕度	relative humidity
相对速度	相對速[率]	relative speed
相对效率	相對效率	relative efficiency
相对运动雷达	相對運動雷達	relative motion radar, RM radar
相对[运动]显示	相對運動顯示	relative motion display
相互接近	互相接近	approaching one another
相容性	相容性	compatibility
箱格导柱	[貨櫃]導槽	cell guide
箱形舱盖	箱形艙蓋	pontoon hatch cover, pontoon cover
箱形机架	箱形機架	box frame
箱形内龙骨	方形內龍骨	box keelson
箱序号	貨櫃序號	container serial number
箱主代号	櫃主碼	container owner code
箱装货	箱裝貨	cased cargo
响应滞后	反應落後	response lag
相位编码	相位編碼	phase coding
相位差	相[位]差	phase difference
相位日变化	相位日變	diurnal phase change
相位调制(＝调相)		
相位突然异常	相位突然異常	sudden phase anomaly, SPA
相序	相序	phase sequence
相移键控	相移鍵移	phase shift keying, PSK
向东航程	東橫距	easting
向后	向後	aback
向流性	向流性	rheotaxis
向前倒缆	向前斜出各纜	forward-leading springs
向前	向前	ahead

祖 国 大 陆 名	台 湾 地 区 名	英 文 名
向上风(=在上风)		
向上拉	向上紧	bowse up
向位换算	方向换算	conversion of directions
向西航行	往西航	westing
向下拉	向下紧	bowse down
向心式涡轮	向心式渦輪機	centripetal turbine
巷	公定航道,巷道	lane
巷号	巷號	lane letter
巷宽	巷寬	lane width
巷设定	巷設定	lane set
巷识别	巷識別	lane identification, LI
巷识别计	巷識別計	lane identification meter
象限自差	象限自差	quadrantal deviation
橡胶垫	橡皮墊	rubber mat
橡胶轴承	橡膠軸承	rubber bearing
橡皮艇	橡皮艇	rubber boat
消磁按钮开关	消磁按鈕開關	degauss push button switch
消磁场	消磁場	degaussing range
消防泵	救火泵	fire pump
消防部署	消防部署	fire fighting station
消防船	救火船,消防艇,消防船	fire boat
消防规范	防火規則	fire protecting rules
消防控制站	火警控制站	fire control station
消防软管(=水龙带)		
消防栓	消防栓,水龍頭	hydrant
消防系统	滅火系統	fire extinguishing system
消防巡逻制度	火警巡邏系統	fire patrol system
消防演习	消防演習	fire fighting drill
消防员装备	消防員裝具	fireman's outfit
消防总管	主消防水管,救火主水管	fire main
小潮升	小潮升	neap rise, NR
小改正	小修正	small correction
小河	溪	creek
小浪(二级)	微浪	smooth sea
小锚	小錨,艉錨,小移船錨	kedge anchor
小水线面双体船	小水線面雙體船	small water plane twin hull ship, SWATH
小艇结	鬆套結	slippery hitch

祖 国 大 陆 名	台 湾 地 区 名	英 文 名
小湾	小灣	creek
小修	小修	minor overhaul
肖氏硬度	反跳硬度	Scleroscope hardness
楔	楔	wedge
协调避碰操纵	協調避碰操縱	coordinated collision avoidance maneuver
协调世界时	協調世界時	coordinated universal time, UTC
协定航线	航路	shipping route
协作横移线搜寻	協調橫移線搜索	coordinated creep line search
斜桁	[縱帆]斜桁	gaff
谐波发射	諧波發射	harmonic emission
谐波分析	諧波分析	harmonic analysis
谐波振动	諧波振動	harmonic vibration
谐摇	同步	synchronous rolling, synchronism
谐振(=共振)		
泄滑下水	洩滑下水	evacuation-slide launching
泄漏	洩漏	leak
卸荷阀	卸載閥	unloading valve
卸货	卸貨	discharging
卸货港	卸貨港	port of discharge
卸扣	卸扣	shackle
卸压阀	洩壓閥,保險閥	relief valve
卸载	卸載	discharge
心环	繩眼襯環	thimble
心宿二(天蝎 α)	心宿二,大火(天蝎 α)	Antarcs
心形[方向]特性图	心形極圖解	cardioid polar diagram
辛氏法则	辛普生法則	Simpson's rules
锌	鋅	zinc
锌基合金	鋅基合金	zinc-base alloy
新版图	新版海圖	new edition chart
新风	新鮮空氣	fresh air
新杰森条款	紐哲遜條款	New Jason clause
新危险物标志	新危險物標誌	new danger mark
新月	新月,朔	new moon
信标	標台	beacon
信标登记	示標登記	beacon register
信标登记国通告	示標登記國通告	notification of country of beacon registration
信标电文	示標信文	beacon message

祖 国 大 陆 名	台 湾 地 区 名	英 文 名
信标定位概率	示標定位概率	beacon location probability
信标发射天线	示標發射天線	beacon transmit antenna
信标检测概率	示標檢測概率	beacon detection probability
信标识别	示標識別	beacon identification
信标识别数据	示標識別數據	beacon identification data
信标台	信標發射台	beacon station
信道存储	頻道存儲	channel storage
信道模型	頻道模型	channel model
信道申请	頻道申請	channel request
信风	信風	trade wind
信号标志	信號標誌	signal mark
信号弹	信號彈	signal shell
信号灯	號誌柱	signal light
信号控制	信號控制	signal control
信号旗	信號旗	signal flag
信号设备	信號設備	signaling appliance
信号绳	信號繩	signal line
信号桅	信號桅	signal mast
信令分组	信號分組	signaling packet
信文标志	信文標誌	message marker
信文范围	信文欄	message field
信文筛选因素	信文篩檢因素	message filtering factor
信文转换时间	信文轉換時間	message transfer time
信息处理	資訊處理	information processing
信息处理机	資訊處理機	information processing machine
信息发送台	資訊發送台	information sending station
信息接收台	資訊接收台	information receiving station
信息流	資訊流	information flow
信用卡电话	信用卡電話	credit card call
信用证	信用狀	letter of credit, L/C
兴波阻力	興波阻力	wave making resistance
星表	星表	star catalogue
星号	星號	star number
星基导航系统	衛星基導航系統	satellite based navigational system
星历	天文曆	ephemeris
星历数据	衛星運行數據表,衛星軌道數據,天文曆數據	ephemeris data

祖 国 大 陆 名	台 湾 地 区 名	英 文 名
星球仪	示星球	star globe
星–三角起动	星角起动	star-delta starting
星宿一(长蛇α)	星宿一(長蛇α)	Alphard
星图(=天象图)		
星位角	星位角	parallactic angle
星形–三角形接法	星形三角連接	star-delta connection
星型发动机	星型發動機	radial engine
星座	星座	constellation
行程,摆幅	動程,曲柄臂	throw
行程缸径比	衝程口徑比	stroke-bore ratio, S/B
行李	行李	baggage, luggage
行李间	行李間	baggage room, luggage room
行李损坏赔款限额	行李損失責任限制	limit of liability for loss of or damage to luggage
行星轨道	行星軌道	planetary orbit
行星视运动	行星視運動	planet apparent motion
行星型齿轮转动机构	太陽行星齒輪	sun and planet gear
A 形机架	A 形托架	A-frame
Z 形试验,标准操纵性试验	蛇航試驗,標準操縱性試驗	standard maneuvering test
形状稳性力臂	形狀穩度力臂	lever of form stability
V 型柴油机	V 型柴油機	vee-type diesel engine
Z 型传动	Z 型傳動	Z transmission, Z drive
U 型管	U 型管	u-pipe
D 型环	D 型環,拉繫環	d-ring
型宽	模寬	molded breadth
MP 型链	多脈搏型鏈	multi-pulse mode chain, MP mode chain
V 型链	V 型鏈	v-mode chain
V 型起重柱	V 型吊桿柱	v-type derrick post
型深	模深	molded depth
型式	型,式	type
型砧	型鐵砧,花砧	swage block
性能标准	性能標準	performance standard
性能监视器	性能監測器	performance monitor
性能验证测试	性能審認試驗	performance verification test
汹涛阻力	狂浪阻力	rough sea resistance
休息室	休息室	lounge
休渔期	休渔期	[fishing] season off

祖国大陆名	台湾地区名	英文名
休止角	(散装货)止倾角,静止角	angle of repose
修理单	修理單	repair list
修理港	修理港	repair port
修正回路	修正迴路	corrective loop
锈	銹	rust
溴化锂吸收式制冷装置	溴化鋰水吸收冷凍裝置	lithium bromide water absorption refrigerating plant
许可舱长	艙區許可長度	permissible length of compartment
许用应力	容許應力	allowable stress, permissible stress
续航力	續航力,持久	endurance, cruising radius
蓄电池	蓄電池	accumulator battery
轩辕十四(狮子 α)	軒轅十四(獅子 α)	Regulus
宣港	港口申報	declaration of port
宣载	載重噸申報	declaration of dead weight tonnage of cargo
悬高杆长比	懸高桿長比	suspension height boom length ratio
悬挂舵	懸舵	hanging rudder, under hung rudder
悬降区标志	懸降區域標誌	winching area mark
悬链锚腿系泊	懸垂法錨泊(鑽油台)	catenary anchor leg mooring
悬伸型艉	懸伸艉	counter stern
旋回半径	迴旋半徑	turning radius, tactical diameter
旋回圈	旋回圈,迴轉圈,迴旋圈	turning circle
旋回试验	迴旋試驗,迴旋試俥	turning trail, turning circle trial
旋回性	迴轉能力	turning ability
旋回性指数	迴旋指數	turning indices
旋回直径	終結直徑	final diameter
旋回周期	迴旋週期	turning period
旋进性	迴轉偏移	gyroscopic precession
旋塞	旋塞	cock
旋网	旋網	round haul net
旋涡	旋渦,渦動,渦流	vortex
旋涡泵	旋流泵	helical flow pump
旋涡式雾化器	漩渦式霧化器	swirl type atomizer
旋圆双半结	繞轉加雙半套結	round turn and two half hitches
旋转磁场	旋轉磁場	rotating magnetic field
旋转环形天线	旋轉環形天線	rotary loop antenna
旋转角速度	迴轉速率	turning rate
旋转失速	旋轉失速	rotating stall

祖 国 大 陆 名	台 湾 地 区 名	英 文 名
旋转视窗	旋轉視窗	clean view screen
漩水	渦流,旋渦,漩渦	eddy
选港货,选卸货	卸地待定貨	optional cargo
选卸货(=选港货)		
选择船位	假定船位	assumed position
选择呼叫	選擇呼叫	selective calling
选择呼叫号码	選擇呼叫號碼	selective calling number
选择经度	經度採用值	assumed longitude
选择可用性	選擇可用性	selective availability, SA
选择前向纠错	選擇前向偵錯	selective forward error correction
选择前向纠错方式	選擇前向偵錯模式	selective error correcting mode
选择纬度	緯度採用值	assumed latitude
选择性保护	選擇性保護	selectivity protection
选择性广播发射台	選擇性廣播發射台	selective broadcast sending station, SBSS
选择性广播接收台	選擇性廣播接收台	selective broadcast receiving station, SBRS
学徒(=练习生)		
雪盖冰	雪蓋冰	snow-covered ice
薰舱	煙燻法,燻蒸法	fumigation
寻位	定位	locating
寻位信号	定位信號	locating signal
寻位与归航信号	定位與導向信號	locating and homing signals
巡航编队与部署	巡航編隊與序列	cruising formation and disposition
巡航工况管理	巡航操作型式管理	cruising operating mode management
巡航机组	巡航機組	cruising engine unit
巡航涡轮机	巡航渦輪機	cruising turbine
巡回监测器	圓轉偵測器	circular monitor
巡逻船	巡邏艇	patrol boat
巡逻艇信号	巡邏艇信號	patrol boat signal
巡洋舰	巡洋艦	cruiser
巡洋舰[型]艉	巡洋艦型艉	cruiser stern
循环泵	循環泵	circulating pump
循环柜	循環櫃	circulating tank
循环检验	連續檢驗	continuous survey
循环润滑	環流潤滑	circulating lubrication
循环水倍率	循環水率	circulating water ratio
循环水槽	環流水槽	circulating water channel

Y

祖 国 大 陆 名	台 湾 地 区 名	英 文 名
压比	壓力比	pressure ratio
压电效应	壓電效應	piezoelectric effect
压舵,整流舵	壓舵,整流舵	counter rudder
压紧板	押板	keep plate
压扩	壓縮擴展	companding
压力级	壓力級	pressure stage
压力壳	壓力容器	pressure vessel
压力控制阀	壓力控制閥	pressure-control valve
压力水柜	壓力水櫃	water pressure tank
压力调节器	制壓器,調壓器,壓力調整裝置	pressure regulator
压力真空切断阀	壓力真空斷路器	pressure and vacuum breaker
压力-重力式滑油系统	重力進給潤滑系統	gravity forced-feed oiling system
压气机喘振试验	壓縮機顫動試驗	compressor surging test
压气机特性	壓縮機特性	compressor characteristics
压气排水打捞	壓縮空氣排水浮升	raising by dewatering with compressed air
压汽式蒸馏装置	蒸汽壓縮蒸餾裝置	vapor compression distillation plant
压燃式发动机	壓燃式引擎	self-ignition engine
压水试验	壓水試驗	water head test
压缩比	壓縮比	compression ratio
压缩环	壓縮脹圈	compression ring
压缩机油	壓縮機油	compressor oil
压缩空气起动系统	壓縮空氣起動系統	compression air starting system
压缩气体	壓縮氣體	compressed gas
压缩室容积	壓縮室容積	compression chamber volume
压缩图	壓縮圖	compression diagram
压缩行程	壓縮衝程	compression stroke
压缩压力	壓縮壓力	compression pressure
压向下风	乘風而駛	drive
压载	壓載	ballasting
压载泵	壓載泵	ballast pump
压载舱	壓載艙	ballast tank
压载舱漆	壓載艙漆	ballast-tank paint
压载水	壓載水,壓艙水	ballast water

祖 国 大 陆 名	台 湾 地 区 名	英 文 名
压载水系统	壓艙水系統	ballast system
压载状态	壓載船況	ballast condition
哑点	電點,無效點,消盡點	null point
哑控	靜音	muting
哑罗经	啞羅經	dumb card compass, pelorus
亚音速	次音速	subsonic velocity
烟囱标记	煙囪標記	funnel mark
烟囱漆	煙囪漆	funnel paint
烟斗形通风筒	煙斗形通風筒	cowl head ventilator
烟灰沉积物	煙灰沈積物	soot deposits
烟灰着火	煙灰著火	soot fires
烟火信号	煙火信號	pyrotechnic signal
烟迹式烟度计	濾紙測煙計	Bosch filter paper smoke meter
烟盔	防煙盔	smoke helmet
烟气分析	氣體分析	gas analysis
延迟交货	延遲交貨	delay in delivery
延迟开航	延遲發航	delay of ship
延期回扣制	延期回扣制度	deferred rebate system
延伸报警	延伸警報	extension alarm
延伸海事声明	遭難證明書	extended protest
延伸轴,中介轴	延伸軸	extension shaft
延绳钓	延繩釣,長繩釣	long line
沿岸标	沿岸標誌	alongshore mark
沿岸冰带	冰繐	ice fringe
沿岸测量	沿岸測量	coastwise survey
沿岸地形	沿岸地形	coastal feature
沿岸航行	沿岸航行	coastal navigation
沿岸警告	沿海警告	coastal warning
沿岸流	沿岸流	coastal current, littoral current
沿岸通航带	沿岸通航區	inshore traffic zone
沿岸图	沿岸海圖,沿海海圖	coastal chart
沿海船	沿海船	coaster
沿海国	沿海國	coastal state
沿海航行	近岸航行	coastal trip, coastwise navigation
沿海航运	沿海航運,沿海貿易	cabotage
沿海航运权	沿海航運權	right cabotage
盐度	鹵水密度	brine density
盐度计	鹽份計,鹽度計	salinometer

祖 国 大 陆 名	台 湾 地 区 名	英 文 名
盐水泵	鹽水泵	brine pump
眼板	眼板	pad eye
眼高	眼高	height of eye
眼高差	眼高差	dip
眼高差改正	眼高差修正	dip correction
眼环[插]接	眼索接琵琶頭	eye splice
演习区	演習區	practice area, exercise area
验船师	驗船師,公證人,檢驗師	surveyor
燕尾槽	鳩屋槽	dovetail groove
扬尘货	粉狀貨	dusty cargo
扬帆结	上桅揚帆結	topsail halyard bend
扬声器通信	揚聲器通信	loud speaker signaling
羊角,系索耳	繫索扣	cleat
阳极防腐	陽極防蝕	anodic protection
杨氏模数	楊氏模數,彈性模數	Young's modulus
洋流图	洋流圖	ocean current chart
洋区	洋區	ocean regions
洋区码	洋區碼	ocean region code
仰极	仰極	elevated pole
仰极高度	仰極高度	polar altitude
氧化安定性	氧化穩定性	oxidation stability
氧化剂	氧化物質	oxidizing substance
氧气电弧切割	氧氣電弧截割	oxy-arc cutting
氧乙炔焊	氧乙炔熔接	oxy-acetylene welding
样品,试样	樣品,試樣	sample
样品供给泵	樣品供給泵	sample feed pump
腰横缆	舯横纜	waist breast
摇摆误差	搖擺誤差	rolling error
遥控	遙控	telecontrol
遥控扫雷航海勤务	遙控掃雷航海勤務	remote control mine-sweeping navigation service
遥控异常报警	遙控異常警報	remote control abnormal alarm
遥控指令	遙控指令	telecommand
咬缸	咬缸,氣缸膠著	piston seizure, cylinder sticking
药剂泵	複式泵	compound pump
业务代码	業務代碼	service code
业务公电	業務通知	service advice
业务衡准数	業務基準數	criterion of service numeral

祖 国 大 陆 名	台 湾 地 区 名	英 文 名
业务信号	業務信號	service signal
叶尖漏泄损失	葉梢漏洩損失	tip-leakage loss
叶轮	葉輪	impeller, blade wheel
叶片泵	轉葉泵	rotary vane pump, vane pump
叶元体理论	葉片元素理論	blade element theory
曳纲	曳網, 曳繩	warp
曳开桥	曳開橋	draw bridge
曳绳钓	曳繩釣	troll line
曳绳钓起线机	曳繩釣起繩機	trolling gurdy
曳网	地曳網	drag net
夜航命令簿	夜航命令簿	night order book
夜间效应	夜間效應	night effect
夜视六分仪	夜視六分儀	night vision sextant
液舱鉴定	液艙櫃檢查	inspection of tank
液浮陀螺仪	液浮迴轉儀	liquid floated gyroscope
液化化学气体	液化化學氣體	liquefied chemical gas
液化气船	液化氣體船	liquefied gas carrier
液化气体	液化氣體	liquefied gas
液化石油气	液化石油氣	liquefied petroleum gas, LPG
液化石油气船	液化石油氣船	liquefied petroleum gas carrier
液化天然气	液化天然氣	liquefied natural gas, LNG
液化天然气船	液化天然氣船	liquefied natural gas carrier
液货	液狀貨, 液體貨	fluid cargo, liquid cargo
液货泵	液貨泵	liquid pump
液货船	液貨船	liquid cargo ship, tanker
液货船管系	液貨船管路系統	tanker piping system
液控单向阀	液控止回閥	hydraulic control non-return valve
液密	液密	liquid-tight, resistant to liquid
液体动力润滑	流體動力潤滑	hydrodynamic lubrication
液体化学品船	化學液體船	liquid chemical tanker
液体静力润滑	流體靜力潤滑	hydrostatic lubrication
液体罗经	濕羅經	liquid compass
液体润滑	液體潤滑	liquid lubrication
液体散货	液體散貨	liquid bulk cargo
液体物质	液體物質	liquid substance
液体阻尼器	液體阻尼器	liquid damping vessel
液压泵	液力泵	hydraulic pump
液压变矩器	液壓變矩器	hydraulic moment converter, hydraulic

祖国大陆名	台湾地区名	英　文　名
		moment variator
液压变速[传动]装置	液力變速驅動裝置	hydraulic variable speed driver
液压舱盖	液壓艙口蓋	hydraulic hatch cover
液压操纵阀	液力操縱閥	hydraulic operated valve
液压操纵货油阀	液力操作液貨閥	hydraulic operated cargo valve
液压传动	液力傳動	hydraulic transmission [drive]
液压传动装置	液力傳動機構	hydraulic [transmission] gear
液压舵机	液壓操舵裝置	hydraulic steering engine
液压发送器	液壓傳動器	hydraulic transmitter
液压放大器	液壓放大器	hydraulic amplifier
液压缸	液壓缸	hydro cylinder
液压缓冲器	液力緩衝器	hydraulic buffer
液压换向阀	液力方向控制閥	hydraulic directional control valve
液压减速[传动]装置	液壓減速裝置	hydraulic reduction gear
液压接头	液壓接頭	hydraulic joint
液压控制阀	液壓控制閥	hydraulic control valve
液压离合器	液壓離合器	hydraulic [friction] clutch
液压起货机	液壓起貨機	hydraulic cargo winch
液压上支撑	液壓上支撐	hydraulic top bracing
液压升压器	液力增壓器	hydraulic booster
液压式排气阀传动机构	液力致動排氣閥機構	hydraulically actuated exhaust valve mech- anism
液压式调速器	液力調速器	hydraulic governor
液压试验	液壓試驗	hydraulic pressure test
液压伺服阀	液力伺服閥	hydraulic servo valve
液压伺服马达	液力伺服馬達	hydraulic servo-motor
液压锁	液壓鎖	hydraulic lock
液压锁闭装置	液力鎖閉設施	hydraulic blocking device
液压系统	液壓系統	hydraulic system
液压蓄能器	液壓儲蓄器	hydraulic accumulator
液压遥控传动装置	液壓遙控裝置	hydraulic telemotor
液压油	液壓油	hydraulic oil
液压油柜	液壓油櫃	hydraulic oil tank
液压[油]马达	液壓馬達	hydraulic motor, fluid motor
液压执行机构	液力致動機構	hydraulic actuating gear
液压制动器	液壓軔,液壓煞俥,液力剎	hydraulic brake
一般闪光	一般閃光	ordinary flash

祖 国 大 陆 名	台 湾 地 区 名	英 文 名
一般通信	一般通信	general communication
一般照明	一般照明	general lighting
一次场	一次場	primary field
一次惯性力		the first order inertia force
一次力矩补偿器	一次力矩補償器	the first order moment compensator
一次配电系统	一次配電系統	primary distribution system
一次屏蔽水系统	一次屏蔽水系統	primary shield water system
一点锚,平行锚	一字力錨	riding one point anchors
一挂	一掛	a hoist
一回路	一次迴路	primary loop
一级水手	一等水手	able-bodied seaman, AB
一级无线电电子证书	第一級無線電電子員證書	first-class radio electronic certificate
一天航程	一日行程	day's run
一字锚泊	一字雙錨泊	moor
一字双锚泊	雙錨繫泊	mooring to two anchors
医疗	醫療	medical care
医疗电文	醫療信文	medical messages
医疗援助	醫療協助	medical assistance
医疗运输	醫療運送,醫療傳送	medical transports
医疗指导	醫療指導	medical advice
医务室	醫務室	medical premises
医院船	醫院船	hospital ship
仪表板(=控制台)		
仪表引航技术	盲目導航技術	blind pilotage techniques
移泊	移位	shifting
移动单元	行動單元	mobile unit
移动地球站	行動地球台,行動衛星電台	mobile earth station
移动地球站信息信道	行動地球台資訊頻道	MES message channel
移动地球站状态	行動地球台狀態	mobile earth station status
移动电台	機動電台	mobile station
移动罐柜	可攜式櫃	portable tank
移动式起重机,滑马	吊運車	traveller
移动	衝程	travel
移动业务	行動業務	mobile service
移动重量式减摇装置	多重式穩定器	moving-weight stabilizer
移频电报	移頻電報	frequency shift telegraphy

祖 国 大 陆 名	台 湾 地 区 名	英 文 名
移线船位	航進定位	running fix, RF
移线定位	航進定位	running fixing
移载法	移重法	shifting weight method
疑存	疑有	existence doubtful, ED
疑位	可疑位置	position doubtful, PD
乙级分隔	"B"級區(防火)	B class division
已充气设备	已充氣設備	inflated appliance
已定位报警	經定位之警報	located alert
已装船提单	已裝船載貨證券,裝船載貨證券	on board bill of lading, shipped bill of lading
异常	異常	anomaly
异常磁区	磁異常區	magnetic anomaly
异常磨损	異常磨損	abnormal wear
异常喷射	異常噴射	abnormal injection
抑制偏摆试验	止擺試驗	yaw checking test
抑制载波发射	遏止載波發射	suppressed carrier emission
译码,解码	譯碼	decoding
译码器	譯碼機	decoder
易爆货	易爆炸貨	explosive cargo
易腐货	易腐貨	perishable cargo
易流态化物质	易液化物質	material which may liquefy
易燃固体	易燃固體	flammable solid
易燃限度	易燃限度	inflammable limit
易燃液体	易燃液體	flammable liquid
易碎货	易碎貨	fragile cargo
易自燃固体	易自燃固體	flammable solid liable to spontaneous combustion
意大利船级社	意大利驗船協會	Registro Italiano, RI
溢出	溢出	spilling
溢流阀	溢流閥	overflow valve
溢流管	溢流管	overflow pipe
溢卸	溢卸	over-landed, over-delivery
溢油柜	溢流櫃	overflow tank
逸出	逸出	escape
因果链	因果鏈	chains of causation
阴	陰	covered
阴极防腐	陰極防蝕	cathodic protection
阴影扇形	扇形陰影	shadow sector

祖国大陆名	台湾地区名	英文名
阴影图	陰影圖	shadow diagram
音响榴弹	音響信號彈	sound signal shell
银河	銀河	Milky Way
银河系	銀河系	Galaxy
引潮力	生潮力,引潮力	tide-generating force
引导	導向	homing
引导滑车	導滑車	leading block
引航	地文導航法	piloting
引航班	導航組	piloting team
引航船	引水船	pilot vessel
引航费	引水費	pilotage
引航锚地	引水錨地	pilot anchorage
引航信号	引領信號	homing signal
引航学	引水術	pilotage
引航员	引水,引水人	pilot
引航员软梯	引水人梯	pilot ladder
引航员升降设备	引水人升降機	pilot hoist
引航站	引水站	pilot station
引缆	傳遞索	messenger
引缆回收索	回收索	messenger return line
饮水泵	飲用水泵	drinking water pump
饮用水臭氧消毒器	飲用水臭氧消毒器	drinking water ozone disinfector
饮用水矿化器	飲用水礦化器	mineralizing equipment of drinking
饮用水系统	飲用水系統	drinking water system
印度洋区	印度洋區域	Indian Ocean Region, IOR
应变部署表	部署表	muster list, station bill
应变仪传感器	應變計測感器	strain-gauge transducer
应答信号	收到信號,確認信號	acknowledge signal
应舵时间	迴轉回應延遲	delay of turning response
应付票据	付款單	bills payable
应付日期	到期日	due date
应急操舵装置	應急操舵裝置,應急舵機	emergency steering gear
应急操纵	應急操縱	emergency maneuvering
应急电池	應急電池	emergency cell
应急电气设备	應急電力設備	emergency electric equipment
应急电源	應急電源	emergency power source, emergency source of electrical power

祖国大陆名	台湾地区名	英 文 名
应急电站	應急動力站	emergency power station
应急舵	應急舵	jury rudder
应急发电柴油机	應急柴油發電機	emergency generator diesel engine
应急发电机	應急發電機	emergency generator
应急鼓风机	應急鼓風機	emergency blower
应急空气压缩机	應急用空氣壓縮機	emergency air compressor
应急罗经	應急羅經	emergency compass
应急配电板	應急配電板	emergency switchboard
应急起动	應急起動	emergency starting
应急说明	應急說明	emergency instruction
应急速闭阀	應急速閉閥	quick-closing emergency valve
应急天线	應急天線	emergency antenna
应急停车	緊急停俥	emergency stop
应急停车装置	緊急關閉設施	emergency shut-down device
应急消防泵	應急救火泵	emergency fire pump
应急业务	應急業務	emergency service
应急照明	應急照明	emergency lighting
应急照明系统	應急照明系統	emergency lighting system
应急准备	應急準備	emergency preparedness
应力腐蚀裂纹	應力腐蝕龜裂	stress corrosion cracking
应力集中	應力集中	concentration of stress, stress concentration
应力疲劳	應力疲勞	stress fatigue
迎风换抢	迎風棹餕	tacking
营救浮环	營救浮環	buoyant rescue quoit
营救器电台	營救器電台	survival station
营运吃水	航務吃水	operating draft
硬搓[绳]	硬搓索法	hard lay
硬度数	硬度數	hardness number
硬盔	硬盔	rigid helmet
佣金	傭金	commission
拥挤人群管理培训	群眾管理訓練	crowd management training
永久船磁	永久船磁	ship permanent magnetism
永久性修理	徹底檢查	permanent repair
涌潮	怒潮	tidal bore
涌级	湧級	swell scale
涌浪	湧	swell
涌浪高度	湧高	swell height

祖 国 大 陆 名	台 湾 地 区 名	英 文 名
用户等效距离误差	用戶等效距離誤差	user equivalent range error, UERE
用户电报电话	用戶電報電話	telex telephony, TEXTEL
用户电报书信业务	電報交換書信業務	telex letter service
用户电报业务	電報交換業務	telex service
用户码	用戶碼	subscriber number
优先权	優先權	right of priority
邮船	郵船	mail ship
邮件舱	郵件室	mail room
邮件及行李间	郵件及行李間	mail and baggage room
邮箱业务	郵箱業務	mailbox service
油泵定时标记	泵定時記號	pump timing mark
油舱涂料	油艙塗料	oil tank coating
油船	運油船	oil tanker
油船锚地	油輪錨地	[oil] tanker anchorage
油底壳	油槽	oil sump
油分计	油含量計	oil content meter
油分离器	油水分離器	oil separator
油分瞬时排放率	油份瞬間排洩率	instantaneous rate of discharge of oil content
油封期	拆封前儲存壽命	storage life before unpack
油迹	油跡	traces of oil
油结胶	油黏結	caking of oil
油类	油類	oil
油类沉积物	油[類]殘留物	oily residues
油类燃料	燃油	oil fuel
油轮清洁海洋指南	油輪海洋清潔指南	clean seas guide for oil tankers
油码头	油終端站	oil terminal
油密	油密	oil-tight, resistant to oil
油泥(=污泥)		
油泥焚化炉	油泥焚化爐	sludge incinerator
油泥柜	油泥櫃,油泥艙櫃	sludge tank
油破布	含油破布	oily rags
油漆	油漆,漆	coating, paint
油漆间	油漆間	paint room
油气分离器	油氣分離器	oil-air separator
油燃烧器	燃油器	oil burning unit
油润滑	油潤滑	oil lubricating
油水分离器	油水分離器	oily water separator

祖 国 大 陆 名	台 湾 地 区 名	英 文 名
油水分离设备	油水分離設備	oily-water separating equipment
油水界面	油水分界面	oil/water interface
油水界面探测仪	油水分界面探測儀	oil water interface detector
油污染	油污	oil pollution
油污水	[含]油污水	oily water
油污水处理船	油污水處理船	oily water disposal boat
油污损害	油污損害	damage from oil pollution
油雾浓度探测器	油霧偵測器	oil mist detector
油压压差控制器	油壓差控制器	oil pressure differential controller
油毡	油氈	linoleum
游览船	遊覽船	excursion boat
游丝	遊絲	hair wire
游艇	遊艇	pleasure yacht, yacht
鱿鱼钓机	釣鱿魚機	squid angling machine
有毒货	毒性貨	poisonous cargo
有毒物质	有毒物質	poisonous substance
有杆锚	有桿錨,普通錨	stock anchor
有功负荷	動力負載	power load
有功功率	瓩功率	active power, kW power
有功功率自动分配装置	有效功率自動分配裝置	automatic distributor of active power
有害干扰	有害干擾	harmful interference
有害货	有害貨物	harmful cargo, noxious cargo
有害物质	有害物質	harmful substance
有害物质报告	有害物質報告	harmful substance report, HS
有害液体物质	有毒液體物質	noxious liquid substance
有机过氧化物	有機過氧化物	organic peroxide
有机酸	有機酸	organic acid
有限元计算	有限元素計算	finite-element calculation
有效的	有效的	useful
有效波高	有義波高	significant wave height
有效辐射功率	有效輻射功率	effective radiated power
有效高度	實際高度	virtual height
有效功率	有效功率	effective power
有效供油行程	有效供油衝程	effective delivery stroke
有效全向辐射功率	有效等方向性輻射電力	effective isotropically radiated power
[有效]燃油消耗率	單位燃料消耗量	specific fuel consumption
有效声号	有效聲號	efficient sound signal
有效效率	有效效率	effective efficiency

祖国大陆名	台湾地区名	英 文 名
有效压缩比	有效壓縮比	effective compression ratio
有效载荷	有效負載	payload
有眼环的短索	末端附眼環短索	lizard
有意搁浅	故意擱淺	voluntary stranding
有用的	有用的	useful
有源卫星	運作中衛星	active satellite
有坐标格网海图	有格海圖	gridded chart
右侧浮标	右舷通過浮標	starboard hand buoy
右满舵	右滿舵	hard starboard
右舷	右舷	starboard side, starboard
右舷发动机	右舷引擎	starboard engine
右旋	右轉	right-handed turning
右旋柴油机	右轉柴油機	right-hand rotation diesel engine
右旋螺旋杆	右旋螺釘,右旋螺桿	right-hand screw
右旋螺旋桨	右旋螺槳	right handed propeller
右转发动机	右轉動力機	right-hand engine
幼鱼栖息场	幼魚棲息場	fish nursery ground
诱导比	感應比	induction ratio
诱导器	感應器	induction unit
淤锚	錨埋入	anchor embedded
余流	剩餘電流	residual current
余面	餘面	lap
鱼粉加工船	捕魚加工船	fish meat factory ship
鱼类回游	魚群洄游	[fishing] mass migration
鱼群	魚群	fish school
鱼群密度	魚群密度	concentration of fish
鱼群探测器(=鱼探仪)		
鱼肉分选机	魚肉採取機	meat separator
鱼探仪,鱼群探测器	魚群探尋器,魚群探測器	fish finder
渔场	漁場	fishing ground
渔场图	漁場圖	fishing chart
渔船	漁船	fishing vessel
渔港	漁港	fishing harbor
渔港规章	漁港規則	regulations of fishery harbor
渔监	漁業監督	fishing supervision
渔礁	漁礁	fish reef
渔具	漁具	fishing gear, fishing tackle

祖 国 大 陆 名	台 湾 地 区 名	英 文 名
渔捞限制	漁撈限制	fishing restriction
渔捞作业	漁撈作業	fishing operation
渔区	漁區	fishing zone, fishing area
渔群指示标	魚群指示浮標	fish group indicating buoy
渔汛	漁期	catching season, fishing season
渔业调查船	漁業研究船	fishery research vessel
渔业法规	漁業法規	fishery rules and regulations
渔业公司	漁業公司	fisheries company
渔业权	漁業權	fishing right
渔业协定	漁業協定	fishery agreement
渔业巡逻船	漁業巡邏船	fishery patrol boat
渔业指导船	漁業指導船	fishery guidance boat
渔栅	漁柵	fishing stake
渔政	漁業行政	fishery administration
渔政船	漁政船	fishery administration vessel
隅点	象限點	intercardinal point
雨层云	雨層雲	nimbo-stratus
雨量	雨量	rainfall
雨雪干扰抑制	抗雨雪干擾	anti-clutter rain
语音/数据群呼	語音/數據群呼	voice/data group call
预报波向	預報波向	forecasted wave direction
预备舵工	預備舵手	lee helmsman
预备航次	預備航程	preliminary voyage
预备间	特別房艙	state room
预测危险区	預測危險區	predicted area of danger, PAD
预防接种证书	預防接種證書	vaccination certificate
预付运费	預付運費,預支運費	advanced freight
预告	預告	preliminary notice
预计到达时间	預計到達時間	estimated time of arrival, ETA
预计开航时间	預計離開時間	estimated time of departure, ETD
预借提单	預借載貨證券	advanced bill of lading
预燃室	預燃室	precombustion chamber
预洗	預洗	prewash
阈限系数	界限係數	threshold factor
阈限值	低限值	threshold limit value
遇水易燃固体	遇濕易燃固體	flammable solid when wet
遇险	遇難,遇險	distress
遇险报告	遇難信文	distress message

祖 国 大 陆 名	台 湾 地 区 名	英 文 名
遇险报警	遇險警報	distress alerting
DSC 遇险报警	數位選擇呼叫遇險警報	DSC distress alerts
遇险报警转发	遇險報中繼	distress alert relay
遇险船员	遇險船員	distressed seaman
遇险电传呼叫	遇險電傳呼叫	distress telex call
遇险电话呼叫	遇險電話呼叫	distress telephone call
遇险和安全通信	遇險與安全通信	distress and safety communications
遇险呼叫	遇難呼叫,求救呼叫	distress call
遇险呼叫程序	遇險呼叫程序	distress call procedure
遇险呼叫格式	遇險呼叫格式	distress call format
遇险火光信号	遇難照明彈	distress flare
遇险阶段	遇險階段	distress phase
遇险求救程序	遇險程序	distress procedures
遇险收妥承认	遇險收到確認	distress acknowledgement
遇险通信	遇險通信	distress communication
遇险通信业务	遇險業務	distress traffic
遇险信道	遇險頻道	distress channel
遇险信号	遇險信號,遇難信號	distress signal
遇险优先等级	遇險優先順序	distress priority
遇险优先呼叫	遇險優先呼叫	distress priority call
遇险优先申请信息	優先請求遇險信文	distress-priority request message
遇险者	遇險人員	person in distress
元件破损事故	元件破損事故	element breakdown accident
元件烧毁事故	元件燒毀事故	element burnout accident
[原地]转向	轉向	cast
原动机自动起动装置	原動機自動起動器	prime mover automatic starter
原油	原油	crude oil
原油船	原油輪	crude oil tanker
原油洗舱	原油洗艙	crude oil washing, COW
原子时	原子時	atomic time, AT
圆材结	木材結	timber hitch
圆度	圓度	roundness, circularity
圆极化	圓極化	circular polarization
圆角区	內圓角區	fillet area
圆木	圓形木材	log
圆频率	圓頻率	circular frequency
圆筒形浮标	筒形浮標	cylindrical buoy
圆形编队	圓形編隊	circular formation

祖 国 大 陆 名	台 湾 地 区 名	英 文 名
圆型尾	圓形艉	round stern
圆柱测径规	柱形塞規	cylindrical plug, cylindrical gauge
圆柱度	圓柱度	cylindricity
圆柱滚子轴承	滾子軸承	roller bearing
圆柱体号型	圓筒形號標	cylinder［shape］
圆柱投影	圓筒投影法	cylindrical projection
圆锥滚子轴承	滾錐軸承	taper roller bearing
圆锥号型	圓錐形號標	conical shape
圆锥投影	圓錐投影法	conical projection
远程扫描	長距程掃描	long-range scanning
远地点	遠地點	apogee
远海测量	遠洋測量	pelagic survey
远距离水位指示计	遠隔水位指示器	remote water level indicator
远离	遠離	away from
远日点	遠日點	aphelion
远星点	遠星點	apastron
远洋船	遠洋船舶	oceangoing vessel, ocean trader
远洋提单	海運載貨證券	ocean bill of lading
远洋渔业	遠洋漁業	distant fishery
远月点	遠月點	apocynthion
约定最大持续功率	約定最大連續輸出功率	contract M. C. R.
约克–安特卫普规则	約克–安特衛普規則	York-Antwerp Rules
月潮	太陰潮	lunar tide
月潮间隙	月中天潮汐間歇	lunitidal interval
月出	月出	moon rise
月龄	月齡	moon's age
月没	月沒	moon set
月球视运动	月視運動	moon's apparent motion
月蚀	月食	lunar eclipse
月相	月相	lunar phases
跃层	銳變層	spring layer
跃水现象	躍水現象	porpoising
云高	雲高	cloud height
云量	雲量	cloud amount
云图［册］	雲圖冊	cloud atlas
云状	雲狀	cloud form
运兵船	運兵船	troopship
运动模拟	動力模擬	kinematic simulation

祖国大陆名	台湾地区名	英 文 名
运动黏度	動力黏度	kinematic viscosity
运动中	運動中	on the fly
运费	運費	freight
运费保险	運費保險	freight insurance
运费吨	運費噸,載貨容積噸,貨物噸	freight tons
运费付讫	運費付訖	carriage paid
运费率	運費率	freight rate
运费清单	運費清單	freight manifest
运费未收	運費由收貨人支付	carriage forward
运河操纵	運河中操縱	maneuvering in canal
运河灯	運河燈	canal light
运河吨位	運河噸位	canal tonnage
运河航标	運河助航標	navigation aids on canal
运价本(=费率本)		
运粮船	糧食船	victualler
运泥船	運泥船	sludge boat
运输船	運輸艦	transport ship
运输合同	運送契約	contract of carriage
运载体	載具	vehicle
晕船药	暈船藥	anti-seasickness medicine

Z

祖国大陆名	台湾地区名	英 文 名
杂货	雜貨	general cargo
杂货船	雜貨船	general cargo ship
杂类危险物质	雜項危險物質	miscellaneous dangerous substance
杂散发射	混附發射	spurious emission
载波功率	載波功率	carrier power
载波频率	載波頻率	carrier frequency
载驳船	子母船,浮貨櫃船	lighter aboard ship, LASH
载货舱位,载货容积	貨艙空間	cargo space
载货清单	載貨單,艙單	manifest
载货容积(=载货舱位)		
载冷剂	冷卻劑	cooling medium, coolant
载损鉴定	貨損檢查	inspection on hatch and/or cargo
载运违禁品	載運違禁品	carriage of contraband

祖 国 大 陆 名	台 湾 地 区 名	英 文 名
载重标尺	载重標尺	dead weight scale
载重线	载重線,载重水線	load line
载重线标志	载重線標誌	load line mark
载重线检验	载重線檢驗	load line survey
载重线勘划	载重線勘劃	load line assignment
载重线区域	载重線區域	load line area
载重线圆圈	载重線圈	load line disc
载重线证书	载重線證書	load line certificate
载装手册	装载手册	loading manual
再热器	重熱器,再熱器	reheater
再热式汽轮机	再熱式蒸汽渦輪機	reheat steam turbine
再热系数(=重热系数)		
再生制动	再生制動	regenerative braking
在航	航行中	underway
[在]后,近艉	在後,近艉,向艉	aft, abaft
在上风,向上风	居上風,頂風	in the wind
在职培训	在職訓練	in service training
在子午线上航行	經線航法	meridian sailing
遭遇周期	遭遇週期	period of encounter
藻海	藻海	sargasso sea
造船厂	造船廠	ship yard
噪声污染	噪音污染	noise pollution
责任级	責任層級	level of responsibility
责任限制	(船東)責任限制	limitation of liability
增黏剂	增黏劑	viscosity index improver
增强群呼接收机	強化群呼接收機	enhanced group calling receiver, EGC receiver
增强群呼系统	強化群呼系統	Enhanced Group Call System
增强群呼业务码	強化群呼業務碼	enhanced group call service code
增强群呼译码器	強化群呼解碼器	EGC decoder
"增速"按钮	"增速"按鈕	push button "up"
增压	增壓	supercharge
增压度	渦輪增壓度	degree of turbocharging
增压器喘振	渦輪增壓器波振	turbocharger surge
增压器涡轮特性	增壓器渦輪特性	turbocharger turbine characteristics
增压式发动机	增壓發動機	supercharged engine
增压系统辅助鼓风机	渦輪增壓器輔助鼓風機	turbocharging auxiliary blower
增压系统应急鼓风机	渦輪增壓器應急鼓風機	turbocharging emergency blower

祖 国 大 陆 名	台 湾 地 区 名	英 文 名
扎绳头	紮頭,繩頭紮束	whipping
渣滓	渣滓	sediment
闸阀	閘閥	gate valve
窄带直接印字	狹頻帶直接印字	narrow band direct print
窄带直接印字电报	狹頻帶直接電報	narrow-band direct-printing telegraphy
窄带直接印字电报设备	狹頻帶直接印字電報設備	narrow-band direct-printing telegraph equipment, NBDP
展期检验	延期檢驗	extension survey
展弦比	寬高比	aspect ratio
占用带宽	佔用頻帶寬	occupied bandwidth
战斗船	戰艦	fighting boat
战斗工况管理	戰鬥操作方式管理	combat operating mode management
战斗航海勤务	戰鬥航海勤務	combating navigation service
战列舰	戰艦,戰鬥艦	battle ship
战争条款	戰時險條款	war risk clause
站到站	貨櫃站到站	container freight station to container freight station, CFS to CFS
栈桥	棧橋	jetty
张力传感器	張力轉換器	tension transducer
张力控制器	張力控制器	tension control assembly
张力轮	拉緊帶輪	straining pulley
张力腿平台	張力腳式鑽油台	tension-leg platform
张索加油装置	張索加油裝置	span wire fuel rig
章动	章動(地軸之章動)	nutation
涨潮	漲潮	flood [tide]
涨潮标志,泛滥标	洪水位,高潮位標誌,泛水標誌	flood mark
涨潮持续时间	漲潮持續期	duration of flood
涨潮顶点	最高潮	top of the flood
涨潮流	漲潮流	flood stream, flood current
丈量吨位	丈量噸位	measurement tonnage
丈人一(天鸽α)	丈人一(天鴿α)	Phact
账务机构	賬務機構	accounting authority
账务机构识别码	賬務機構識別碼	accounting authority identification code, AAIC
障碍物探测	障礙物探測	obstruction sounding
招引注意信号	引起注意信號	signal to attract attention
照距(=光力射程)		

祖 国 大 陆 名	台 湾 地 区 名	英 文 名
遮蔽光弧	遮光弧	obscured sector
遮蔽甲板	遮蔽甲板	shelter deck
遮蔽甲板船	遮蔽甲板船	sheltered deck vessel
折,卷	收,捲	furl
折叠式舱盖	摺式艙蓋	folding hatchcover
折叠式减摇鳍装置	摺鰭穩定器	folding fin stabilizer
折叠式箱型货盘	摺疊式箱形托貨板	folding box pallet
折断	折斷	carry away
折光差(=蒙气差)		
折射角	折射角	angle of refraction
针阀	針閥	needle valve
针状星云	針狀星雲	acicular nebula
帧频	圓框頻率	frame frequency
真北	真北	true north
真出没	真出沒	true rise and set
真地平	真地平線,真水平線	true horizon
真地平圈	天球水平線圈	celestial horizon
真方位	真方位	true bearing, TB
真风	真風	true wind
真高度	真高度	true altitude
真航向	真航向	true course, TC
真近点角	真近點角	true anomaly
真空	真空	vacuum
真空安全阀	呼吸閥	pressure vacuum relief valve
真空泵	真空泵	evacuation pump, vacuum pump
真空压力表	真空壓力器	vacuum manometer
真空蒸馏	真空蒸餾	vacuum distillation
真太阳	真太陽	true sun
真误差	真誤差	true error
真运动雷达	真運動雷達	true motion radar, TM radar
真[运动]显示	真運動顯示	true motion display
真蒸汽压力	真蒸汽壓力	true vapour pressure
阵风	風陣	gust
阵雨	陣雨	shower
振动	振動	vibration
振型	振動模式	vibration mode
镇浪油	鎮浪油	wave quelling oil
争议	爭端	dispute

祖国大陆名	台湾地区名	英 文 名
蒸发管束	蒸發管排	evaporator tube bank
蒸发盘管	蒸發盤管	evaporating coil
蒸发压力调节阀(背压调节阀)	背壓調整器	evaporator pressure regulator, back pressure regulator
蒸发压缩制冷	蒸汽壓縮冷凍	vapor compression refrigeration
蒸馏法	蒸餾法	distillation method
蒸馏器	蒸餾器	distiller
蒸馏温度	蒸餾溫度	distillation temperature
蒸馏装置	蒸餾設備	distillation plant
蒸汽	蒸汽	steam
蒸汽舵机	蒸汽舵機	steam steering engine
蒸汽供暖系统	暖汽系統	steam heating system
蒸汽机船	汽船,輪船	steam ship, SS
蒸汽机–废汽汽轮机联合装置	蒸汽機與排汽渦輪機複合機	combined steam engine and exhaust turbine installation
蒸汽截止阀	停汽閥	steam stop valve
蒸汽喷射油气抽除装置	噴汽清除[有害]氣體系統	steam ejector gas-freeing system
蒸汽喷射制冷	蒸汽噴射冷凍	steam jet refrigeration
蒸汽起货机	蒸汽吊貨絞機	steam cargo winch
蒸汽熏舱管系	蒸汽窒火管路系統	tank steaming-out piping system
蒸汽直接作用泵	直聯蒸汽泵	direct acting steam pump
整车货	整車貨物	full truck load, FTL
整锻曲轴	實心曲柄軸	solid crankshaft
整理水池	整理水池	unravel water tank
整流舵(=压舵)		
整流叶片	整流葉片	straightening vane
整箱货	全貨櫃之貨物	full container load, FCL
正常起动程序	正常起動程序	normal starting sequence
正常使用功率	常用[定額]出力	normal service rating
正常损耗	正常損耗	normal loss
正车	正俥	ahead
正车操纵阀	正俥操縱閥	ahead maneuvring valve
正车汽轮机	正俥蒸汽渦輪機	ahead steam turbine
正车燃气轮机	正俥燃氣渦輪機	ahead gas turbine
正舵	正舵	amidships
正反馈	正反饋	positive feedback
正浮	正浮	upright

祖 国 大 陆 名	台 湾 地 区 名	英 文 名
正浮位置	正浮位置	upright position
正庚烷不溶物	正庚烷不溶物	n-heptane insoluble
正横	正横	abeam
正横接近法	正横接近法	abeam approaching method
正横距离	正横距離	distance abeam
正后方	正後方	dead astern
正螺旋试验	渦漩試驗	direct spiral test
正前方	正前方	dead ahead
正时图	正時圖	timing diagram
正蒸汽分配	正蒸汽分配	positive steam distribution
政委	政工官	political officer
政务电报	公務電報	government telegram
支撑	支撐	shoring
支承液体	支持液	supporting liquid
支持级	助理級	support level
支付	給付	payment
支架	支座	support
支流	支流	side stream，tributary
支索,拉条	牽索,支桿,拉條	stay
支线	枝線,枝繩	branch line
支线传送装置	枝繩輸送裝置	branch line conveyer
支线起线机	枝繩捲揚機	branch line winder
支线运输	輔助運務	feeder service
支柱	支柱	pillar
知识表达技术	知識表達術	knowledge presentation technique
知识库	知識庫	knowledge base
织女一(天琴 α)	織女一(天琴 α)	Vega
执行机构	執行機構	routine organization
执行器	引動器	actuator
直达港	直達港	direct port
直达货	直達貨	direct cargo
直达提单	直達載貨證券	direct bill of lading
直航船	直航船舶	stand-on vessel
直接传动	直接傳動	direct transmission
直接换装	直接換裝	direct transshipment
直接印字电报	直接印字電報	direct-printing telegraphy
直接原因	直接原因	immediate cause
直接蒸发式空气冷却器	直接蒸發空氣冷卻器	direct evaporating air cooler

祖国大陆名	台湾地区名	英文名
直接装注油管	直接注入管路	direct filling line, direct loading pipe line
直立	直柱	upright
直立[型]艏	直立艏	vertical bow, straight stem
直列式柴油机	直列式柴油機	in-line diesel engine
直列式发动机	直列型動力機	straight type engine
直流电力推进装置	直流電力推進裝置	D.C. electric propulsion plant
直流电站	直流電站	DC power station
直流扫气	單向流驅氣	uniflow scavenging
直流式锅炉	直通式鍋爐	straight-through boiler
直馏矿物油	純礦油	straight mineral oil
直升机救生套	直升機救生環索	helicopter rescue strop
直升机援助	直升機援助	assistant by helicopter
直升机作业	直升機作業	helicopter operation
直叶片	直葉片	straight blade
值班	當值	watch keeping
植物纤维绳	天然纖維索	natural fiber rope
止荡锚	止擺錨	yaw checking anchor
止回阀	止回閥	back pressure valve, non-return valve
止脱结	止脫繩, 止脫結	mousing
止移板	防動板	shifting board
纸制品	紙製品	paper products
指标差	器差, 指標誤差	index error
指标改正量	指標改正	index correction
指标镜	[六分儀]器鏡	index-mirror
指导船长	駐埠船長	port captain
指定仲裁员	指定仲裁人	appointed arbitrator
指挥舰	指揮艦	commanding ship
指令和数据获取	指令和數據擷取	command and data acquisition
指令和数据获取站	指令和數據擷取站	command and data acquisition station
指令译码器	指令譯碼器	instruction decoder
指南	指南	guide book
指配频带	指配頻帶	assigned frequency band
指配频率	指配頻率	assigned frequency
指示标志	指示標誌	instructive mark
指示功率	指示功率	indicated power
指示器	指示器	indicator
指示燃油消耗率	指示燃油消耗率	indicated specific fuel oil consumption
指示热效率	指示熱效率	indicated thermal efficiency

祖国大陆名	台湾地区名	英文名
指示提单	指示載貨證券	order bill of lading
指向力	指向力	directive force
指向力矩	尋子午線力矩	meridian-seeking torque, meridian seeking moment
制荡板	制水板	swash plate
制荡舱壁	制水艙壁	swash bulkhead
制动平均有效压力	制動平均有效壓力	brake mean effective pressure
制动索	制索繩	rope stopper
制舵器	制舵器	rudder stopper
制冷吨	冷凍噸	refrigerating ton
制冷剂	冷凍劑,冷媒	refrigeration agent, refrigerant
制冷剂计量装置	冷凍劑計算裝置	refrigerant metering device
制冷量	冷凍能量	refrigerating capacity
制冷系数	冷凍性能係數	coefficient of refrigerating performance
制冷系统	冷凍系統	refrigeration system
制冷循环	冷凍循環	refrigeration cycle
制链器	錨鏈扣	chain stopper
制链爪	止鏈爪,吊鏈鉤	devil's claw
制索结	絆索套結	stopper hitch
制索器	制鏈器	stopper
制止危及海上航行安全非法行为公约	制止危及海上航行安全非法行爲公約	Convention for the Suppression of Unlawful Acts against the Safety of Maritime Navigation
质量保证	品質保證	quality assurance
质量管理	品[質]管[制]	quality control
质量惯性矩	質量慣性矩	mass moment of inertia
质询条款	解釋條款	interpellation clause
窒息法	窒息法	smothering method
智能避碰系统	智慧型避碰系統	intelligent collision avoidance system
滞航	頂風緩航	heave to
滞留费(=停泊费)		
滞期	延滯費,滯船費	demurrage
滞燃期	延遲期間	delay period
滞止	停滯	stagnation
滞止蒸汽参数	停滯蒸汽參變數	stagnation steam parameter
置信度	可信度	confidence factor
中舱	中心艙	center tank
中层拖网	中層拖網	mid-water trawl

祖国大陆名	台湾地区名	英 文 名
中垂	舯垂[现象]	sagging
中断服务程序	中斷服務常規	interrupt service routine
中断系统	中斷系統	interrupt system
中分纬度	中間緯度	middle latitude
中分纬度改正量	中緯修正量	correction of middle latitude
中分纬度算法	中緯航法	mid-latitude sailing
中/高频无线电设备	中/高頻無線電裝置	MF/HF radio installation
中国船级社	中國船級社	Zhongguo Shipping Register, ZG
中国帆船	中國帆船	junk
中桁材	內龍骨	keelson, center girder
中机型船	舯機艙船	amidships engined ship
中级压紧配合	中級壓緊配合	medium force fit
中介轴(=延伸轴)		
中间壳体	中間殼體	intermediate casing
中间轮(=惰轮)		
中间燃料油	中間燃油	intermediate fuel oil
中间轴	中間軸	intermediate shaft
中间轴承	中間軸承	intermediate bearing
中浪	和浪	moderate sea
中锚	流錨,中錨	stream anchor
中能见度(6 级)	能見度中等(6 級)	visibility moderate
中频通信	中頻通信	MF communication
中频无线电设备	中頻無線電裝置	MF radio installation
中期检验	中期檢驗	intermediate survey
中期天气预报	中期預報	medium-range forecast
中速柴油机	中速柴油機	medium speed diesel engine
中碳钢	中碳鋼	medium carbon steel
中天	中天	transit, meridian passage
中天顶距	[過]子午圈天頂距	meridian zenith distance
中天高度	中天高度	meridian altitude
中天观测	中天高度觀測	meridian observation
中途港	中途港	intermediate port
中心扩大显示	中心擴大顯示	center-expand display
中心偏移	偏心	off-centering
中心线	中心線	center line
中性点	中性點	neutral point
中性流	中線電流	neutral current
中央处理单元	中央處理單元	central processor unit

祖 国 大 陆 名	台 湾 地 区 名	英 文 名
中央浮标	航道中流浮標	mid-channel buoy
中央空调器	中央空調	central air conditioner
中央冷却系统	中央冷卻系統	central cooling system
中涌	中湧	moderate swell
中游	中游	middle reach
中子功率表	中子功率表	neutron power meter
终端	終端	terminal
终航向	終結航向	final course
终结信号	終止信號	finishing signal
钟锤拉索	拉鈴把手	bell pull
钟浮标	鐘浮標	bell buoy
舯	舯	midship
舯拱	舯拱	hogging
舯剖面积	舯剖面積	area of midship section
舯剖面模数	舯剖面模數	modulus of midship section
舯剖面系数	舯剖面係數	midship section coefficient
舯突出部,舭	舯膨出部	bulge
仲裁	仲裁	arbitration
仲裁裁决	仲裁判斷	arbitration award
仲裁条款	仲裁條款	arbitration clause
仲裁庭	仲裁法庭	arbitration tribunal
仲裁委员会	仲裁委員會	board of arbitration
仲裁员	仲裁人	arbitrator
重大改装	重大改裝	major conversion
重大件运输船	重大件貨運輸船	heavy and lengthy cargo carrier
重吊杆	重吊桿	heavy derrick, jumbo boom
重吊起货机	重貨起重機	heavy lift derrick cargo winch
重力柜	重力櫃	gravity tank
重力式吊艇架	重力吊架(小艇)	gravity davit
重力式供给系统	重力供給系統	gravity feed system
重力式平台	重力式鑽油台	gravity platform
重力系泊塔	重力繫泊塔	gravity mooring tower
重力异常图	重力異常圖	gravity anomaly chart
重力油柜	重力油櫃	gravity oil tank
重量	重量	weight
重量货	過秤貨	deadweight cargo
重量鉴定	重量檢定	inspection of weight
重量稳性力臂	重量穩度力臂	lever of weight stability

祖 国 大 陆 名	台 湾 地 区 名	英 文 名
重心	重心	center of gravity
重心高度	重心高	height of center of gravity
重心距中距离	縱向重心與舯距離	longitudinal distance of center of gravity from midship
重要负载	重要負載	important load
重质燃料油	重燃油	heavy fuel oil
周波重合	週波匹配	cycle matching
周计时	以週計時	time of week，TOW
周年光行差	週年光行差	annual aberration
周年日	週年日	anniversary date
周期测量系统	定期測量系統	period measurement system
周期风	週期風	periodic wind
周期性变化	週期性變化	cyclic variation
周期性检查	定期檢查	periodical inspection
周日视运动	每日視運動(天體)	diurnal [apparent] motion
X 轴	X 軸線	x-axis
Y 轴	Y 軸線	y-axis
Z 轴	Z 軸線	z-axis
轴承	軸承	bearing
轴承刮削	軸承刮削	scraping of bearing
轴承间隙	軸承間隙	bearing clearance
轴带发电机	軸驅動發電機	shaft-driven generator
X 轴分量	X 軸分量	x-component
轴功率	軸功率	shaft power
轴毂	軸轂	shaft bossing
轴流泵	軸流泵	axial-flow pump
轴流式汽轮机	軸流蒸汽渦輪機	axial flow steam turbine
轴流式涡轮机	軸流渦輪機	axial-flow turbine
轴流式压气机	軸流壓縮機	axial-flow compressor
轴隧	軸道	shaft tunnel
轴头销(＝活塞销)		
轴系校中	軸系校準	shafting alignment
轴系制动器	軸系制動器	shafting brake
轴向减振器	縱向振動阻尼器	longitudinal vibration damper
轴向位移保护装置	軸向位移保護設施	axial displacement protective device
轴向振动	軸向振動	axial vibration
轴向柱塞式液压马达	軸向活塞式液力馬達	axial-piston hydraulic motor
轴制动器	軸靭,軸制動器	shaft brake

祖 国 大 陆 名	台 湾 地 区 名	英 文 名
轴墜	軸道	shaft alley
肘板	腋板	bracket
竹排	竹筏	bamboo raft
逐级,步进式	逐步	step-by-step
主标志	主標誌	main mark
主车钟	主俥鐘	main engine telegraph
主电源	主電源	main source of electrical power
主动齿轮	主動齒輪	driving gear
主动舵	主動舵	active rudder
主动水舱式减摇装置	主動減搖水艙穩定器	activated anti-rolling tank stabilization system
主发电柴油机	主柴油發電機	main generator diesel engine
主钢缆(**撑杆**)	撐桿	jackstay
主管机关	主管機關,主管官署	administration
主锅炉	主鍋爐	main boiler
主航道	主航道	main channel, trunk line routes
主机工况监测器	主機狀況偵測器	condition monitor of main engine
主机故障应急处理	主機故障應急操縱	main engine fault emergency maneuver
主机航程	主機(每時轉數計)航程	distance by engine's RPH
主机航速	機速	engine speed
主机遥控屏	主機遙控屏	main engine remote control panel
主机转速表	主機轉速表	main engine revolution speedometer
主甲板	主甲板	main deck
主控线路	主控線路	master control circuit
主控站	主控站	master control station
主冷凝器循环泵	主冷凝器循環泵	main condenser circulating pump
主冷却剂系统	主冷卻劑系統	main coolant system
主令控制器	主控制器	master controller
主流	主流	main stream
主螺栓	大螺栓	king bolt
主配电板	主配電盤	main switchboard
主起动阀	主起動閥	main starting valve
主汽轮机	主蒸汽渦輪機	main steam turbine
主燃气轮机	主燃氣渦輪機	main gas turbine
主竖区	主要垂直區域	main vertical zone
主台	羅遠主台	master station
主台信号	主台信號	master signal
主台座	主托架	master pedestal

祖国大陆名	台湾地区名	英 文 名
主题标识符	主題標識符	subject indicator character
主题表示类型	主題顯示之形式	subject indication type
主推进机舱	主推進機器	main propelling machinery room
主推进装置	主推進裝置	main propulsion unit
主拖缆	主拖纜	main towing line
主陀螺	子午迴轉儀	meridian gyro
主循环泵	主循環泵	main circulating pump
主要原因	主要原因	major cause
主应力	主應力	primary stress
主用发信机	主發射機	main transmitter
主用收信机	主接收機	main receiver
主用天线	主天線	main antenna
主轴承	主軸承	main bearing
主轴颈	曲柄軸頸	crankshaft journal
主转子	主轉子	main rotor
住舱	住艙	living quarter
助航标志	導航設備	aids to navigation
助艏缆	前艏纜	forward bow spring
助艉缆	艏向後倒纜	head aft spring
注册	註冊,登記	register
注入管	注入管	filling pipe, filling line
注入站	注入站	injection station
注销登记	撤銷登記	registration of withdrawal
注意标志	注意標誌	notice mark
驻波	定波,駐波	standing wave
柱塞泵	唧子泵,柱塞泵	plunger pump
柱塞偶件	柱塞偶合件	plunger matching parts
柱塞套筒组件	柱塞與套筒組件	plunger and sleeve assembly
柱形浮标	柱形浮標	pillar buoy
柱状桅	柱狀桅	pole mast
抓到就打住	打住,拉平各纜	hold what you've got
抓斗	抓斗	grab
专业救助	專業救助業務	specialized salvage service
专用标志	特殊標誌	special mark
专用灯,特别工作灯	整補作業燈	task light
专用航道	專用航道	special purpose channel
专用清洁压载水舱	清潔壓艙水專用艙	dedicated clean ballast tank
专用压载舱	隔離壓載水艙	segregated ballast tank, SBT

祖 国 大 陆 名	台 湾 地 区 名	英 文 名
专用压载舱保护位置	隔離壓艙水艙保護位置	protective location of segregated tank ballast
专用压载系统	隔離壓載系統	segregated ballast system
专属经济区	專屬經濟區	exclusive economic zone
专属渔区	專屬漁業區	exclusive fishery zone
转车机	迴轉裝置,盤俥裝置	turning gear
转船货	轉船貨	transshipment cargo
转船力矩	船舶迴轉力矩	moment of turning ship
转船提单	轉口載貨證券	transshipment bill of lading
转船条款	轉船條款	transshipment clause
转动	迴轉	turn
转动配合	轉動配合	running fit
转舵阶段	轉舵階段	maneuvering period
转舵时间	轉舵時間	time of rudder movement
转发无线电话遇险信号	轉發遇險通報(無線電話用語)	mayday relay
转发信标	轉發信標	responder beacon
转环	轉環	swivel
转换阀	變換閥	change-over valve
转换机构	變換機構	change-over mechanism
$p\text{-}V$ 转角示功图	失相圖	out-of-phase diagram
转矩,扭矩	扭矩	torque
转口港	轉口港	port of transshipment
转口货(=过境货)		
转口提单(=过境提单)		
转流	潮流旋轉	turn of tidal current
转鳍机构	傾轉鰭機構	fin-tilting gear
转数计数器	轉數計	revolution counter
转速表	轉速計	tachometer
转速波动率	速率波動係數	coefficient of speed fluctuation
转速禁区	速限區	barred-speed range
转塔式系泊系统	轉塔式繫泊系統	turret mooring system
转向	轉向	alter course, a/c
转向点	轉向點	turning point
转向点浮标	轉向點浮標	turning buoy
转向工况管理	迴轉操作形式管理	turning operating mode management
转向上风	轉向上風	bear down
转向下风	轉向下風	bear up

祖国大陆名	台湾地区名	英 文 名
转叶式转舵机构	轉葉式操舵裝置	rotary vane steering gear
转移位置线	轉移位置線	position line transferred
转子相对位移	轉子相對位移	relative rotor displacement
转租	轉租	subletting, subchartering
桩基码头	碼頭	pier
桩基系泊塔	樁基繫泊塔	piled mooring tower
装船通知单	下貨單	shipping note
装货单	裝貨通知單	shipping order
装货港	裝貨港	port of loading
装货清单	裝貨清單	cargo list, loading list
装货准单(海关)	載貨單	shipping-bill
装配附件	裝配件	mounting fittings
装燃料	裝載燃料	bunkering
装哨浮标	鳴笛浮標	whistling buoy
装箱单	裝箱單	packing list
装卸货天数	許可裝卸日數	laydays
装卸工,码头工人	裝卸工,碼頭裝卸工人	stevedore
装卸期限	約定裝卸時間	laytime
装卸长,领班	領班,工頭	foreman
装油港站	裝油終端站	oil loading terminal
装有电报或电话通信的 系泊浮筒	裝有電報或電話通信之 繫泊浮筒	mooring buoy with telephonic or telegraph- ic communications
装载	裝載	loading
装载和污底工况管理	裝載與污底作業形式管 理	load and fouling hull operating mode man- agement
状态记录系统	衛星報況系統	states recording system
状态询问	狀況詢問	status enquiry
撞锤张力器	撞鎚張力機	ram tensioner
追过	追越	out-foot
追越	追越	overtaking
追越船	超越船,追趕船	overtaking vessel
追越声号	追越信號	overtaking sound signal
追踪	追蹤	homing
锥度规	斜度計,推拔規	taper gauge
锥形	楔形,推拔	taper
锥形导槽	錐形導槽	conic guide
锥形浮标	圓錐形浮標	conical buoy
坠落试验	墜落試驗	drop test

祖国大陆名	台湾地区名	英　文　名
准备就绪通知书	(装卸)準備完成通知書	notice of readiness
准备开动(=准备开航)		
准备开航,准备开动	備便放鬆,準備開航	clear for running
浊点	雲點,濁點	cloud point
姿态测定系统	姿態測定系統	attitude determination system, ADS
姿态角	姿態角	attitude angle
资费表	價目表	tariff
资源量	資源量	population size
子程序	子程序	sub-program
子母钟	子母鐘	primary-secondary clocks
子圈	子午線下半部	lower branch of meridian
子午角	子午線角	meridian angle
子午圈	子午圈	meridian circle
子午线	子午線,經線	meridian
子午仪	子午儀,中星儀	meridian instrument
子午仪系统(=海军导航卫星系统)		
8 字结	8 字結	figure of eight knot
字母电报	文字電報	alphabetical telegraph
字母拼读法	字母拼音	letter pronunciation
字母旗	字母旗	alphabetical flag
自闭式安全门	自閉式安全門	automatic self-closing safety door
自差	自差	deviation
自差表	自差表	deviation table
自差补偿装置	自差補償設施	deviation compensation device
自差校正场	迴旋場所	swinging ground
自差曲线	自差曲線	deviation curve
自差系数	自差係數	coefficient of deviation
自带行李	隨身行李	cabin luggage
自动报警	自動警報器	auto alarm
自动避碰系统(**避碰系统**)	避碰系統	collision avoiding system, CAS
自动并联运行	自動並聯運轉	automatic parallel operation
自动拨号双向电话	自動撥號雙向電話	automatic dial-up two-way telephony
自动操舵仪	自動駕駛儀	autopilot
自动测量仪	自動測量儀	automatic measuring instrument
自动测向仪	自動測向儀	automatic radio direction finder, ADF

祖 国 大 陆 名	台 湾 地 区 名	英 文 名
自动车辆定位	自動車輛定位	automatic vehicle location, AVL
自动除渣分杂机	自動除渣澄清器	automatic dislodging clarifier
自动吹灰器	自動吹灰器	automatic soot blower
自动电压调节器	自動電壓調整器	automatic voltage regulator, AVR
自动扶正部分封闭救生艇	自行扶正部分圍蔽救生艇	self-right partially enclosed lifeboat
自动负荷控制	自動負載控制	automatic load control, ALC
自动航海通告系统	自動航船布告系統	automatic notice to mariners system, ANMS
自动呼叫	自動呼叫	automatic call
自动减速	自動減速	automatic slow down
自动校平装置	自動準平裝置	autolevelling assembly
自动解列	自動解除並聯	automatic parallel off
自动纠错	自動改錯	automatic error correction
自动空气断路器	自動空氣斷路器	automatic air circuit breaker
自动雷达标绘仪	雷達自動測繪設備	automatic radar plotting aids, ARPA
自动拍发器	自動鍵發設施	automatic keying device
自动喷水、报警和探测系统	自動噴水器與火警警報及滅火系統	automatic sprinkler and fire alarm and fire detection system
自动膨胀阀	自動膨脹閥	automatic expansion valve
自动平舱煤船	自動均載煤船	self trimming collier
自动起动空气压缩机	自動起動空氣壓縮機	auto-starting air compressor
自动清洗滤器	自動清洗過濾器	auto-clean strainer
自动请求重发方式	自動複傳申請	automatic repetition request mode, ARQ
自动扫描	自動掃瞄	auto scanning
自动式主起动阀	自動起動空氣關斷閥	automatic starting air shut-off valve
自动疏水	自動排洩裝置	automatic drainage
自动洒水探火系统(**自动探火喷水系统**)	自動探火噴水系統	automatic sprinkler fire detection system
自动调谐	自動調諧	automatic tuning
自动停车	自動停俥	auto-stop
自动同步装置	自動同步設施	automatic synchronizing device
自动系泊绞车	自動繫泊絞車	automatic mooring winch
自动显示仪表	自動顯示儀	automatic display instrument
自动延绳机	自動延繩機	autolongline machine
自动验潮仪	自動測潮計	automatic tide gauge
自动业务	自動業務	automatic service
自动用户电报试验	自動電傳試驗	automatic telex test

祖国大陆名	台湾地区名	英文名
自动止回阀	自動止回閥	automatic non-return valve
自给空气支持系统	自供空氣支援系統	self-contained air support system
自供空气救生艇	自供空氣系統救生艇	lifeboat with self-contained air support system
自航试验	自[行]推[動]試驗	self propulsion test
自激(=自励)		
自记式转速表	紀錄轉速計	recording tachometer
自检功能	自核功能	self-checking function
自励,自激	自激,自勵	self-excitation
自励交流发电机	自激交流發電機	self-excited AC generator
自亮浮灯	[救生圈]自燃燈	self igniting buoy light
自耦变压器起动	自耦變壓器起動	auto-transformer starting
自清洗粗滤器	自淨過濾器	self-cleaning strainer
自清洗分油机	自清分離器	self-cleaning separator
自清洗式滤器	自清洗濾器	self-cleaning strainer
自然地貌	自然地貌	natural feature
自然减速	標稱速降	nominal speed loss
自然磨损	自然耗損	ordinary wear and tear
自然行诉	自然源繪地,自然管轄地	natural forum
自然循环锅炉	自然循環鍋爐	natural circulation boiler
自燃点	自燃點	spontaneous combustion point
自燃烟雾信号	自動煙號	self-activating smoke signal
自扫舱装置	自行收艙裝置	self stripping unit
自身标识	自船識別碼	self-identification
自适应操舵仪	適應自動操舵儀	adaptive autopilot
自适应控制	適應控制	adaptive control
自吸式离心泵	自注離心泵	self-priming centrifugal pump
自修	自行檢修	self repair
自由活塞燃气轮机	自由活塞燃氣渦輪機	free piston gas turbine
自由降落下水	自由降落下水	free-fall launching
自由陀螺仪	自由迴轉儀	free gyroscope
自由液面	自由液面	free surface
自由液面效应	自由液面效應	free surface effect
自由液面修正值	自由液面修正值	free surface correction
自由振动	自由振動	free vibration
自整位式轴承	自動對正軸承	self-aligning bearing
自主式导航设备	自備助航設備	self-contained navigational aids

祖国大陆名	台湾地区名	英文名
总布置图	總佈置圖	general arrangement
总长	全長	length overall, LOA
总吨位	總噸位	gross tonnage, GT
总付运费	[總額]包載運貨	lumpsum freight
总碱值	總鹼值	total base number, TBN
总容量	總容量	aggregate capacity
总酸值	總酸值	total acid number, TAN
总图	總海圖,沿岸通用海圖	general chart
总效率	總效率	total efficiency
总压头	總落差	total head
总载重量	載重量	dead weight, DW
总装式锅炉	組合鍋爐	packaged boiler
总阻力	總阻力	total resistance
纵舱壁	縱艙壁	longitudinal bulkhead
纵荡	激變	surging
纵队	縱隊	column formation
纵帆	縱帆	fore-and-aft sail, fore and after sail
纵骨架式	縱肋系統	longitudinal frame system
纵距	進距,前進距離(迴旋圈)	advance
纵剖面图	縱剖面圖	longitudinal section plan
纵倾角	俯仰角	trimming angle
纵倾力距	俯仰力矩	trimming moment
纵稳心	縱定傾中心	longitudinal metacenter
纵稳心半径	縱定傾半徑	longitudinal metacentric radius
纵稳心高度	縱定傾中心在基線以上高度	longitudinal metacentric height above baseline
纵稳性	縱穩度	longitudinal stability
纵稳性高度	縱定傾中心高	longitudinal metacentric height
纵稳性力臂	縱穩度力臂	longitudinal stability lever
纵向补给装置	艉向補給裝置	astern replenishing rig
纵向冲距	縱向衝距	head reach
纵向磁棒	縱向磁棒	fore-and-aft magnet
纵[向]强度	縱向強度	longitudinal strength
纵向止移板	縱防動板	longitudinal shifting board
纵摇	縱搖	pitching
纵摇周期	縱搖週期	pitching period, period of pitching
纵摇阻尼控制器	縱搖阻尼控制	pitch damping control

祖 国 大 陆 名	台 湾 地 区 名	英 文 名
走廊	走廊	corridor
走锚(＝拖锚)		
走私	走私	smuggling
租船费	租傭費	charterage
租船合同	租傭契約	charter party
租船提单	傭船載貨證券	charter party bill of lading
租船运费	傭船運費	charter freight
租金支付	租金支付	payment of hire
租期	租期	period of hire
阻火器	防焰網	flame trap
阻力	拖曳力	drag force
阻尼器	阻尼板	damper
阻尼系数	阻尼因素	damping factor
阻尼振动	阻尼振動	damped vibration
阻尼重物	鎮偏重物	damping weight
阻汽器	蒸汽除水閘	steam trap
阻塞	阻流	choking
组重复周期	組重複週期	group repetition interval，GRI
组合报警(**基群报警**)	群警報	group alarm
组合导航系统	組合導航系統	integrated navigation system
组合模式	組合模式	integrated mode
组合式锅炉	複合鍋爐	composite boiler
组合[式]曲轴	組合曲柄軸	built-up crankshaft
组合体	組合體	composite unit
组合与测试	整合及測試	integration and test
祖母结(＝错平结)		
钻井平台	鑽油平台	drilling platform
钻孔冷却	鏇孔冷卻	bore-colded
钻探船	鑽探船	drilling vessel
钻探设备	鑽探設備	drilling rig
最大测量深度	最大測量深度	maximum measuring depth
最大持续功率	額定最大連續出力	maximum continuous rating
最大舵角	滿舵角	hard-over angle
最大复原力矩	最大扶正力矩	maximum righting moment
最大高度	最大高度	maximum height
最大宽度	最大寬度	maximum breadth
最大起升高度	最大提升高度	maximum height of lift
最大燃油量限制螺钉	最大燃油量限制螺釘	maximum fuel limit screw

祖 国 大 陆 名	台 湾 地 区 名	英 文 名
最大容许稳定运行功率	容許最大穩定運轉功率	maximum permissible stable operation power
最大容许压力	最大容許壓力	maximum allowable pressure
最大稳性力臂	最大穩度力臂	maximum stability lever
最大稳性力臂角	最大穩度力臂角	angle of maximum stability lever
最低安全配员	船員最低安全配額	minimum safe manning
最低潮	最低潮	dead tide
最低发光强度	最低照明強度	minimum luminous intensity
最低起动压力	最低起動壓力	minimum starting pressure
最低稳定转速	最低穩定轉速	minimum stable engine speed
最低有效位	最低有效位元	least significant bit
最低运费	起碼運費	minimum freight
最低运费吨	起碼運費噸	minimum freight ton
最低运费提单	起碼運費載貨證券	minimum freight bill of lading
最概率船位	最可能船位	most probable position, MPP
最高爆发压力	最高爆發壓力	maximum explosive pressure
最高爆发压力表	最高爆發壓力錶	maximum explosion pressure gauge
最高空载转速	最大無負載速度	maximum no-load speed
最高速度	飛速[率],強速	flank speed
最后不合法航次	最後違法航程	illegitimate last voyage
最后的	最後的	ultimate
最后合法航次	最後合法航程	legitimate last voyage
最后机会原则	最後機會原則	last opportunity rule
最后文件	蔵事文件	final act
最惠国待遇	最惠國待遇	most favored nation treatment, MFNT
最佳负荷分配	最佳負載分配	optimum load sharing
最佳航迹定线	最佳船舶航路	optimum track ship routing
最佳航速	最佳航速	optimum speed
最佳航线	最佳航路	optimum route
最佳航线拟定	最佳航線擬定	optimum routing
最佳决策	最佳決定	optimum decision
最佳匹配点	最佳吻合點,最佳匹配點	optimum match point
最佳性能	最佳性能	optimum performance
最接近时间	最接近之時刻	time of closet approach
最近会遇点	最接近點	closest point of approach, CPA
最近会遇距离	最接近點距離	distance to closest point of approach, DCPA

祖国大陆名	台湾地区名	英 文 名
最近会遇时间	最接近點時間	time to closest point of approach, TCPA
最近距离	最小距離	minimum distance
最近陆地	最近陸地	nearest land
最亮(金星)	最亮(金星)	greatest brilliancy
最小测量深度	最小測量深度	minimum measuring depth
最小倾覆力矩	最小翻覆力矩	minimum capsizing moment
最优控制	最佳控制	optimal control
最有效位	最高有效位元	most significant bit
最终报告	最終報告	final report
左侧浮标	左舷通過浮標	port hand buoy
左邻舰	左側船艦	next ship on the left
左满舵	左滿舵	hard port
左舷	左舷	port side
左舷发动机	左舷引擎	port engine
左向旋转,逆时针旋转	左向旋轉,逆時針旋餞	left-hand rotation
左旋	左轉	left-hand turning
左旋柴油机	左旋柴油機	left-hand rotation diesel engine
左旋圆极化	左旋圓形極化	left-hand circular polarization
左翼舰	左翼船艦	left flank ship
左右舵停船,蛇航制动	循環操舵停船法	rudder circling stop
左右通航标	分道航行標	separate channel mark
作业废弃物	操作所生廢棄物	operational wastes
作业控制中心	作業控制中心	operations control center
[作业]跳板	跳板	plank stage
作用区	接觸區	zone of contact
坐板升降结	工作吊板套結	bosun's chair hitch
坐标变换器	坐標變換器	coordinate conversion device

副 篇

A

英 文 名	祖 国 大 陆 名	台 湾 地 区 名
AAIC (= accounting authority identifica-tion code)	账务机构识别码	賬務機構識別碼
a. a. r(= against all risk)	承保一切险	全險
AB(= able-bodied seaman)	一级水手	一等水手
AB(= American Bureau of Shipping)	美国船级社	美國船級協會
aback	向后	向後
abaft	在后	在後
abandonment	委付	委付
abandonment of ship	弃船	棄船
abandon ship drill	弃船救生演习(弃船演习)	棄船演習
abandon ship signal	弃船信号	棄船信號
abeam	正横	正横
abeam approaching method	正横接近法	正横接近法
abeam replenishing rig	横向补给装置	横向補給裝置
abeam replenishment at sea	航行横向补给	航行中横向補給
aberration	光行差	光行差
able-bodied seaman(AB)	一级水手	一等水手
abnormal injection	异常喷射	異常噴射
abnormal wear	异常磨损	異常磨損
above deck equipment(ADE)	舱外设备	艙面設備
abrasive wear	磨料磨损	磨料磨損
absolute bolometric magnitude	绝对辐射热星等	絕對熱星等
absolute delay	绝对延迟	絕對遲延
absolute humidity	绝对湿度	絕對濕度
absolute log	绝对计程仪	絕對測程儀
absolute parallax	绝对视差	絕對視差
absolute pressure	绝对压力	絕對壓力

英　文　名	祖国大陆名	台湾地区名
absolute temperature	绝对温度	絕對溫度
absolute velocity	绝对速度	絕對速度
absorbent material	吸收材料	吸收材
absorption agent	吸收剂	吸收劑
absorption refrigeration	吸收制冷	吸收冷凍
a/c(=alter course)	转向	轉向
acceleration of ship	船舶加速度	船舶加速度
acceleration performance	加速性能	加速性能
acceleration turn	加速旋回	加速迴旋
accelerometer	加速度计(**加速计**)	加速計
access opening	出入口	出入口
accessorial charge	附加费用	附收費用
accidental pollution	事故污染	事故污染,意外污染
accommodation	居住舱,起居设备	起居廳房,起居設備
accommodation deck	起居甲板	住艙甲板
accommodation ladder	舷梯	舷梯
accommodation ladder winch	舷梯绞车	舷梯絞機
accommodation space	起居处所	起居艙空間
accounting authority	账务机构	賬務機構
accounting authority identification code (AAIC)	账务机构识别码	賬務機構識別碼
accumulated rate	积差[率]	[累]積差率
accumulator battery	蓄电池	蓄電池
accuracy of position	船位精度	船位精度
A. C. electric propulsion plant	交流电力推进装置	交流電力推進裝置
Achernar	水委一(波江 α)	水委一(屬銀河星座)
acicular nebula	针状星云	針狀星雲
acknowledge	收妥	收到
acknowledgement of distress alert	确认遇险报警收妥	確認收妥遇險警報
acknowledge signal	应答信号	收到信號,確認信號
A class division	甲级分隔	"A"級區(防火)
acoustic depth finder(=echo sounder)	回声测深仪	回聲測深儀
AC power station	交流电站	交流電站
acquisition	捕获	截獲
Acrux	十字架二(南十字 α)	十字架二(南十字 α)
AC three-phase three-wire system	交流三相三线制	交流三相三線制
activated anti-rolling tank stabilization system	主动水舱式减摇装置	主動減搖水艙穩定器

英　文　名	祖国大陆名	台湾地区名
active power(=kW power)	有功功率	瓩功率
active rudder	主动舵	主動舵
active satellite	有源卫星	運作中衛星
actual carrier	实际承运人	實際運送人
actual cycle	实际循环	實際循環
actual total loss	实际全损	實際全損
actual track	实际航迹向	實際航跡
actuator	执行器	引動器
adaptive autopilot	自适应操舵仪	適應自動操舵儀
adaptive control	自适应控制	適應控制
Adcook antenna	爱德考克天线	亞德考克天線
additional charge	附加运费(**附加费**)	附加費
additional secondary phase factor(ASF)	附加二次相位因子	附加二次相位因數
additional survey	附加检验	額外檢驗
address	收报人名址	地址
addressee	收报人	收信人
ADE(=above deck equipment)	舱外设备	艙面設備
ADF (=automatic radio direction finder)	自动测向仪	自動測向儀
adiabatic process	绝热过程	絕熱過程
adjacent angle	邻角	鄰角
adjustable vane	可调叶片	可調葉片
administration	主管机关	主管機關,主管官署
admissible error	容许误差	容許誤差
ADS(=attitude determination system)	姿态测定系统	姿態測定系統
ad valorem	从价	從價
ad valorem rate	从价运费	從價運費
advance	纵距	進距,前進距離(迴旋圈)
advanced bill of lading	预借提单	預借載貨證券
advanced freight	预付运费	預付運費,預支運費
advance of periastron	近星点前移	近星點前移
advance of perihelion	近日点前移	近日點前移
advance to captain	船长借支	船長借支
aerial scouting	空中探鱼	空中探測魚群,飛行探測魚群
aerofoil boat	气翼船	氣翼艇
AF(=astronomical fix)	天文船位	天文定位
A-frame	A 形机架	A 形托架

英　文　名	祖 国 大 陆 名	台 湾 地 区 名
aft	[在]后,近艉	在後,近艉,向艉
after body	后部	船尾部
after-burning	后燃	後燃
after burning period	后燃期	後燃期
after mast	后桅	後桅
after peak bulkhead	艉尖舱壁	艉艙壁
after-sales service	售后服务	售後服務
aft peak tank	艉尖舱	艉[尖]艙
aft perpendicular	艉垂线	後垂標
against all risk（a. a. r）	承保一切险	全險
ageing of lubricating oil	机油老化	潤滑油老化
agent	代理人	代理人
age of correction	校正龄期	校正齡期
aggregate capacity	总容量	總容量
aging	老化	老化
agonic line	零磁差线	無磁偏線
agricultural vessel	农用船	農用船
aground	搁浅	擱淺
ahead	①向前 ②前方 ③正车	①向前 ②前方 ③正俥
ahead gas turbine	正车燃气轮机	正俥燃氣渦輪機
ahead maneuvring valve	正车操纵阀	正俥操縱閥
ahead steam turbine	正车汽轮机	正俥蒸汽渦輪機
a hoist	一挂	一掛
aids to navigation	助航标志	導航設備
aiming circle	标圈	標圈
air clutch	气力离合器	氣力離合器
air compressor	空气压缩机	空氣壓縮機
air compressor auto-control	空压机自动控制	空壓機自動控制
air conditioning evaporator	空气调节装置蒸发器（**空调蒸发器**）	空調器蒸發器
air container	通气箱（**储气箱**）	空氣儲蓄器
air-cooled condenser	风冷式冷凝器	氣冷式冷凝器
air cooler	空气冷却器	空氣冷卻器
air cooling machine	冷风机	冷氣機
aircraft carrier	航空母舰	航空母艦
aircraft station	航空器电台（**航空电台**）	航空電台
air-cushion vehicle（＝hovercraft）	气垫船	氣墊船

英　文　名	祖国大陆名	台湾地区名
air defense formation and disposition	防空编队与部署	防空編隊與序列
air distributor	空气分配器	空氣分配器
air draft	净空高度	淨空高度
air flask	储气瓶	儲氣瓶
air floatation correction coefficient	空气浮力修正系数	空氣浮力修正係數
air-fuel ratio	空燃比	空氣燃油比
air heater	空气加热器	空氣加熱器
air injection type	空气喷射式	空氣噴射式
air inlet unit	进气装置	進氣裝置
air intake casing	进气壳体	進氣殼體
air jet diffuser	集散式布风器	噴氣擴散器
air jet ship	喷气推进船	噴氣推進船
airless injection	无气喷射	無空氣噴射
air lift mud pump	吸泥泵	吸泥泵
air mass	气团	氣團
air motor	空气马达	氣動馬達
air pipe	通气管	通氣管
air preheater	空气预热器	空氣預熱器
air resistance	空气阻力	空氣阻力
air temperature	气温	氣溫
airtight	气密	氣密
airtight test	气密试验	氣密試驗
alarm monitoring system	报警监视系统	警報監視系統
alarm printer	报警打印	警報打印
ALC(=automatic load control)	自动负荷控制	自動負載控制
Aldebaran	毕宿五(金牛α)	畢宿五(金牛α)
alert data	报警数据	警報數據
alert message	报警报文	警報信文
alert phase	告警阶段	警戒階段
aligning angle	看齐角，校准角	校準角
all direction propeller(=Z propeller)	全向推进器	全向推進器
alley way	通道	通道
allowable stress	许用应力	容許應力
all risks	[船舶]一切险	一切險,全險
all-round light	环照灯	環照燈
all told	全部容量	總括數
almanac	历书	曆書,航海曆,天文曆
almucantar	高度圈	等高圈

英　文　名	祖国大陆名	台湾地区名
aloft	高处	在上,在高處
aloft work	高空作业	高空作業
alongshore mark	沿岸标	沿岸標誌
alongside	靠泊	靠泊
alongside berth wharf	靠码头	靠[泊]碼頭
alongside delivery(=free alongside ship)	船边交货	船邊交貨
along-side method	傍靠补给法	傍靠傳遞法
alphabetical flag	字母旗	字母旗
alphabetical telegraph	字母电报	文字電報
Alphard	星宿一(长蛇 α)	星宿一(長蛇 α)
Alphecca	贯索四(北冕 α)	貫索四(北冕 α)
Alpheratz	壁宿二(仙女 α)	壁宿二(仙女 α)
Altair	河鼓二(天鹰 α)	河鼓二,牽牛[星]牛郎 [星](天鷹 α)
altazimuth	经纬仪	經緯儀
alter course(a/c)	转向	轉向
alternating current	交流电	交流電,直線流
alternating light	互光	變色燈
alternating stress	交变应力	交變應力
altitude difference	高度差	高度差
altitude difference method	高度差法	高度差法
alto-cumulus	高积云	高積雲
alto-starts	高层云	高層雲
aluminum base bearing metals	铝基轴承合金	鋁基軸承合金
aluminum paint	铝粉漆	鋁漆
ambient temperature	环境温度	環境溫度
ambiguity	多值性(**模糊度**)	模糊度
ambiguity resolution	模糊度解析	不明確訊息解析
Amendments to the 1974 SOLAS Convention	1974 年 SOLAS 公约修正案	一九七四年海上人命安全國際公約修正案
American Bureau of Shipping（AB）	美国船级社	美國船級協會
American Petroleum Institute Classification	美国石油协会分级	美國石油學會分級
amidships	正舵	正舵
amidships engined ship	中机型船	舯機艙船
amphidromic point	无潮点	無潮點
amphidromic region	无潮区	無潮區
amplitude shift keying(ASK)	幅移键控	移幅按鍵

英　文　名	祖国大陆名	台湾地区名
AMVER(=automated mutual-assistance vessel rescue system)	船舶自动互救系统	自動互助船舶救助系統
anabatic	上升风	上升風, 飆, 上坡風
analog-to-digital converter	模–数转换器	模擬–數位轉換器
analytic inertial navigation system	解析式惯性导航系统	解析慣性航行系統
anchor	锚	錨
anchorage	锚地	錨地
anchorage chart	锚地海图	泊地海圖
anchor at short stay	锚链收短	錨鏈收短
anchor aweigh	锚离底	錨離地
anchor ball	锚球	錨球
anchor buoy	锚标	錨標
anchor by the stern	抛艉锚	抛後錨
anchor capstan	起锚系缆绞盘(**起锚绞盘**)	起錨絞盤, 起錨機
anchor chain	锚链	錨鏈
anchor davit(=cat davit)	吊锚杆	吊錨桿
anchored vessel	锚泊船	錨泊船
anchor embedded	淤锚	錨埋入
anchor gear	起锚设备(**锚设备**)	錨具
anchor holding power to weight ratio	锚抓重比	錨抓著力與重量比
anchor ice	锚冰	錨冰
anchoring	锚泊	錨泊
anchoring in fog	雾泊(**雾中抛描**)	霧中抛錨
anchoring orders	抛起锚口令(**锚令**)	錨令
anchoring test	抛锚试验	錨泊試驗
anchor in sight(=clear of water)	锚出水	錨出水
anchor is up	起锚完毕	起錨完畢
anchor light	锚灯	錨燈
anchor penetration	锚啮入性	錨嵌入性
anchor position(AP)	锚位	錨位
anchor prohibited mark	禁止锚泊标	禁泊標誌
anchor recess	锚穴	嵌錨穴, 錨龕
anchor rope	锚缆	錨索
anchor up and down	锚垂直	錨垂直
anchor watch	锚更	錨更
anchor windlass(=windlass)	起锚机	起錨機
anemograph	风速计	風速紀錄計

英 文 名	祖 国 大 陆 名	台 湾 地 区 名
anemometer	风速表	風速計
angle of attack	冲角	切水角,攻角
angle of deck immersion	甲板进水角	甲板浸水角
angle of heel	横倾角	橫傾角
angle of maximum stability lever	最大稳性力臂角	最大穩度力臂角
angle of refraction	折射角	折射角
angle of repose	休止角	(散裝貨)止傾角,静止角
angle of sight	视线角	視[線]角
angle rod	钓竿	釣竿
angle valve	角阀	[折]角閥,肘閥
angular misalignment	角偏差	角偏差
angular position sensor	角度传感器	角度感應器
animal or plant quarantine	动植物检疫	動植物檢疫
ANMS(=automatic notice to mariners system)	自动航海通告系统	自動航船布告系統
annealing	退火	退火
anniversary date	周年日	週年日
annual aberration	周年光行差	週年光行差
annual parallax	[恒星]周年视差	週年視差
annual survey	年度检验	歲驗
annular combustor	环形燃烧室	環形燃燒室
annular eclipse	[日]环食	[日]環食
annular nebula	环状星云	環狀星雲
anodic protection	阳极防腐	陽極防蝕
anomalistic month	近点月	近點月
anomaly	异常	異常
anomaly	近点角	近點角
answering pendant	回答旗	回答旗,答應旗
Antarcs	心宿二(天蝎α)	心宿二,大火(天蝎α)
antenna effect	天线效应	天線效應
antenna switch	天线开关	天線開關
antenna tuning	天线调谐	天線調諧
anthelion	幻日	幻日
anti-clutter rain	雨雪干扰抑制	抗雨雪干擾
anti-clutter sea	海浪干扰抑制	抗海浪干擾
anti-corrosion treatment	防腐处理	防蝕處理
anticyclone	反气旋	反氣旋

英 文 名	祖 国 大 陆 名	台 湾 地 区 名
anti-dated bill of lading	倒签提单	倒簽載貨證券
anti-explosion bulkhead stuffing box	舱壁防爆填料函	艙壁防爆填料函
anti-fouling paint	防污漆	防污漆
anti-friction composition	减磨剂	抗磨劑
anti-icing equipment	防冰装置	防冰設備
anti-oxidant anti-corrosion additive	抗氧化抗腐蚀剂	抗氧化抗腐蝕劑
anti-rolling tank	减摇水舱	減搖水艙
anti-rolling tank stabilization system	水舱式减摇装置	減搖水艙穩定系統
anti-roll pump	减摇泵	減搖泵
anti-rust(=rust proof)	防锈	防銹
antirust paint	防锈漆	防銹漆
anti-seasickness medicine	晕船药	暈船藥
antiselena	幻月	幻月
antiskid deck paint	防滑甲板涂料	防滑甲板漆
anti-slack device	防松装置	抗鬆裝置
anti-toppling device	防倾装置	抗搖裝置
anti-wear agent	抗磨剂	抗磨劑
AOR(=Atlantic Ocean Region)	大西洋区	大西洋區域
AP(=anchor position)	锚位	錨位
apastron	远星点	遠星點
aperiodic compass	非周期罗经	無週期羅經,立復羅經, 安定羅經
aperiodic transitional condition	非周期过渡条件	非週期過渡條件
apex downwards	尖端向下	錐尖向下
apexes together	尖端对接	錐尖相連
apex upwards	尖端向上	錐尖向上
aphelion	远日点	遠日點
apocynthion	远月点	遠月點
apogee	远地点	遠地點
apparatus for on-board communication	船内通信设备	船上通信設備
apparent altitude	视高度	視高度
apparent declination	视赤纬	視赤緯
apparent equatorial coordinates	视赤道坐标	視赤道坐標
apparent horizon	视地平	視海平線
apparent period	视周期	視週期
apparent position	视位置	視位
apparent power	视功率	視[在]功率
apparent rise and set	视出没	視出沒

英　文　名	祖国大陆名	台湾地区名
apparent [solar] time	视[太阳]时	視時,視太陽時
apparent sun	视太阳	視太陽,真太陽
apparent wind	视风	視風向,似風
apparent zenith distance	视顶距	視天頂距
appendage resistance	附体阻力	附屬物阻力
applicable law of ship collision	船舶碰撞准据法	船舶碰撞適用法
appointed arbitrator	指定仲裁员	指定仲裁人
appraisal of traffic safety	交通安全评估	交通安全評估
apprentice	练习生(**实习生**),学徒	實習生,學徒
approaching one another	相互接近	互相接近
approximate contour line	未精测的等高线	近似等高線
approximate depth contour	未精测的等深线	近似等深線
apron	副艏材	艏護木,艏牆
aquaculture	水产养殖	水産養殖
arbitration	仲裁	仲裁
arbitration award	仲裁裁决	仲裁判斷
arbitration clause	仲裁条款	仲裁條款
arbitration tribunal	仲裁庭	仲裁法庭
arbitrator	仲裁员	仲裁人
archipelago sea area	群岛水域	群島海域
arctic air mass	北极气团	北極氣團
arctic circle	北极圈	北極圈
arctic current	北冰洋海流	北極海流
arctic pack	北极冰	北極流動水田
Arcturus	大角(牧夫 α)	大角(牧夫 α)
arc welding	电弧焊	電[弧]焊[接]
area	区域	區域
area code	区域码	地區碼
area co-ordinator	区域协调人	區域協調人
area of midship section	舯剖面积	舯剖面積
area of propeller disc	螺旋桨盘面积	螺槳盤面積
area of rudder	舵面积	舵面積
area of water plane	水线面积	水線面積
areas dangerous due to mines	水雷危险区	水雷危險區
arena	动界	界
armature reaction	电枢反应	電樞反應
ARPA(=automatic radar plotting aids)	自动雷达标绘仪	雷達自動測繪設備
ARQ(=automatic repetition request	自动请求重发方式	自動複傳申請

英　文　名	祖国大陆名	台湾地区名
mode)		
arrest of ship	扣押船舶	扣押船舶
arrival ballast	到港压载水	抵港壓艙水
arrival notice	到货通知	到貨通知
arrival point	到达点	到達點
arrived ship	到达船	到港船
articulated loading tower	铰接式装油塔	活節裝載塔
articulated tower mooring system	铰接塔系泊系统	活節塔繫泊系統
artificial antenna	仿真天线	假天線
artificial harbor	人工港	人工港
artificial horizon	人工地平	人工水平儀
artificial intelligent language	人工智能语言	人工智慧語言
artificial island	人工岛	人工島
artificial planet	人造行星	人造行星
artificial satellite	人造卫星	人造衛星
asbestos	石棉	石棉
asbestos cord	石棉绳	石棉繩
ascending node	升交点	昇交點
ASF(= additional secondary phase factor)	附加二次相位因子	附加二次相位因數
ash content	灰分	灰分含量
ASK(= amplitude shift keying)	幅移键控	移幅按鍵
ASO(= auxiliary ship observation)	船舶辅助观测	輔助船舶觀測
aspect ratio	展弦比	寬高比
asphaltenes content	沥青分	瀝青含量
aspirated engine	非增压发动机	吸氣式發動機
assigned frequency	指配频率	指配頻率
assigned frequency band	指配频带	指配頻帶
assistant by helicopter	直升机援助	直升機援助
assistant engineer	轮助	助理工程師,助理輪機員
assistant officer	驾助	助理船副
associated rescue coordination center	联合救助协调中心	聯合救助協調中心
assumed latitude	选择纬度	緯度採用值
assumed longitude	选择经度	經度採用值
assumed position	选择船位	假定船位
assurer	保险人	保險人
astern	后方,向后,倒车	後方,向後,在……後,倒俥,後退

英　文　名	祖国大陆名	台湾地区名
astern approaching method	艉部接近法(**艉接近法**)	艉接近法
astern exhaust chest sprayer	倒车排汽室喷雾器	倒俥排氣室噴霧器
astern gas turbine	倒车燃气轮机	倒俥燃氣渦輪機
astern guarding valve	倒车隔离阀	倒俥護閥
astern maneuving valve	倒车操纵阀	倒俥操縱閥
astern power(=backing power)	倒车功率	倒俥動力
astern replenishing rig	纵向补给装置	艉向補給裝置
astern replenishment at sea	航行纵向补给	海上艉向補給
astern running operating mode management	倒航工况管理	倒航操作形式管理
astern steam turbine	倒车汽轮机	倒俥蒸汽渦輪機
astern trial	倒车试验	倒俥試航
astronomical coordinate	天文坐标	天文坐標
astronomical fix(AF)	天文船位	天文定位
astronomical latitude	天文纬度	天文緯度
astronomical longitude	天文经度	天文經度
astronomical meridian	①天文子午线 ②天文子午圈	①天文子午線 ②天文子午圈
astronomical orientation	天文定向	天文定向
astronomical tide	天文潮	天文潮汐
astronomical triangle	天文三角形	天文三角形
astronomical twilight	天文晨昏朦影	天文曦光
AT(=atomic time)	原子时	原子時
Athens Convention Relating to the Carriage of Passengers and Their Luggages by Sea	海上旅客及其行李运输雅典公约	海上運送旅客及其行李雅典公約
athwartships	横向	橫切,橫向(與龍骨成直角之方向)
athwartships magnet	横向磁棒	橫向磁棒
Atlantic Ocean Region (AOR)	大西洋区	大西洋區域
Atlantic Ocean Region East	大西洋东区	大西洋東區
Atlantic Ocean Region West	大西洋西区	大西洋西區
Atlantic polar front	大西洋极锋	大西洋極鋒
atlas of tidal stream	潮流图	潮流圖
atlas of tides	潮汐[海]图	潮汐圖
atmosphere	大气	大氣
atmospheric distillation	常压蒸馏	常壓蒸餾

英 文 名	祖国大陆名	台湾地区名
atmospheric pressure	气压(**大气压**)	大氣壓力
atmospheric transmissivity	大气透射率	大氣傳達量
atomic time(AT)	原子时	原子時
atomization	雾化	霧化
atomizer	雾化器	霧化器
Atria	三角形三(南三角 α)	三角形三(南三角 α)
at sea maintenance	海上维修	海上維修
attemperator	减温器	減溫器
attitude angle	姿态角	姿態角
attitude determination system(ADS)	姿态测定系统	姿態測定系統
auction of ship	拍卖船舶	拍賣船舶
aurora	极光	極光
aurora australis	南极光	南極光
aurora borealis(=northern light)	北极光	北極光
auto alarm	自动报警	自動警報器
auto-clean strainer	自动清洗滤器	自動清洗過濾器
auto crane	汽车起重机	汽伸式起重機
autolevelling assembly	自动校平装置	自動準平裝置
autolongline machine	自动延绳机	自動延繩機
automated control system of dredging process	挖泥船自控程序	挖泥船自控程序
automated mutual-assistance vessel rescue system(AMVER)	船舶自动互救系统	自動互助船舶救助系統
automatic air circuit breaker	自动空气断路器	自動空氣斷路器
automatic call	自动呼叫	自動呼叫
automatic collision avoidance system	自动避碰系统	自動避碰系統
automatic combustion control	燃烧自动控制	自動燃燒控制
automatic constant tension mooring winch	恒张力带缆绞车	自動張力繫船絞車
automatic constant tension towing winch	恒张力拖缆机	自動等拉力拖攬機
automatic dial-up two-way telephony	自动拨号双向电话	自動撥號雙向電話
automatic direction finder	自动测向仪	自動測向儀
automatic dislodging clarifier	自动除渣分杂机	自動除渣澄清器
automatic display instrument	自动显示仪表	自動顯示儀
automatic distributor of active power	有功功率自动分配装置	有效功率自動分配裝置
automatic distributor of reactive power	无功功率自动分配装置	無功功率自動分配裝置
automatic drainage	自动疏水	自動排洩裝置
automatic error correction	自动纠错	自動改錯
automatic expansion valve	自动膨胀阀	自動膨脹閥

英　文　名	祖国大陆名	台湾地区名
automatic filter cleaning	滤器自动清洗	過濾器自動清洗
automatic fire alarm and fire detection system	失火自动报警与探测系统	自動火警警報與探火系統
automatic fire alarm system	失火自动报警系统	自動火警警報系統
automatic keying device	自动拍发器	自動鍵發設施
automatic load control(ALC)	自动负荷控制	自動負載控制
automatic measuring instrument	自动测量仪	自動測量儀
automatic mooring winch	自动系泊绞车	自動繫泊絞車
automatic non-return valve	自动止回阀	自動止回閥
automatic notice to mariners system (ANMS)	自动航海通告系统	自動航船布告系統
automatic parallel off	自动解列	自動解除並聯
automatic parallel operation	自动并联运行	自動並聯運轉
automatic radar plotting aids(ARPA)	自动雷达标绘仪	雷達自動測繪設備
automatic radio direction finder (ADF)	自动测向仪	自動無線電測向儀
automatic repetition request mode	自动请求重发方式	自動複傳申請
automatic self-closing safety door	自闭式安全门	自閉式安全門
automatic service	自动业务	自動業務
automatic slow down	自动减速	自動減速
automatic soot blower	自动吹灰器	自動吹灰器
automatic sprinkler and fire alarm and fire detection system	自动喷水、报警和探测系统	自動噴水器與火警警報及滅火系統
automatic sprinkler fire detection system	自动洒水探火系统(**自动探火喷水系统**)	自動探火噴水系統
automatic starting air shut-off valve	自动式主起动阀	自動起動空氣關斷閥
automatic synchronizing device	自动同步装置	自動同步設施
automatic telex test	自动用户电报试验	自動電傳試驗
automatic tide gauge	自动验潮仪	自動測潮計
automatic tuning	自动调谐	自動調諧
automatic vehicle location(AVL)	自动车辆定位	自動車輛定位
automatic voltage regulator(AVR)	自动电压调节器	自動電壓調整器
automobile ferry(=car ferry)	汽车轮渡	車輛渡輪
autopilot	自动操舵仪	自動駕駛儀
auto scanning	自动扫描	自動掃瞄
auto-starting air compressor	自动起动空气压缩机	自動起動空氣壓縮機
auto-stop	自动停车	自動停俥
auto-transformer starting	自耦变压器起动	自耦變壓器起動
autumnal equinoctial spring tide	秋分大潮	秋分大潮

英　文　名	祖国大陆名	台湾地区名
autumnal equinox	秋分点	秋分
auxiliary boiler automatic control system	辅锅炉自动控制系统	輔鍋爐自動控制系統
auxiliary condenser circulating pump	辅冷凝器循环泵	輔機冷凝器循環泵
auxiliary device	辅助装置	輔助設施
auxiliary engine log book	副机日志	副機日誌
auxiliary generator diesel engine	辅助发电柴油机	輔助柴油發電機
auxiliary gyro	副陀螺	輔迴轉儀
auxiliary hose rigs	辅助软管设备	輔助軟管設備
auxiliary machinery diesel engine	辅助柴油机	輔助柴油機
auxiliary ship observation	船舶辅助观测	輔助船舶觀測
auxiliary steam turbine	辅汽轮机	輔蒸汽渦輪機
auxiliary steering gear	辅助操舵装置	輔助操舵裝置
auxiliary towing line	副拖缆	副拖纜
auxiliary weather chart	辅助天气图	輔助天氣圖
avast hauling	停拉	停曳
average adjuster	海损理算人	海損清算人
average bale capacity	平均包装容积	平均包裝容積
average bond	海损分担保证书	共同海損承諾書
average guarantee	海损担保函	海損擔保
average per ton	平均每吨容积	平均每噸容積
average statement	海损理算书	海損理算書
average void depth	平均空档深度	平均空檔深度
AVL(=automatic vehicle location)	自动车辆定位	自動車輛定位
AVR(=automatic voltage regulator)	自动电压调节器	自動電壓調整器
away from	远离	遠離
awkward cargo	笨重货	笨重貨
awning	天幕	天遮
axe	①斧头 ②太平斧	①斧頭 ②太平斧
axial displacement protective device	轴向位移保护装置	軸向位移保護設施
axial-flow compressor	轴流式压气机	軸流壓縮機
axial-flow pump	轴流泵	軸流泵
axial flow steam turbine	轴流式汽轮机	軸流蒸汽渦輪機
axial-flow turbine	轴流式涡轮机	軸流渦輪機
axial-piston hydraulic motor	轴向柱塞式液压马达	軸向活塞式液力馬達
axial vibration	轴向振动	軸向振動
azimuth	①[天体]方位角 ②地平经度	①方位角 ②地平經度
azimuthal equidistant projection	方位等距投影	方位等距投影

英　文　名	祖国大陆名	台湾地区名
azimuthal projection	方位投影	方位投影
azimuth difference	方位角差	方位角差
azimuth gyro	方位陀螺	方位迴轉儀

B

英　文　名	祖国大陆名	台湾地区名
Babbitt metal	巴氏合金	巴比合金
back	倒车	倒俥
back emergency	紧急倒车	紧急倒俥
back full	全速后退	全速後退
background light	背景亮光	背景亮光
backing power	倒车功率	倒俥動力
back pressure regulator	蒸发压力调节阀(**背压调节阀**)	背壓調整器
back pressure steam turbine	背压式汽轮机	背壓蒸汽渦輪機
back pressure valve(=non-return valve)	止回阀	止回閥
backsplice	绳头插接	反插接
back track	艉迹	艉跡
back water!	退桨!	反划!
baggage	行李	行李
baggage room	行李间	行李間
bagged cargo	袋装货	袋裝貨
bait	钓饵	釣餌
balanced gyroscope	平衡陀螺仪	平衡迴轉機
balanced valve	平衡阀	均壓閥,均衡閥
balance rudder	平衡舵	平衡舵
bale capacity	包装容积	包裝容積
baled cargo	捆包货	細包貨
ball and socket joint	球窝节	球接頭
ballast condition	压载状态	壓載船況
ballasting	压载	壓載
ballast pump	压载泵	壓載泵
ballast system	压载水系统	壓艙水系統
ballast tank	压载舱	壓載艙
ballast-tank paint	压载舱漆	壓載艙漆
ballast water	压载水	壓載水,壓艙水
ballistic error	冲击误差	衝擊誤差

英　文　名	祖国大陆名	台湾地区名
ball［shape］	球体［号型］	球形號標
ball thrust bearing	滚球止推轴承	滾球止推軸承
bamboo raft	竹排	竹筏
bank clearance	船岸间距	船岸間距
bank effect	岸壁效应	岸壁效應
bank suction	岸吸	岸吸力
bank the oars！	平桨！	平槳！
bar draft	过滩吃水	淺灘低潮水深
bare boat	光船	空船
bareboat charter	光船租赁	光船租賃
bareboat charter registration	光船租赁登记	光船租賃登記
bareboat charter with hire purchase	船舶租购(**光船租购**)	分期付款之光船租約
barge	驳船	駁船
barge train	驳船队	駁船隊
barge train formation	驳船队编组	駁船隊編組
barkentine	多桅帆船	多桅帆船
barnacle	藤壶	海生介
barograph	气压计	氣壓計
barometer	气压表	氣壓表,氣壓計,晴雨計
barometric tendency	气压趋势	氣壓趨勢
bar port	候潮港	淺灘港
barred-speed range	转速禁区	速限區
barrel	桶	桶(燃油之單位)
barrel buoy	桶形浮标	桶狀浮標
bar the engine	盘车	盤俥
base cargo	垫底货	底艙貨
baseline	基线	基線
baseline delay	基线延迟	基線遲延(羅遠)
baseline extension	基线延长线	基線延長(羅遠)
baseline of territorial sea	领海基线	領海基線
base oil	基油	基油
base point	基点	基點
base station	基地电台	基［地］台
basic freight	基本运费	基本運費
basic port(BP)	基本港	母港
basic repetition frequency	基本重复频率	基本重複頻率
basket［shape］	篮子［号型］	籃形號標
bathymetric map	等深线图	水深圖,海底地形圖,等

英　文　名	祖国大陆名	台湾地区名
		深線圖
bathymetric navigation	测深导航	測深航法
bathymetric survey	水深测量	水深測量
battle ship	战列舰	戰艦，戰鬥艦
B class division	乙级分隔	"B"級區（防火）
BDC（=bottom dead center）	下止点	下死點
BDE（=below deck equipment）	舱内设备	艙內設備
beach	海滩	海灘
beaching	抢滩	搶灘
beacon	①立标 ②信标	①標桿 ②標台
beacon buoy	带顶标浮标	桿形浮標
beacon detection probability	信标检测概率	示標檢測概率
beacon identification	信标识别	示標識別
beacon identification data	信标识别数据	示標識別數據
beacon location probability	信标定位概率	示標定位概率
beacon message	信标电文	示標信文
beacon register	信标登记	示標登記
beacon station	信标台	信標發射台
beacon tower	立标塔	指標塔
beacon transmit antenna	信标发射天线	示標發射天線
beam distance between ships	舰间间隔（船[舰]间横距）	船艦間橫距
beam draft ratio	船宽吃水比	船寬吃水比
beam sea	横浪	橫浪
beam trawl	桁拖网	桁拖網
beam trawler	舷侧拖网船	舷側拖網船
bear down	转向上风	轉向上風
bearing	①方位 ②轴承	①方位 ②軸承
bearing beam	承推架	軸承架
bearing circle	方位圈	方位圈
bearing clearance	轴承间隙	軸承間隙
bearing line	方位线	方位線
bearing resolution	方位分辨力	方位分析度
bear up	转向下风	轉向下風
Beaufort［wind］scale	蒲福风级	蒲福風級
bedplate	机座	座板
beginning of morning twilight	晨光始	黎明之開始
Beijing coordinate system	北京坐标系	北京座標系統

英 文 名	祖 国 大 陆 名	台 湾 地 区 名
bell	号钟	號鐘
bell book	车钟记录簿	俥鐘紀錄簿
bell buoy	钟浮标	鐘浮標
bellows	①波纹管 ②风箱	①伸縮囊 ②風箱,皮老虎
bell pull	钟锤拉索	拉鈴把手
below deck equipment(BDE)	舱内设备	艙內設備
bench mark	水准点	基準標誌
bending fatigue cracks	弯曲疲劳破裂	彎曲疲勞破裂
bending strength	抗弯强度	彎曲強度
bend of channel	航道弯曲度	航道彎曲度
bend on	系住	繫住
bends	潜水病	潛水病
bends and hitches	绳结	繩結
benthic division	深海分区	深海分區
berg	冰山	冰山
berg-bit	碎冰山	碎冰山
berm	冲积岸堤	堤岸
berthage	码头费(港口费)	碼頭費,船席費
berth cargo	填舱货	補充貨載
berth charter party	泊位租船合同	泊位傭船契約
berthing	系泊	舷緣殼板
berth note	订舱单	訂運單
berth term	泊位条款	碼頭收交貨條件
Betelgeuse	参宿四(猎户 α)	參宿四(獵戶 α)
bight	①绳的弯曲部 ②绳扣	①繩[索]套 ②回頭索
bilge automatic discharging device	污水自动排除装置	舭水自動排除設施
bilge bracket	舭肘板	舭腋板
bilge keel	舭龙骨	舭龍骨
bilge pump	舱底泵	舭泵
bilge pumping arrangement	污水排吸装置	舭水抽排裝置
bilge system	污水系统	舭水系統
bilge tank	污水柜	舭櫃,舭水艙
bilge water	污水	舭水,艙底污水
bilge well	污水井	污水井
bill board	锚床	錨床
bill of clearance	结关清单	出口結關單,出港證書
bill of entry	进口关单	進口[報]關單,進港證

英　文　名	祖国大陆名	台湾地区名
	書	
bill of lading (BL)	提单	提單,載貨證券
bills payable	应付票据	付款單
binnacle	罗经座	羅經座
binoculars	双筒望远镜	雙筒遠望鏡
bipod mast	人字桅,双脚桅	雙腳桅
bitter end	锚链内端	錨鏈内端
bitts(=bollard)	缆桩	雙繫纜樁
bituminous solution	沥青清漆	瀝青溶液
BL(=bill of lading)	提单	提單,載貨證券
black box	黑匣子	黑盒子
Black stream(=kuroshio)	黑潮	黑潮
blade element theory	叶元体理论	葉片元素理論
blade wheel	叶轮	葉輪
blade wheel rotor	轮型转子	葉輪轉子
blank bill of lading	不记名提单	不記名載貨證券
blank flange	盲板法兰	管口蓋板
bleed air rate	抽气量	抽氣率
bleeding steam turbine	抽汽式汽轮机	抽汽渦輪機
blind pilotage techniques	仪表引航技术	盲目導航技術
blind sending	盲发	盲發
blind zone	盲区	盲區
blinker yardarm	横桁闪光信号灯	横桁閃光信號燈
blip	尖头信号	跳波
blister	气泡	舯膨出部
block	①滑车 ②封锁	①滑車 ②封鎖
block brake	滑块制动器	塊狀靭,塊狀煞車
block coefficient	方形系数	方塊係數
block diagram	方块图	方塊圖
blow-by	漏气	漏氣
blow down pump	排盐泵	衝放泵
blower	鼓风机	鼓風機
blue peter	开船旗	開船旗
blurring effect	模糊效应	模糊效應
board measure	板材量尺	板材量法
board of arbitration	仲裁委员会	仲裁委員會
board of inquiry	调查委员会	調查委員會
board of investigation(=board of inquiry)	调查委员会	調查委員會

英　文　名	祖国大陆名	台湾地区名
boat	艇	小艇
boat boom	系艇杆	繫艇桿
boat cover	艇罩	艇罩
boat davit	吊艇架	小艇吊架
boat fall	吊艇索	小艇吊索
boat gripe	艇系紧带	小艇扣帶
boat handling gear	吊艇装置	小艇吊放裝置
boat hook	钩篙	鉤篙
boating	操艇	操艇
boat launching appliance	艇下水装置	小艇下水設施
boat lifting hook	吊艇钩	吊艇鉤
boat plug	艇底塞	艇底塞
boat sailing	驶风	小艇駛帆
boat skate	艇滑架	艇滑道
boat station bill	救生部署表	救生部署表
boatswain(=bosun)	水手长	水手長
boatswain's chest	帆缆箱	帆纜箱
boatswain's store	帆缆间	帆纜庫
boat the oars	收桨	收槳
boat winch	吊艇机	小艇吊機
bobstay	艏斜桅支索	艏斜桅拉索
boiler automatic control system	锅炉自动控制系统	鍋爐自動控制系統
boiler auxiliary steam system	锅炉辅助蒸汽系统	鍋爐輔助蒸汽系統
boiler blow down valve	锅炉排污阀	鍋爐放水閥
boiler blower	锅炉鼓风机	鍋爐鼓風機
boiler body	锅炉本体	鍋爐體
boiler casing	锅炉外壳	鍋爐殼,鍋爐襯套
boiler clothing(=boiler casing)	锅炉外壳	鍋爐殼,鍋爐襯套
boiler dry pipe	锅炉干汽管	鍋爐乾汽管
boiler feed check valve	锅炉给水止回阀	鍋爐給水止回閥
boiler feed pump	锅炉给水泵	鍋爐給水泵
boiler feed system	锅炉给水系统	鍋爐給水系統
boiler firing equipment	锅炉点火设备	鍋爐點火設備
boiler fittings	锅炉附件	鍋爐附件,鍋爐裝具
boiler forced-circulating pump	锅炉强制循环泵	鍋爐強力循環泵
boiler fuel oil pump	锅炉燃油泵	鍋爐燃油泵
boiler fuel oil system	锅炉燃油系统	鍋爐燃油系統
boiler heating surface	锅炉受热面	鍋爐受熱面

英 文 名	祖国大陆名	台湾地区名
boiler ignition oil pump	锅炉点火泵	鍋爐點火泵
boiler induced-draft fan	锅炉引风机	鍋爐誘導通風扇
boiler lighting up	锅炉点火	鍋爐點火
boiler main steam system	锅炉主蒸汽系统	鍋爐主蒸汽系統
boiler room	锅炉舱	鍋爐艙,鍋爐間
boiler safety valve	锅炉安全阀	鍋爐安全閥
boiler secondary air blower	锅炉二次鼓风机	鍋爐二次鼓風機
boiler stay	锅炉牵条	鍋爐牽條,鍋爐拉條
boiler stay tube	锅炉牵条管	鍋爐拉條管
boiler uptake	锅炉烟箱	鍋爐煙道
boiler water gauge	锅炉水位表	鍋爐水位計
boiler water level regulator	锅炉水位调节器	鍋爐水位調整器
boiler water wall	锅炉水冷壁	鍋爐水管壁
boiling evaporation	沸腾蒸发	沸騰蒸發
bollard	缆桩	雙繫纜樁
bollard pull	系桩拉力	繫纜樁拖力
bolt rope	帆缘索	帆[帳]緣索
bonded store	保税库	保稅倉庫
bond room (=bond store)	保税库	保稅倉庫
booking	订舱	訂載
boom	帆脚杆	帆桁
boom cradle	吊杆架	吊桿架
boom outstretch	吊杆跨距	吊桿伸出舷外距離
boom topping angle	吊杆仰角	吊桿俯仰角
booster gas turbine	加速燃气轮机	加力燃氣渦輪機
booster pump	升压泵	增壓泵,加力泵
booster servomotor	升压伺服器	增壓伺服馬達
boot	桅套	靴,桅跟帆套
boot-topping paint	水线漆	水線漆
bore-colded	钻孔冷却	鏇孔冷卻
bore maximum wear	缸径最大磨损	缸徑最大磨損
Bosch filter paper smoke meter	烟迹式烟度计	濾紙測煙計
Bosch injection pump	回油孔式喷油泵	布氏噴油泵
bossing	艉膨出部	艉膨出部
bosun	水手长	水手長
bosun's chair	单人坐板	工作吊板
bosun's chair hitch	坐板升降结	工作吊板套結
both to blame collision	双方责任碰撞	雙邊過失碰撞

英　文　名	祖国大陆名	台湾地区名
both to blame collision clause	互有责任碰撞条款	雙邊過失碰撞條款
bottom current	底层流	底層流
bottom dead center (BDC)	下止点	下死點
bottom longitudinal	船底纵骨	船底縱肋
bottom paint	船底漆	船底漆
bottom plate	船底板	船底殼板
bottomry	船舶抵押贷款	船舶押款
bottomry bond	船舶抵押合同	船舶抵押契約,船舶貸款保證書
bottomry bondholder	船舶抵押合同持有人	船舶抵押債券持有人
bottom tracking	底迹	底跡
bottom trawl	底拖网	底曳網
boundary lubrication	边界润滑	邊界潤滑
bound from	下行(**来自**)	由…來
bound to	上行(**开往**)	開往…
bow	艏	艏
bow	艏舷	
bow anchor(=bower)	艏锚	艏錨
bow and beam bearing	四点方位法	四點方位法
bow breast	艏横缆	頭腰纜
bow chock	艏导缆孔	艏導索椿,分水艏板
bower	艏锚	艏錨
bow hook	艏钩篙	掌篙手
bowline	单套结	單套結
bowline on the bight	双套结	兜腰稱人結,腰結
bow painter(=painter)	艇首缆	艇首索
bow pudding	艏碰垫	艏碰墊
bow ramp	艏门跳板	艏大門跳板
bow rudder	艏舵	艏舵
bow sea	艏舷浪	艏側浪
bowse down	向下拉	向下緊
bowser boat	加油艇	加油艇
bowse up	向上拉	向上緊
bowsing tackle	收拉绞辘	收繫轆轤
bowsprit	艏斜桅	艏斜桅
bow wave	艏波	艏波
box frame	箱形机架	箱形機架
box keelson	箱形内龙骨	方形內龍骨

英　文　名	祖国大陆名	台湾地区名
BP(=base point)	基点	基點
bracket	肘板	腋板
brails	卷帆索	捲帆索
brake mean effective pressure	制动平均有效压力	制動平均有效壓力
branch line	支线	枝線,枝繩
branch line conveyer	支线传送装置	枝繩輸送裝置
branch line winder	支线起线机	枝繩捲揚機
brash ice	碎冰	碎冰
breadth depth ratio	宽深比	寬深比
break	船楼端	艛端
breakaway emergency	紧急脱离	緊急脫離
break bulk cargo	件杂货	零散雜貨
breakdown lights	故障灯	故障號燈
breakers	浪花	碎浪
breaking strain	断裂应变	裂斷應變
breaking strength(BS)	破断强度	裂斷強度
breakwater	①防波堤 ②挡浪板	①防波堤 ②擋浪板
breathing apparatus	呼吸器	呼吸器
breasting the ship apart	将船撑开	使船橫著離開
breast line	横缆	橫纜
breast off	撑开	撐開
breather valve	呼吸阀	呼吸閥
breech buoy	围裙救生圈	圍裙救生圈
breeding ground(=nursery ground)	繁殖场	繁殖場
breeding season	繁殖季节	繁殖期
bridge	驾驶台	駕駛台
bridge control	驾驶台控制	指揮台操縱
bridge deck	驾驶台甲板	駕駛台甲板
bridge fitting	桥式联结器	橋式聯結器
bridge gauge	桥规	橋形規,橋形軸規,軸規
bridge gauge value	桥规值	橋形規值
bridge opening mark	桥涵标	橋通路標
bridge pier light	桥柱灯	橋柱燈
bridge remote control system	驾驶台遥控系统	駕駛台遙控系統
bridge-to-bridge communication	驾驶台间通信	船橋間通信
bridle	缰绳,系船索	叉索
brig (=brigantine)	二桅帆船	雙桅帆船
brigantine	二桅帆船	雙桅帆船

英 文 名	祖国大陆名	台湾地区名
brine density	盐度	卤水密度
Brinell figure	布氏[硬]度数	勃式硬度數
brine pump	盐水泵	鹽水泵
brine rate	排盐量	排鹽率
bring home	归位,收紧	歸位
bring to	使船停住	拋錨停航
broach to	打横	船身突橫(順風駛帆時)
broadcast ephemeris	广播星历	廣播天文曆
broadcasting-satellite service	卫星广播业务	衛星廣播業務
broadcasting service	广播业务	廣播業務
broadside	舷侧	舷側
broken coast	断续海岸	斷續海岸
broken space	亏舱	堆貨餘隙,貨載空隙
broken stowage(=broken space)	亏舱	堆貨餘隙,貨載空隙
broken water	乱水	亂水
bronze strip	青铜条	青銅條
brow	①跳板 ②窗楣	①跳板 ②窗楣
brow landing	跳板台	跳板著陸架
brushless AC generator	无刷交流发电机	無刷交流發電機
BS(=breaking strength)	破断强度	裂斷強度
bubble sextant	气泡六分仪	氣泡六分儀
bucket arrangement	链斗装置	鏈斗裝置
buckler	锚链孔盖	錨鏈孔蓋
buffer	缓冲器	緩衝器
buffer spring	缓冲弹簧	緩衝彈簧
built-up crankshaft	组合[式]曲轴	組合曲柄軸
bulbous bow	球鼻[型]艏	球形艏
bulb rudder	导流罩舵	球形舵
bulb stern	球形艉	球形艉
bulge	舯突出部,舭	舯膨出部
bulk capacity(=grain capacity)	散装容积	散裝容積
bulk cargo	散货	散裝貨
bulk cargo clause	散装货条款	散裝貨條款
bulk-cargo ship(=bulk carrier)	散货船	散裝貨船
bulk carrier	散货船	散裝貨船
bulkhead	舱壁	艙壁
bulkhead deck	舱壁甲板	艙壁甲板

英 文 名	祖 国 大 陆 名	台 湾 地 区 名
bulkhead plan	舱壁图	艙壁圖
bulkhead resistant to water(=watertight bulkhead)	水密舱壁	水密艙壁
bulkhead stuffing box	隔舱填料函	艙壁填料函
bulky cargo	轻泡货	輕笨貨(泡貨)
bulldog grip	绳头卸扣	鋼絲索扣
bull eye ring	牛眼环	牛眼環
bull-rope	拦索	攔索,吊貨控索
bulwark	舷墙	舷牆
bulwark freeing port	舷墙排水口	舷牆排水口
bulwark gripper	舷墙系索器	舷牆繫索器
bundle of bulk grain	散装谷物捆包	散裝穀類捆包
bundling of bulk	散货捆包	大包捆
bunk	床铺	床[鋪]
bunker	燃料舱	燃料艙,煤艙
bunkering	装燃料	裝載燃料
bunker oil	燃料油	重油
buoy	浮标	浮標,浮筒
buoyage system	水上助航标志系统	浮標系統
buoyancy	浮力	浮力
buoyancy regulating system	浮力调节系统	浮力調節系統
buoyancy test for buoyant apparatus	救生浮具浮力试验	救生浮具浮力試驗
buoyancy test for lifebuoy	救生圈试验	救生圈浮力試驗
buoyancy test for life-jacket	救生衣试验	救生衣浮力試驗
buoyant bailer	浮水杓	浮水杓
buoyant lifeline	救生浮索	救生浮索
buoyant paddle	浮桨	浮槳
buoyant rescue quoit	营救浮环	營救浮環
buoyant smoke signal	漂浮烟雾信号	浮煙信號
buoy hook	浮筒系钩	浮筒繫鉤
buoy tender	航标船	浮標母船,浮標管理船
burden of proof(=onus of proof)	举证责任	舉證責任
Bureau International de'l Heure	国际时间局	國際時間局
Bureau Veritas (BV)	法国船级社	法國驗船協會
burnable poison element	可燃毒物元件	可燃有毒元素
burner	燃烧器	燃燒器
burn-out	燃料烧毁(**烧尽**)	燒壞,燒毀
burn-out heat flux	烧毁热负荷(**烧尽热负**	燒盡熱通量

英 文 名	祖 国 大 陆 名	台 湾 地 区 名
	荷)	
burn-up	燃耗	燒盡,燒光,燒完
burton method of transfer	双索吊送传递法	雙索吊送傳遞法
burton rig	联杆操作补给装置	聯桿操作補給裝置
busbar	汇流排	匯流排
butchery	肉库	肉庫
butterfly valve	蝶阀	蝶形閥
Butterworth pump	洗舱泵	洗艙泵,巴特華斯泵（洗艙用）
Butterworth tank cleaning system	巴氏货油舱清洗系统	巴氏貨油艙清洗系統
butt joint	对接	對接
BV(=Bureau Veritas)	法国船级社	法國驗船協會
by-pass governing	旁通调节	旁路調節
by-pass valve	旁通阀	旁通閥
by the board	跌落水	落水
by the run	全松	鬆脱
by the wind	逆风航驶	逆戧

C

英 文 名	祖 国 大 陆 名	台 湾 地 区 名
C&F(=cost and freight)	成本加运费价格	成本與運費
CA(=course of advance)	计划航迹向	預期航向
cab.	链(长度单位 =1/10 海里)	鏈(長度單位)
cabin	居住舱,房舱	房艙,艙間
cabin luggage	自带行李	隨身行李
cabin passenger	客舱旅客	房艙乘客
cabin plan	房舱布置图	房艙佈置圖
cabin ventilator	舱室通风机	房艙通風機
cable	①粗缆 ②电缆	①粗纜 ②電纜
cable buoy	电缆浮标	電纜浮標
cable burying machine	埋缆机	埋纜機
cable cutter	切缆机	切纜機
cable gram	海底电报	海底電報
cable grapnel	捞缆钩	撈纜機
cable layer	布缆船	佈纜船
cable laying machine	布缆机	佈纜機

英 文 名	祖 国 大 陆 名	台 湾 地 区 名
cable mark	锚缆标记	錨纜標誌
cable position sensor	海底电缆传感器	電纜位置測定器
cable releaser	弃链器	釋纜器
cable slack meter	电缆松紧指示器	電纜鬆緊指示器
cabotage	沿海航运	沿海航運,沿海貿易
C/A code(=coarse/acquisition code)	C/A 码	C/A 碼
caking of oil	油结胶	油黏結
calcium grease	钙基润滑脂	鈣基滑脂
calculated altitude	计算高度	計算高度
calculated azimuth(=computed azimuth)	计算方位	計算方位
calculated wind pressure lever	计算风力力臂	計算風壓力臂
calculated wind pressure moment	计算风力力矩	計算風壓力矩
calendar line	日界线	日界線
calibrating lever	校准杆	校準桿
calibration	校准	校準
calk	捻缝	捻縫
call	呼叫,通话	呼叫,通話
call attempt	呼叫尝试	嘗試性呼叫
called party	被呼方	被呼用戶,受話方
calling	呼叫	呼叫
calling-in-point(CIP)	呼叫点	呼叫點
calling party	呼叫方	主叫用戶,發話方
call sign(CS)	呼号	呼號
calm	无浪	無風,浪靜
calm(rippled) sea	微浪(1 级)	靜海,浪靜(1 級)
calorific value	热值	發熱量
calving	裂冰	冰崩
cam	凸轮	凸輪
camber	梁拱	弧高,拱高
cam controller	凸轮控制器	凸輪控制器
camel	码头护木	碼頭護木
camshaft	凸轮轴	凸輪軸
canal light	运河灯	運河燈
canal tonnage	运河吨位	運河噸位
can annular type combustor	环管形燃烧室	環筒形燃燒器
can buoy	罐形浮标	罐形浮標
cancel	撤销	取消
canceling	解约	解約

英　文　名	祖国大陆名	台湾地区名
canceling date	解约日	解約日
candidate	候选人,报考人	申請發證者
can hook	筒钩	筒鉤
canning factory ship	罐头加工船	罐頭工作船
Canopus	老人(船底 α)	老人(船底 α)
canopy	艇天幕	小艇天遮
canvas	帆布	帆布
canvas and rope work	帆缆作业	帆纜作業
canvas cargo bag	盛货帆布袋	載貨帆布袋
capacity	容量	容量
capacity adjusting valve	能量调节阀	能量調節閥
capacity plan	舱容图	容量圖
cape	岬角(岬)	岬,岬角
Capella	五车二(御夫 α)	五車二(御夫 α)
Caph	王良一(仙后 β)	王良一(仙后 β)
capsize	倾覆	傾覆
capstan	绞盘	絞盤,起錨機
captain	船长	船長,艦長
carbon arc welding	碳弧焊	碳[極電]弧熔焊
carbon deposit	积碳	積碳
carbon residue	残碳值	殘留碳
carbon ring gland	碳环式密封	碳精迫緊
cardinal mark	方位标志	方位標誌
cardinal point	罗经基点	四向基點
cardinal winds	基点风	四方位風
cardioid polar diagram	心形[方向]特性图	心形極圖解
car ferry	汽车轮渡	車輛渡船
cargo-associated waste	货物伴生废弃物	貨物所生相關廢棄物
cargo boom	吊货杆	吊貨桿
cargo drop reel	吊货索卷筒	吊貨索捲軸
cargofall	吊货索	吊貨索
cargo gear	货物装卸设备	貨物裝卸設備
cargo-handling and stowage	货物装载	貨物裝載
cargo hold	货舱	貨艙
cargo hold dehumidification system	货舱空气干燥系统	貨艙空氣乾燥系統
cargo hook	吊货钩	吊貨鉤
cargo lifting equipment	起货设备	吊貨設備
cargo list	装货清单	裝貨清單

英　文　名	祖 国 大 陆 名	台 湾 地 区 名
cargo net	吊货网	吊貨網
cargo oil control room	货油控制室	貨油控制室
cargo oil deck pipe line	甲板货油管系	甲板貨油管路
cargo oil heating system	货油加热系统	貨油加熱系統
cargo oil hose	货油软管	貨油軟管
cargo oil pump	货油泵	貨油泵
cargo oil pumping system	货油装卸系统	貨油裝卸系統
cargo oil pump room pipe line	货油泵舱管系	貨油泵室管路
cargo oil suction heating coil	吸油口加热盘管	貨油吸入加熱盤管
cargo oil tank	货油舱	貨油艙
cargo oil tank cleaning installation	货油舱洗舱设备	貨油艙清洗裝置
cargo oil tank gas-freeing installation	货油舱油气驱除装置	貨油艙清除有害氣體裝置
cargo oil tank gas pressure indicator	货油舱气压指示器	貨油艙氣壓指示器
cargo oil tank pipe line	货油舱管系	貨油艙管路
cargo oil tank stripping system	货油舱扫舱系统	貨油艙收艙系統
cargo oil tank venting system	货油舱透气系统	貨油艙通氣系統
cargo oil valve	货油阀(**液货阀**)	貨油閥
cargo pallet	货盘	托貨板
cargo plan	配载图	載貨圖,貨物裝載圖
cargo pump room	货泵舱	貨泵室
cargo purchase eye	吊货眼板	吊貨眼板
cargo record book	货物记录簿	液貨紀錄簿
cargo runner(=cargofall)	吊货索	吊貨索
cargo safety hook	安全吊货钩	安全吊貨鉤
cargo shifting	货物移位	貨物移位
cargo ship(=freighter)	货船	貨船
cargo ship safety construction certificate	货船构造安全证书	貨船安全構造證書
cargo ship safety radio certificate	货船无线电安全证书	貨船安全無線電話證書
cargo ship safety radiotelegraphy certificate	货船无线电报安全证书	貨船安全無線電報證書
cargo ship safety radiotelephony certificate	货船无线电话安全证书	貨船安全無線電話證書
cargo space	载货舱位,载货容积	貨艙空間
cargo tracer	货物查询单	貨物追查單
cargo war risk	货物战争险	貨物戰爭險
cargo winch	起货机	吊貨機
cargo worthiness	适货	適於運貨之(船舶)
carpenter	木匠	木匠
carriage forward	运费未收	運費由收貨人支付

英　文　名	祖国大陆名	台湾地区名
carriage free	免费运送	免費運送
carriage of contraband	载运违禁品	載運違禁品
Carriage of Goods by Sea（COGSA）	海上货物运输法	海上貨物運輸法
carriage of life saving appliances on board	救生设备配备	救生設備配備
carriage of passenger	旅客运输	旅客載運
carriage paid	运费付讫	運費付訖
carrier	①承运人 ②承运船	①運送人 ②運送船
carrier frequency	载波频率	載波頻率
carrier power	载波功率	載波功率
carry away	折断	折斷
carrying capacity of craft	艇筏乘员定额	艇筏乘載量
carrying capacity of lifeboat	救生艇乘员定额	救生艇容載量
carry on	继续	繼續操作
CAS（=collision avoiding system）	自动避碰系统（**避碰系统**）	避碰系統
cased cargo	箱装货	箱裝貨
cast	①抛投 ②[原地]转向	①抛,投 ②轉向
cast loose	松掉	鬆掉
cast net	投网	投網
cast of all lines	全部缆绳松掉	各纜鬆開
cast off	解开	解開
Castor	北河二(双子α)	北河二(雙子α)
casualty report （CASREP）	事故报告	損傷報告
cat	吊锚	起錨
catalog of charts and publications	航海图书目录	航海圖書目錄
catalytic cracking	催化热裂	觸媒裂解
catamaran	双体船	雙體或三體船
cat block	吊锚滑车	吊錨滑車
catch	捕获量	漁獲物
catch a crab	桨入水过深(出不了水面)	划空槳,槳入水過深
catch a turn	绕住	繞住
catching season	渔汛	漁期
cat davit	吊锚杆	吊錨桿
category A noxious liquid substance	A 类有毒液体物质	A 類有毒液體物質
category B noxious liquid substance	B 类有毒液体物质	B 類有毒液體物質
category C noxious liquid substance	C 类有毒液体物质	C 類有毒液體物質
category D noxious liquid substance	D 类有毒液体物质	D 類有毒液體物質

英 文 名	祖 国 大 陆 名	台 湾 地 区 名
catenary anchor leg mooring	悬链锚腿系泊	懸垂法錨泊(鑽油台)
cathode-ray tube display	CRT 显示器	陰極射線管顯示器
cathodic protection	阴极防腐	陰極防蝕
cattle container	动物箱(**牲口箱**)	牲口櫃
cat walk	步桥	窄道,步橋
cat's paw	猫爪结	貓爪結
caulking(=calk)	捻缝	捻縫
cause of pollution	污染源	污染源
cavitation erosion	气蚀,空泡腐蚀	孔蝕
CB(=compass bearing)	罗方位	羅經方位
CBRS(=collective broadcast receiving station)	通播接收台	通播接收台
CBSS(=collective broadcast sending station)	通播发射台	通播發射台
CBT(=clean ballast tank)	清洁压载舱	清潔壓艙水艙
CC(=compass course)	罗航向	羅經航向
CE(=chronometer error)	天文钟误差	天文鐘誤差
celestial altitude	天体高度	天體高度
celestial axis	天轴	天軸
celestial body	天体	天體(星)
celestial body apparent motion	天体视运动	天體視運動
celestial equator	天赤道	天球赤道
celestial fixing	天文定位	天體定位
celestial horizon	真地平圈	天球水平線圈
celestial meridian	[测者]子午圈	天子午線
celestial navigation	天文航海	天文航海
celestial observation	天文观测	測天
celestial pole	天极	天極
celestial sphere	天球	天球
cell guide	箱格导柱	[貨櫃]導槽
cementation	渗碳	滲碳
cement box	堵漏水泥箱	水泥堵
center-expand display	中心扩大显示	中心擴大顯示
center girder(=keel son)	中桁材	中線縱梁,内龍骨
center line	中心线	中心線
center of buoyancy	浮心	浮[力中]心
center of gravity	重心	重心
center tank	中舱	中心艙

英　文　名	祖国大陆名	台湾地区名
central air conditioner	中央空调器	中央空調
central air conditioning system	集中式空气调节系统（**中央空调系统**）	中央空調系統
central cooling system	中央冷却系统	中央冷卻系統
central island	江心洲	江心洲
centralized monitor	集中监测器	集中偵測器
centralized monitoring system	集中监视系统	中央監視系統
centralized operation cargo oil pumping system	集中操纵货油装卸系统	集中操縱貨油裝卸系統
central processor unit	中央处理单元	中央處理單元
centrifugal compressor	离心式压气机	離心壓縮機
centrifugal inertia moment	离心惯性力矩	離心慣性力矩
centrifugal oil separator	[离心]分油机	油水離心分離器
centrifugal pump	离心泵	離心泵
centrifugal refrigerating compressor	离心式制冷压缩机	離心冷凍壓縮機
centripetal turbine	向心式涡轮	向心式渦輪機
ceramic insulation	陶瓷绝缘	陶瓷絕緣
certificated lifeboat person	持证艇员	持證救生艇員
certificate of seafarer	船员证书	船員證書,航海人員證書
certified safe type apparatus	合格安全型设备	合格安全型設備
cesser clause of charterer's liability	承租人责任终止条款	租傭人責任留置條款
cetane number	十六烷值	十六烷值
CF(=combined fix)	联合船位	混合定位
CFS(=container freight station)	集装箱货运站	貨櫃集散站
CFS to CFS(=container freight station to container freight station)	站到站	貨櫃站到站
chafing gear	耐磨装置	耐擦器
chain	台链	
chain block	机械滑车,链条滑车	鏈滑車
chain cable fairlead	导链轮	導鏈器
chain drive	链传动	鏈傳動
chain hook	锚链钩	錨鉤
chain lashing device	绑扎链扣	拉繫鏈扣
chain locker	锚链舱	錨鏈艙
chain scope	出链长度	放鏈長度
chains of causation	因果链	因果鏈
chain stopper	制链器	錨鏈扣

英　文　名	祖国大陆名	台湾地区名
chain tightener	链条张紧机构	緊鏈器
change-over mechanism	转换机构	變換機構
change-over valve	转换阀	變換閥
changing formation	队形变换	隊形變換
changing plate	换板	换板
channel	①水道 ②频道	①水道,航道 ②頻道
channel buoy	航道浮标	航道浮標
channel light	航道灯标	航道燈標
channel marker	航道标志	航道標誌
channel model	信道模型	頻道模型
channel navigation	狭水道航行	狭水道航行
channel request	信道申请	頻道申請
channel ship	海峡船	海峽船
channel storage	信道存储	頻道存儲
character	品质	品質
characteristic curve at constant speed	恒速特性曲线	等速特性曲線
characteristic curve of propeller	螺旋桨特性曲线	推進器特性曲線
characteristic number	特性数	示性數
charge	充电	充電
chargeable time	计费时间	計費時間
charging rate	充电率	充電率
chart	海图	海圖
chart card	海图卡片	海圖卡
chart datum	海图基准面	海圖深度基準面
chart datum for inland navigation	内河航行基准面	內水航行海圖深度基準面
charterage	租船费	租傭費
chartered ship	出租船	出租船
charterer	承租人	租傭人
charter freight	租船运费	傭船運費
charter hire	船舶租金	船舶租金
charter party	租船合同	租傭契約
charter party bill of lading	租船提单	傭船載貨證券
chart legend	海图标题栏	海圖圖例
chart number	图号	圖號
chart of inland waterway	内河航道图	內水航道圖
chart room	海图室	海圖室
chart scale	海图比例尺	海圖比例尺

英　文　名	祖国大陆名	台湾地区名
chart table	海图桌	海圖桌
chart work	航迹绘算,海图作业	海圖作業
chart work tools	海图作业工具	海圖作業工具
chassis	底盘车	車底盤,底盤
check	打住,稍松	打住,稍鬆
check digit	核对数字	核對數位
check-off list	检查清单	檢查表
check valve	单向止回阀(**止回阀**)	止回閥
chemical addition system	化学物添加系统	化學品添加系統
chemical cargo ship	化学品船	化學品船
chemical corrosion	化学腐蚀	化學腐蝕
chemical fire extinguisher	化学灭火器	化學滅火器
chemical light	化学灯	指距用化學燈
chief engineer	轮机长	輪機長
chief mate	大副	大副
chief officer (=chief-mate)	大副	大副
chief steward	事务长	勤務長
chip log	测速板	測速板
chock	垫木	墊木
choking	阻塞	阻流
chromel	铬镍合金	克鉻美
chrome plated liner	镀铬缸套	鍍鉻缸套
chrome-plate ring	镀铬环	鍍鉻環
chrome plating	镀铬	鍍鉻
chronometer	天文钟	天文鐘,船鐘
chronometer error(CE)	天文钟误差	天文鐘誤差
chronometer rate	[天文钟]日差	天文鐘日差率
CIF(=cost insurance and freight)	到岸价格	含保險費與運費之貨價
CIP(=calling-in-point)	呼叫点	呼叫點
cipher	密码	密碼
cipher device	密码器	密碼器
ciphony	密码电话学	密語電話
circle of equal altitude	等高圈	等高圈
circle of position	船位圆	位置圈
circle of uncertainty	[船位]误差圆	未定的船位圈
circular error probable	概率误差圆	圓形概差
circular formation	圆形编队	圓形編隊
circular frequency	圆频率	圓頻率

英 文 名	祖国大陆名	台湾地区名
circularity(=roundness)	圆度	圓度
circular monitor	巡回监测器	圓轉偵測器
circular polarization	圆极化	圓極化
circulating lubrication	循环润滑	環流潤滑
circulating pump	循环泵	循環泵
circulating tank	循环柜	循環櫃
circulating water channel	循环水槽	環流水槽
circulating water ratio	循环水倍率	循環水率
circulation theory	环流理论	環流理論
circumference of rope	缆绳周径	纜繩週長
circum-navigation	环球航行	環航
cirrocumulus	卷积云	卷積雲
cirrostratus	卷层云	卷層雲
cirrus	卷云	卷雲
civil embargo	民船禁航	禁止民船出港
civil ship	民船	民用船
civil twilight	民用晨昏朦影	民用朦光
CL(=course line)	航向线	航向線
claim for salvage	救助报酬请求	救助求償
clarifier	分杂机	淨油機
classification	入级	船級
classification certificate	入级证书	船級證書
class of emission	发射类别	發射等級
class of ship	船级	船級
clean ballast	清洁压载水	清潔壓艙水
clean ballast pump	清洁压载泵	清潔壓艙水泵
clean ballast tank(CBT)	清洁压载舱	清潔壓艙水艙
clean ballast tank operation manual	清洁压载舱操作手册	清潔壓艙水艙操作手冊
clean bill of lading	清洁提单	無批註載貨證券,清潔 提單
clean cargo	清洁货	潔淨貨
clean hull propeller curve	清洁船体螺旋桨特性曲 线	潔淨船體螺槳特性曲線
clean seas guide for oil tankers	油轮清洁海洋指南	油輪海洋清潔指南
clean view screen	旋转视窗	旋轉視窗
clearance	结关	結關,出港許可
clearance certificate	结关单	出口結關證書,出航許 可證

英　文　名	祖国大陆名	台湾地区名
clearance gauge	量隙规,塞尺	餘隙規
clear for running	准备开航,准备开动	備便放鬆,準備開航
clearing from alongside(=unberthing)	离泊	離泊
clearing from buoy	离浮筒	離浮筒
clearing hawse	清解锚链	解錨鏈
clear of water	锚出水	錨出水
clear zone	净区	淨區
cleat	羊角,系索耳	繫索扣
clew	帆下角	縱帆踵,吊鋪攀
climate	气候	氣候
climate routing	气候航线	氣候航路
clinometer	倾斜仪	傾斜儀
clipper bow	飞剪[型]艏	飛剪式艏
clipper stem(=clipper bow)	飞剪[型]艏	飛剪式艏
clock	船钟	船鐘
clockwise rotation	顺时针旋转	順時針旋轉
closed and pressured fuel system	加压式燃油系统	封密加壓燃油系統
closed container	封闭箱(**封闭集装箱**)	封閉貨櫃
closed cooling system	闭式冷却系统	閉式冷卻系統
closed cooling water system	闭式冷却水系统	閉式冷卻水系統
closed cup test	闭杯试验	閉杯法試驗
closed-loop system	闭环系统	閉環系統
closed network	闭路网络	閉路網路
closed type fuel valve	闭式喷油器	閉式噴油閥
closed-type hydraulic system	闭式液压系统	閉式液壓系統
closed user group	闭路用户组	閉路用戶組
close haul	抢风	迎風
close-in fueling rig	靠近加油装置	靠近加油裝置
close quarters situation	紧迫局面	彼此接近
close running fit	紧转配合	緊轉配合
closest point of approach(CPA)	最近会遇点	最接近點
close up	拉到顶	滿懸
closing cylinder	封缸	封缸
cloud amount	云量	雲量
cloud atlas	云图[册]	雲圖冊
cloud form	云状	雲狀
cloud height	云高	雲高
cloud point	浊点	雲點,濁點

英 文 名	祖 国 大 陆 名	台 湾 地 区 名
clove hitch	丁香结	丁香結
clubbing	拖锚滑行	拖錨溜行
CMI Rules of Electronic Bills of Lading	国际海事委员会电子提单规则	國際海事委員會電子載貨證券規則
CMI Uniform Rules for Sea Waybills	国际海事委员会海运单统一规则	國際海事委員會海運單統一規則
coarse/acquisition code(C/A code)	C/A 码	C/A 碼
coarse alignment	粗对准	粗校準
coarse synchronizing method	粗同步法	粗同步法
coast	[海]岸	海岸
coastal chart	沿岸图	沿岸海圖,沿海海圖
coastal current	沿岸流	沿岸流
coastal effect	海岸效应	海岸效應
coastal feature	沿岸地形	沿岸地形
coastal navigation(=coastal trip)	沿岸航行	沿岸航行
coastal state	沿海国	沿海國
coastal trip	沿海航行	近岸航行
coastal warning	沿岸警告	沿海警告
coastal zone	海岸带	海岸地帶
coast earth station	①海岸地球站 ②海岸无线电台	①海岸衛星電台 ②海岸無線電台
coast earth station identification	海岸地球站识别码	海岸衛星電台識別碼
coaster	沿海船	沿海船
coastline	岸线	海岸線
coast radar station	海岸雷达站	海岸雷達站
coast station	海岸电台	海岸電台
coast station charge	岸台费	岸台費
coast station identity	海岸电台识别	海岸電台識別
coastwise navigation	沿海航行	近岸航行
coastwise survey	沿岸测量	沿岸測量
coating(=paint)	①涂料 ②涂层 ③油漆	①塗料 ②塗層 ③油漆,漆
cock	旋塞	旋塞
cocked hat	[船位]误差三角形	誤差三角形
code	码	電碼
coded information	编码信息	編碼數據
code division system	码分隔制	碼分隔制
Code of Safe Practice for Solid Bulk Car-	固体散装货物安全操作	散裝固體貨物安全實務

英　文　名	祖国大陆名	台湾地区名
goes	规则	章程
code phase	码相位	碼相位
coding delay	编码延迟	密碼遲延
coefficient of cargo handling	货物操作系数	貨物操縱係數
coefficient of deviation	自差系数	自差係數
coefficient of hold	舱容系数	貨艙係數
coefficient of refrigerating performance	制冷系数	冷凍性能係數
coefficient of speed fluctuation	转速波动率	速率波動係數
cofferdam	围堰,空隔舱	堰艙,圍堰
COGSA(=Carriage of Goods by Sea)	海上货物运输法	海上貨物運輸法
cold advection	冷平流	冷平流
cold air mass	冷气团	冷氣團
cold blow-off operation	冷吹运行	冷吹操作
cold-coolant accident	冷水事故	冷水事故
cold current	寒流	寒流
cold front	冷锋	冷鋒
cold high	冷高压	冷高壓
cold shut-down	冷停堆	冷關閉
cold starting	冷态起动	冷溫起動
cold wave	寒潮	寒潮,寒流
collect call	受话人付费电话	受話人付費電話
collective broadcast receiving station (CBRS)	通播接收台	通播接收台
collective broadcast sending station (CBSS)	通播发射台	通播發射台
collision angle	碰角	碰撞角
collision avoidance behavior	避碰行为	避碰行爲
collision avoidance expert system	避碰专家系统	避碰專家系統
collision avoiding system(CCAS)	自动避碰系统(**避碰系统**)	避碰系統
collision bulkhead	防撞舱壁	防撞艙壁,防碰艙壁,碰撞隔堵
collision insurance	碰撞保险	碰撞保險
collision mat	堵漏毯	堵漏毯,防水墊
collision speed	碰撞速度	碰撞速度
collision warning	碰撞警报	碰撞警報
column formation	纵队	縱隊
column open order	散开纵队命令	疏開編隊

英 文 名	祖 国 大 陆 名	台 湾 地 区 名
combating navigation service	战斗航海勤务	戰鬥航海勤務
combat operating mode management	战斗工况管理	戰鬥操作方式管理
combination carrier	混装船	混載船
combined bill of lading	并装提单	併裝載貨證券
combined diesel and gas turbine power plant	柴油机和燃气轮机联合动力装置	柴油燃氣渦輪機複合動力設備
combined fix(CF)	联合船位	混合定位
combined frame system(=mixed frame system)	混合骨架式	混合肋骨系統
combined lantern	合并舷灯	合併燈,聯合燈
combined steam engine and exhaust turbine installation	蒸汽机–废汽汽轮机联合装置	蒸汽機與排汽渦輪機複合機
combined steam-gas turbine [propulsion] plant	汽轮机–燃气轮机联合装置	蒸汽與燃氣渦輪複合推進裝置
combustion chamber	燃烧室	燃燒室
combustion efficiency	燃烧效率	燃燒效率
combustor outer casing	燃烧室外壳	燃燒室外殼
come about	掉抢	掉餞
come home	锚向船来	收錨
come up	[缆]回松,回升	緩緩放鬆
coming to a single anchor	抛单锚	抛單錨
command and data acquisition	指令和数据获取	指令和數據擷取
command and data acquisition station	指令和数据获取站	指令和數據擷取站
commanding ship	指挥舰	指揮艦
commence search point	起始搜寻点	起始搜索點
comminuter	粉碎设备	粉碎設備
commission	佣金	傭金
commissioning test	启用试验	啓用試驗
commixture and unidentifiable cargo	混杂不清货	混雜不清貨
commodity freight	分货种运费	商品運費
common calling channel	公共呼叫频道	公共呼叫頻道
common danger(=common peril)	共同危险	共同危險
common peril	共同危险	共同危險
common safety	共同安全	共同安全
common text message	同文电报	同文電文
communication	①通信 ②交通 ③沟通	①通信 ②交通 ③溝通
communication log	通信记录	通信紀錄簿
communication protocol	通信协议	通信協定

英　文　名	祖国大陆名	台湾地区名
communication satellite	通信卫星	通信衛星
companding	压扩	壓縮擴展
companion ladder	升降梯	升降梯
companion way	升降口	升降口
comparator	比较器	比測儀
comparing unit	比较单元	比較單位
compartment loaded in combination	混装舱间	共同裝載艙間
compass	罗经	羅經
compass bearing（CB）	罗方位	羅經方位
compass binnacle	罗经柜	羅經針箱
compass bowl	罗经盆	羅經碗
compass buoy	罗经校正浮标	校正羅經浮標
compass card	罗经盘	羅經盤
compass course（CC）	罗航向	羅經航向
compass deck	罗经甲板	羅經甲板
compass error	[磁]罗差	羅經誤差，羅經差
compass liquid	罗经液体	羅經液體
compass north	罗北	羅經北
compass point	罗经点	羅經點
compass repeater	分罗经	羅經複示儀
compass rose	罗经花	圖上羅經
compatibility	相容性	相容性
compensation	补偿	補償金
compensation adjusting pointer	补偿调节指针	補償調節指針
compensation current	补偿流	補償流
compensation for damage	损害赔偿	損害賠償
compensation needle valve	补偿针阀	補償針閥
complement	船员定员	船員配額
complement code	补充码	補充碼
complete combustion	完全燃烧	完全燃燒
completing cargo	完整货	混載貨
complex cycle gas turbine	复杂循环燃气轮机	複合循環燃氣渦輪機
compliant structures or systems	顺应式结构或系统	順應式鑽油台
composite boiler	组合式锅炉	複合鍋爐
composite sailing	混合航线算法	混合航法
composite stress	复合应力	複應力
composite unit	组合体	組合體
compound expansion steam engine	双胀式蒸汽机	複膨脹式蒸汽機

英 文 名	祖 国 大 陆 名	台 湾 地 区 名
compound formation	混合编队	複列編隊
compound generator	复励发电机	複激發電機
compound impulse turbine	复式冲动涡轮机	複式衝動渦輪機
compounding impedance	复励阻抗	複激阻抗
compound pump	药剂泵	複式泵
compound stress(=composite stress)	复合应力	複應力
compound supercharging	复合增压	複合增壓
compressed gas	压缩气体	壓縮氣體
compressibility	可压缩性	可壓性
compression air starting system	压缩空气起动系统	壓縮空氣起動系統
compression chamber volume	压缩室容积	壓縮室容積
compression diagram	压缩图	壓縮圖
compression pressure	压缩压力	壓縮壓力
compression ratio	压缩比	壓縮比
compression ring	压缩环	壓縮脹圈
compression stroke	压缩行程	壓縮衝程
compressor characteristics	压气机特性	壓縮機特性
compressor oil	压缩机油	壓縮機油
compressor surging test	压气机喘振试验	壓縮機顫動試驗
compulsory pilotage	强制引航	強制引水
compulsory removal of wreck	强制打捞	強制打撈沉船
computed altitude	计算高度	計算高度
computer assisted collision avoidance	计算机辅助避碰	電腦輔助避碰
concentration of fish	鱼群密度	魚群密度
concentration of stress(=stress concentration)	应力集中	應力集中
conclusive evidence	确定证据	確定證據
condensate pump	凝水泵	凝水泵
condensate recirculating pipe line	凝水再循环管路	凝水再循環泵
condensate system	凝水系统	凝水系統
condenser vacuum	冷凝器真空度	冷凝真空
condensing steam turbine	凝汽式汽轮机	凝水式蒸汽渦輪機
condition alarm	工况报警	狀況警報
condition indicator	工况显示器	狀況指示器
condition monitor	工况监视器	狀況偵測器
condition monitor of main engine	主机工况监测器	主機狀況偵測器
conditions of discharge	排放条件	排洩條件
conducting liquid	导电液体	導電液體

英　文　名	祖国大陆名	台湾地区名
conferencing call	会议电话	會議電話
confidence factor	置信度	可信度
conical buoy	锥形浮标	圓錐形浮標
conical projection	圆锥投影	圓錐投影法
conical shape	圆锥号型	圓錐形號標
conic guide	锥形导槽	錐形導槽
connected replenishment	连接补给	連接整補
connected with the high seas	连接公海	與公海相通
connecting bridge	天桥	連橋
connecting fitting	连接件	連接裝具
connecting line	连接缆	連接線
connecting link(=joining link)	连接链环	連接鏈環
connecting-rod	连杆	連桿
connecting shackle(=joining shackle)	连接卸扣	連接接環
consecutive days(=running days)	连续日	連續自然日
consignee	收货人	受貨人,收貨人,受貨單 　　位
consignor	发货人	發貨人
consistency	稠度	稠度
Consol	康索尔	康蘇(電子航海儀)
Consolan	康索兰	康蘇蘭
console	控制台,仪表板	控制台,電子儀器座
constant deviation	固定自差	固定自差
constant pressure cycle	定压循环	定壓循環
constant pressure turbo-charging	定压涡轮增压	定壓渦輪增壓
constant volume cycle	定容循环	定容循環
constellation	星座	星座
constructive classification	建造入级	建造入級
constructive total loss	推定全损	推定全損
consular invoice	领事签证发票	領事簽證發票
contact damage	触损	接觸損害
container	集装箱	貨櫃
container country code	集装箱国家代号	貨櫃國碼
container freight station(CFS)	集装箱货运站	貨櫃集散站
container freight station to container freight 　station(CFS to CFS)	站到站	貨櫃站到站
container guide fitting	集装箱导具	貨櫃導具
containerized cargo	集装货	貨櫃裝載貨物

英　文　名	祖国大陆名	台湾地区名
container lifting spreader	集装箱吊架	貨櫃吊架
container load plan	集装箱装箱单	貨櫃裝櫃圖
container owner code	箱主代号	櫃主碼
container serial number	箱序号	貨櫃序號
container service charge	集装箱服务费	貨櫃服務費
container ship	集装箱船	貨櫃船
container terminal	集装箱[装卸]作业区	貨櫃終站基地
container yard(CY)	集装箱堆场	貨櫃場
container yard to container yard(CY to CY)	场到场	貨櫃場到場
contamination index	污染指数	污染指數
contiguous zone	毗连区	鄰接區
continental shelf	大陆架	大陸架,陸棚
continental shelf boundary	大陆架界线	大陸架界線
continental slope	大陆坡	大陸斜坡
continuous deck	统长甲板	連續甲板
continuous service rating	连续输出功率	額定連續常用出力
continuous serving rating	持续使用功率	額定連續常用出力
continuous survey	循环检验	連續檢驗
continuous watch	连续值守	連續守值
contour chart	等压面图	等值圖
contour light	轮廓灯	整補輪廓燈
contour lines	等高线	等高線
contract freight system	合同费率制	合約運費制
contract government	缔约国政府	締約國政府
contract M. C. R.	约定最大持续功率	約定最大連續輸出功率
contract of affreightment	货运合同	貨運契約
contract of carriage	运输合同	運送契約
contract of carrier	定约承运人	契約運送人
contributory value of general average	共同海损分摊价值	共同海損分攤價值
control air compressor	控制用空气压缩机	控制用空[氣]壓[縮]機
controllable passive tank stabilization system	可控被动水舱式减摇装置	可控被動水艙穩定系統
controllable phase compensation compound excited system	可控相复励磁系统	可控相之補償複激系統
controllable pitch propeller	可调螺距桨	可控距螺槳
controllable pitch propeller control system	可调螺距桨控制系统	可控距螺槳控制系統

英　文　名	祖国大陆名	台湾地区名
(CPP control system)		
controllable pitch propeller transmission	调距桨传动	可控距螺槳傳動
controllable self-excited constant voltage device	可控自励恒压装置	可控自激等電壓設施
controlled object	被控对象	控制對象
controlled swirl scavenging	控制涡流式[直流]扫气	控制漩流趨氣
controlling operator	管制值机员	管制值機員
control of discharge of oil	排油控制	排油管制
control of on scene communication	现场通信管制	現場通信管制
control point	控制点	控制點
control rod	①控制棒 ②操纵杆	①控制桿 ②操縱桿
control rod drive mechanism	控制棒驱动机构	操縱桿驅動機構
control rod guide tube	控制棒导管	操縱桿導管
control room maneuvering panel	控制室操纵屏	控制室操縱板
control station	控制站	控制站
control station change-over switch	控制部位转换开关	控制站換向開關
control tower	控制塔	控制塔
conventional radio service	常规无线电业务	一般無線電業務
conventional ship	普通货船	傳統式船
conventional transfer rig	传统式输送设备	傳統式輸送設備
Convention and Stature on Freedom of Transit	过境自由公约与规约	過境自由公約與規約
Convention and Stature on the International Regime of Maritime Ports	国际海港制度公约与规约	國際海港制度公約與規約
Convention for the Protection of Submarine Cables	保护海底电缆公约	保護海底電纜公約
Convention for the Suppression of Unlawful Acts against the Safety of Maritime Navigation	制止危及海上航行安全非法行为公约	制止危及海上航行安全非法行爲公約
Convention for Unification of Certain Rules of Law Relating to Assistance and Salvage at Sea	统一海难援助和救助某些法律规定公约	海上救助及撈救統一規定公約
Convention on a Code of Conduct for Liner Conference	班轮公会行动守则公约	定期船同盟行動章程公約
Convention on Facilitation of International Maritime Traffic	便利国际海上运输公约	便利國際海上運輸公約
Convention on Limitation of Liability for	海事索赔责任限制公约	海事求償責任限制公約

英　文　名	祖　国　大　陆　名	台　湾　地　区　名
Maritime Claims		
Convention on the International Hydrographic Organization	国际水道测量组织公约	國際海道測量組織公約
Convention on the International Maritime Organization	国际海事组织公约	國際海事組織公約
Convention on the International Maritime Satellite Organization	国际海事卫星组织公约	國際海事衛星組織公約
Convention on the Liability of Operators on Nuclear Ships	核动力船舶经营人责任公约	核子船舶營運人責任公約
Convention on the Prevention of Marine Pollution by Dumping of Wastes and other Matters	防止倾倒废物及其他物质污染海洋公约	防止傾倒廢棄物及其他物質污染海洋公約
Convention Relating to Civil Liability in the Field of Maritime Carriage of Nuclear Materials	海上核材料运输民事责任公约	海上運載核子物質民事責任公約
Convention Relating to Registration of Rights in Respect of Vessels under Construction	建造中船舶权利登记公约	建造中船舶權利登記公約
convergence	辐合	輻合
convergence line	辐合线	輻合線
convergent area of main and branch	干支流交汇水域	主支流匯流區
convergent nozzle	渐缩喷嘴	漸縮噴嘴
converging shafting	内斜轴系	漸縮軸系
conversion chart	换算图表	換算圖表
conversion of directions	向位换算	方向換算
converter set	变流机组	換流機組
convertible container ship	可变换的集装箱船	可變換貨櫃船
convoy	护航	護航
convoy escort	护送	船團護航
convoy in ice	冰中护航	冰中護航
cooling medium(=coolant)	载冷剂	冷卻劑
cooling method	冷却法	冷卻法
cooling system	冷却系统	冷卻系統
cooling water ratio	冷却水倍率	冷卻水率
cooperative international GPS network	国际合作 GPS 跟踪网	國際合作 GPS 網路
coordinate conversion device	坐标变换器	坐標變換器
coordinated collision avoidance maneuver	协调避碰操纵	協調避碰操縱
coordinated creep line search	协作横移线搜寻	協調橫移線搜索

英 文 名	祖 国 大 陆 名	台 湾 地 区 名
coordinated universal time(UTC)	协调世界时	協調世界時
co-owner of ship	船舶共有人	船舶共有人
copper base bearing metals	铜基轴承合金	銅基軸承合金
copper-lead bearing	铜铅轴承	銅鉛軸承
corange line	等潮差线	等潮差線
Cor Carole	常陈一(猎犬α)	常陳一(獵犬α)
cordage	绳索	繩索
core	堆芯	索芯
Coriolis force	科里奥利力	科氏力,自轉偏向力
corner fitting	角件	櫃角裝置
corner reflector	角形反射器	角形反射器,雷達波反射器
corona	日冕	日冕
correction of geocentric latitude	地心纬度改正量	地心緯度修正
correction of middle latitude	中分纬度改正量	中緯修正量
corrective loop	修正回路	修正迴路
corridor	走廊	走廊
corroded limit	蚀耗极限	蝕耗極限
corrosion	腐蚀	腐蝕
corrosion fatigue	腐蚀疲劳	銹蝕疲勞,腐蝕疲勞
corrosion-resisting steel	耐腐蚀钢	耐蝕鋼
corrosion wear	腐蚀磨损	腐蝕耗損
corrosives	腐蚀性物质	腐蝕性物質
corrugated bulkhead	波形舱壁	波形艙壁
corrugated expansion pipe	波形膨胀管	波形膨脹管
COSPAs-SARSAT council	低极轨道卫星搜救组织理事会	國際衛星輔助搜救組織理事會
COSPAs-SARSAT message	低极轨道卫星搜救系统信文	衛星輔助搜救系統信文
COSPAs-SARSAT system	低极轨道卫星搜救系统	衛星輔助搜救系統
cost and freight(C&F)	成本加运费价格	成本與運費
cost insurance and freight(CIF)	到岸价格	含保險費與運費之貨價
cotidal chart	等潮时图	等潮圖
cotidal hour	等潮时	等潮時
cotidal line	等潮时线	等潮線
cotter pin	开尾销	開口銷
counter-acting force	反作用力	反作用力
counter-clockwise rotation	逆时针转动	逆時針旋轉

英　文　名	祖国大陆名	台湾地区名
counter-current braking	反接制动	逆流制動
counter flood	对称灌水	平衡泛水
counter mark	副标志	副標誌
counter rudder	压舵,整流舵	壓舵,整流舵
counter rudder angle	反舵角	反舵角
counter stern	悬伸型艉	懸伸艉
counterweigh	平衡重	配重,衡重
coupled vibration	耦合振动	偶合振動
course	航向	航向
course autopilot	航向自动操舵仪	航向自動操舵裝置
course changing ability	改向性	變向能力
course changing ability test	改向性试验	變向能力試驗
course keeping quality	保向性	航向保持性
course line(CL)	航向线	航向線
course of advance(CA)	计划航迹向	預期航向
course recorder	航向记录器	航向記錄儀
course recording machine(=course re-corder)	航向记录器	航向記錄儀
course stability	航向稳定性	航向穩定性
course up	航向向上	航向向上
court of survey	调查庭	調查庭
coverage	覆盖区	覆蓋區
covered	阴	陰
COW(=crude oil washing)	原油洗舱	原油洗艙
cowl head ventilator	烟斗形通风筒	煙斗形通風筒
coxswain	艇长	司艇
CPA(=closest point of approach)	最近会遇点	最接近點
CPP(=controllable pitch propeller)	可调螺距桨	可控距螺槳
CPP control system(=controllable pitch propeller control system)	可调螺距桨控制系统	可控距螺槳控制系統
CQ	呼叫各电台(无线电话用语)	呼叫各電台(無線電話用語)
CQD(=customary quick despatch)	习惯装卸速度	依慣例快速處理
crab angle	漂流角,风流压差	偏流修正角
crabber	捕蟹船	捕蟹船
crabbing	横漂	橫漂
crack detection	探伤	探傷
cracked fuel oil	裂化燃料油	裂化燃料油

英　文　名	祖国大陆名	台湾地区名
cradle	①托架 ②船架	①托架 ②船架
craft	船筏	船艇,载具
cranage	吊车费	起重機使用費
crane	起重机	起重機
crane boom	起重机起重臂(**起重机臂**)	起重機臂
crane radius	起重机伸距	起重機伸距
crank angle	曲轴转角	曲柄角
crank arm	曲柄臂	曲柄臂
crankcase	曲轴箱	曲柄軸箱
crankcase explosion	曲轴箱爆炸	曲柄軸箱爆炸
crankcase explosion relief door	曲轴箱防爆门	曲柄軸箱防爆門
crankcase vent pipe	曲轴箱透气管路	曲[柄]軸箱通氣管
crank pin	曲柄销	曲柄軸銷
crank radius-connecting rod length ratio	曲柄连杆比	曲柄半徑與連桿長比
crankshaft	曲轴	曲柄軸
crankshaft counterweight	曲轴平衡重	曲柄軸衡重
crankshaft deflection dial gauge	臂距千分表	曲柄軸撓曲針盤量規
crankshaft fatigue fracture	曲轴疲劳断裂	曲柄軸疲勞破壞
crankshaft journal	主轴颈	曲柄軸頸
crankshaft shrinkage slip-off	曲轴红套滑移	曲柄軸短縮滑移
crank spread	曲柄臂间距(开档)	曲臂間距
crank throw	曲柄半径	曲柄推程
crank web(=crank arm)	曲柄臂	曲柄臂
crash maneuvering	特急操纵	緊急操縱
crash stopping distance	紧急倒车冲程	緊急停俥距離
credit card call	信用卡电话	信用卡電話
creek	①小湾 ②小河	①小灣 ②溪
creep of metal	金属蠕变	金屬潛變
crevice corrosion	缝隙腐蚀	間隙腐蝕
crew	船员	船員
crew list	船员名单	船員名單,船員名冊
crew's customs declaration	船员自用物品报关	船員自用物品申請單
crisis management and human behaviors	危机处理和人的行为	危機處理及行爲管理
criteria of maneuverability	操纵性衡准	操縱性標準
criterion of service	服务标准	航務標準
criterion of service numeral	业务衡准数	業務基準數
critical capsizing lever	临界倾覆力臂	臨界翻覆力臂

英 文 名	祖 国 大 陆 名	台 湾 地 区 名
critical capsizing moment	临界倾覆力矩	臨界翻覆力矩
critical height of center of gravity	极限重心高度	臨界重心高度
critical initial metacentric height	临界初稳性高度	臨界初定傾[中心]高度
criticality test	临界实验	臨界試驗
critical pressure ratio	临界压力比	臨界壓力比
critical relative bearing	临界舷角	臨界相對方位
critical speed	临界速度	臨界速度
crockery	陶器	陶器
cross bitt	十字缆桩	十字形繫樁
cross-compound steam turbine	并联复式汽轮机	並列複式蒸汽渦輪機
cross current	横流	横流,側流
cross-current mark	横流标	横流標
cross curve	交叉曲线	交叉曲線
cross flood	对称进水	對稱浸水
crosshead	十字头	十字頭
crosshead shoe	十字头滑块	十字頭履
crosshead slipper(= crosshead shoe)	十字头滑块	十字頭履
crosshead type diesel engine	十字头式柴油机	十字頭型柴油機
crossing	穿越	横越
crossing ahead	横越	横越
crossing area	横驶区	横越區
crossing mark	过河标	横越標
crossing river point	过河点	渡河點
crossing situation	交叉相遇局面	交叉相遇情況
cross scavenging	横流扫气	横驅氣
cross section	横剖面	横截面,横剖面
cross track angle	交叉轨迹角	交叉軌跡角
crosstree	桅顶横杆	桅頂横桿
cross wind	横风	横風
crowd management training	拥挤人群管理培训	群衆管理訓練
crown knot	绳头结	倒紐結
crude oil	原油	原油
crude oil tanker	原油船	原油輪
crude oil washing(COW)	原油洗舱	原油洗艙
cruiser	巡洋舰	巡洋艦
cruiser stern	巡洋舰[型]艉	巡洋艦型艉
cruising engine unit	巡航机组	巡航機組

英　文　名	祖国大陆名	台湾地区名
cruising formation and disposition	巡航编队与部署	巡航編隊與序列
cruising operating mode management	巡航工况管理	巡航操作型式管理
cruising radius（＝endurance）	续航力	續航力,持久
cruising turbine	巡航涡轮机	巡航渦輪機
CS（＝call sign）	呼号	呼號
culmination	中天	中天
cumulus	积云	積雲
current-limiting starter	限流起动器	限流起動器
current rose	海流花	旋潮流圖
curve of statical stability	静稳性曲线	靜穩度曲線
cushioning effect	缓冲作用	緩衝作用
customary quick despatch（CQD）	习惯装卸速度	依慣例快速處理
custom broker	报关行	報關行
custom of port	港口习惯	港口慣例
customs	海关	海關
customs duties	关税	關稅
customs seal	关封	海關封條
cut-out position	停油位置	停供位置
cutting	切割	切割
cutting away wreck	切除残损物	切除殘骸
CY（＝container yard）	集装箱堆场	貨櫃場
cycle irregularity	回转不均匀	迴轉不規率度
cycle matching	周波重合	週波匹配
cycle skipping	跳周	跳週
cyclic variation	周期性变化	週期性變化
cyclone	气旋	旋風,氣旋
cyclonic wave	气旋波	氣旋波
cyclonic wind	气旋风	氣旋風
cylinder block	气缸体	氣缸體
cylinder blow-by	气缸窜气	氣缸漏氣
cylinder bore	气缸直径	氣缸直徑,氣缸内徑
cylinder constant	气缸常数	氣缸常數
cylinder cover	气缸盖	缸蓋
cylinder jacket	气缸[冷却]水套	缸套
cylinder liner	气缸套	氣缸内襯套,[氣]缸 [襯]套
cylinder lubricator	气缸注油器	氣缸潤滑器
cylinder oil	气缸油	氣缸油

英　文　名	祖 国 大 陆 名	台 湾 地 区 名
cylinder oil dosage	气缸油注油量	氣缸油注量
cylinder oil transfer pump	气缸油输送泵	氣缸油輸送泵
cylinder scraping	拉缸	氣缸刮削
cylinder [shape]	圆柱体号型	圓筒形號標
cylinder starting valve	气缸起动阀	氣缸起動閥
cylinder sticking(=piston seizure)	咬缸	咬缸,氣缸膠著
cylinder total volume	气缸总容积	氣缸總容積
cylindrical buoy	圆筒形浮标	筒形浮標
cylindrical gauge	圆柱测径规	柱形塞規
cylindrical plug(=cylindrical gauge)	圆柱测径规	柱形塞規
cylindrical projection	圆柱投影	圓筒投影法
cylindricity	圆柱度	圓柱度
CY to CY(=container yard to container yard)	场到场	貨櫃場到場

D

英　文　名	祖 国 大 陆 名	台 湾 地 区 名
daily rate	[天文钟]日差	天文鐘日差率
daily tank	日用柜	日用櫃
damage cargo list	货物残损单	貨損清單
damage caused by waves	浪损	浪損
damage control equipment	损管器材	損[害]管[制]設備
damage control plan	海损管制示意图	損害管制圖
damaged stability	破舱稳性	破損穩度,受損穩度
damage from oil pollution	油污损害	油污損害
damage repair	事故修理	損害修理
damped method of horizontal axis	水平轴阻尼法	水平軸阻尼法
damped method of vertical axis	垂直轴阻尼法	垂直軸阻尼法
damped vibration	阻尼振动	阻尼振動
damper	①风门 ②阻尼器	①擋板 ②阻尼板
damping factor	阻尼系数	阻尼因素
damping pollution	倾倒污染	傾倒污染
damping weight	阻尼重物	鎮偏重物
danger	危险[物]	危險
danger buoy	碍航浮标	危險浮標
danger message	危险电文	危險消息
danger money	危险津贴	危險加給

英　文　名	祖国大陆名	台湾地区名
dangerous cargo	危险货[物],危险品	危险货物
dangerous cargo anchorage	危险货物锚地	危险货物锚地
dangerous cargo list	危险品清单	危险货物清单
dangerous coast	危险海岸	危险海岸
dangerous goods in limited quantity	限量危险品	限量危险品
dangerous goods report	危险货物报告	危险货物报告
dangerous mark	危险标志	危险物品标志
dangerous quadrant	危险象限	危险象限
dangerous semicircle	危险半圆	危险半圈
database availability ratio	数据库使用率	数据库使用率
database effectiveness ratio	数据库有效率	数据库有效比率
data circuit terminating equipment	数据线路终端	数据线路终端设备
data communication	数据通信	资讯,数据通信
Data Distribution Plan	数据分配计划	数据配送计划
data logger	数据记录器	数据记录器
dataphone	数据电话机	数据电话
data recovery unit	数据复原单元	数据复原单元
data report	数据报告	数据报告
data terminal equipment	数据终端设备	数据终端设备
date line(=calendar line)	日界线	日界线
date of arrival	到港日期	到港日期
date of built	建造日期	建造日期
date of departure	开航日期	启航日期
date of inspection	检查日期	检查日期
date of issue	签发日期	签发日期
date of keel laid	安放龙骨日期	安放龙骨日期
date of launching	下水日期	下水日期
datum ship	基准舰	基准舰
davit launching liftraft	吊放式救生筏	吊杆下水救生筏
daylight signaling lamp	白天信号灯	日间信号灯
daylight signaling mirror	日光信号镜	日光信号镜
day mark	日标	日间助航标志,昼标
days of grace	宽限日期	宽限日期
day's run	一天航程	一日行程
D. C. electric propulsion plant	直流电力推进装置	直流电力推进装置
DC power station	直流电站	直流电站
dead ahead	正前方	正前方
dead astern	正后方	正后方

英　文　名	祖国大陆名	台湾地区名
deadfreight	空舱运费	空艙運費
deadlight	舷窗内盖	舷窗内蓋
dead load	固定负荷	静載負荷
dead reckoning（DR）	积算	推算
dead reckoning position	积算船位	推算船位
dead-reckoning tracer（DRT）	航迹积算仪	推算航跡儀
dead tide	最低潮	最低潮
dead weight（DW）	总载重量	載重量
deadweight cargo	重量货	過秤貨
dead weight scale	载重标尺	載重標尺
deaerator	除氧器	除氧器
Dec（=declination）	赤纬	赤緯
Decca	台卡	迪凱(電子航儀)
Decca chain	台卡链	迪凱鏈
Decca chart	台卡海图	迪凱海圖
Decca data sheet	台卡活页资料	迪凱活頁數據
Decca fix	台卡船位	迪凱船位
Decca navigator	台卡导航仪	迪凱導航儀
Decca position line	台卡位置线	迪凱位置線
decision making of collision avoidance	避碰决策	避碰决策
deck	甲板	甲板
deck beam	横梁	甲板梁
deck cargo	甲板货	艙面貨
deck cleat	甲板羊角,甲板系索耳	艙面繫索扣
deck covering	甲板敷面	甲板被覆
deck department	甲板部	艙面部門
deck girder	甲板纵桁	甲板縱梁
deck house	甲板室	甲板房艙
deck lighting system	甲板照明系统	甲板照明系統
deck line	甲板线	甲板線
deck longitudinal	甲板纵骨	甲板縱材
deck machinery	甲板机械	艙面機械
deck officer	驾驶员	航行員
deck paint	甲板漆	甲板漆
deck passenger	无铺位旅客	艙面旅客
deck plan	甲板布置图	甲板位置圖,甲板佈置圖
deck plate	甲板板	甲板板

英　文　名	祖国大陆名	台湾地区名
deck sprinkler system	甲板洒水系统	甲板灑水系統
deck strake	甲板列板	甲板列
deck stringer	甲板边板	甲板緣板
deck washing piping system	甲板冲洗管系	甲板衝洗管路系統
deck water piping system	甲板水排泄管系	甲板水排洩管路系統
deck water seal	甲板水封	甲板水封
declaration	报关,申报	陳報,申報
declaration of dead weight tonnage of cargo	宣载	載重噸申報
declaration of port	宣港	港口申報
declination(Dec)	赤纬	赤緯
decoder	译码器	譯碼機
decoding	译码,解码	譯碼
decometer	台卡计	相位計
decomposition	分解	分解
decompression of diving	潜水减压	潛水減壓
decontamination system	去污系统	除污系統
decoy ship(=P-ship)	伪装商船	偽裝商船
dedicated clean ballast tank	专用清洁压载水舱	清潔壓艙水專用艙
deductible	绝对免赔额	自負額
deep current	深层流	深水流
deep diving submersible, bathyscaphe	深潜器	深潛器
deep sea anchoring	深水抛锚	深海抛錨
deep submersible rescue vehicle	深潜救生艇	深潛救生艇
deep tank	深舱	深艙
deep water trawl	深水拖网	深水拖網
deep water way	深水航路	深水航路
default in management of the ship	管理船舶过失	管理船舶過失
default in navigation of the ship	驾驶船舶过失	駕駛船舶過失
deferred rebate system	延期回扣制	延期回扣制度
deflectometer	挠度计	撓度計
deflector	偏转仪	磁向偏差測算儀
deformation gauge	变形测量表	變形量規
defrost receiver	融霜储液器	除霜接受器
degaussing range	消磁场	消磁場
degauss push button switch	消磁按钮开关	消磁按鈕開關
degenerate	①退化 ②变质	①退化 ②變質
degree of admission	进汽度	進汽度
degree of partial admission	部分进汽度	部分進汽度

英 文 名	祖 国 大 陆 名	台 湾 地 区 名
degree of reaction	反动度(**反应度**)	反應程度
degree of turbocharging	增压度	渦輪增壓度
dehumidifier	除湿器	消濕器
deicer	除冰装置	除冰裝置
delay in delivery	延迟交货	延遲交貨
delay of ship	延迟开航	延遲發航
delay of turning response	应舵时间	迴轉回應延遲
delay period	滞燃期	延遲期間
delicate cargo	精致货	精細貨
delivery	交付	交貨
delivery ex-warehouse	仓库交货	倉庫交貨
delivery of vessel	交船	交船
delivery order	提货单	提貨通知單,提貨單
delivery ship	补给船	補給船
delivery valve	输出阀	輸出閥
delivery valve line retraction	排油阀卸载作用	
delta	三角洲	三角洲
Delta Echo, DE	来自信号,我是……	本台,是(無線電話用語)
demarcated fishery	划界渔业	區劃漁業
demodulation	解调	解調變
demulsification number	抗乳化度	脫乳化數
demurrage	滞期	延滯費,滯船費
Deneb	天津四(天鹅 α)	天津四(天鵝 α)
dense fog	大雾(0 级)	濃霧(0 級)
density	①密度 ②浓度	①密度 ②濃度
Dep(＝departure)	东西距	東西距
Departure(Dep)	东西距	東西距
departure ballast	出港压载水	出港壓艙水
departure from these rules	背离规则	背離規則
departure point	推算始点	出航點,出發點
depressed pole	俯极	下天極
depth autopilot	深度自动操舵仪	深度自動操舵裝置
depth contour	等深线	等深線
depth floor	高肋板	深肋板
depth indicator	深度指示器	深度指示錶
depth recorder	深度记录器	測深計
depth signal mark	水深信号标	水深信號標誌

英 文 名	祖 国 大 陆 名	台 湾 地 区 名
derating	减额功率	降[低]额[定]馬力
derating certificate	灭鼠证书	除鼠證明書
derating exemption certificate	免予灭鼠证书	除鼠豁免證書
deratting	除鼠	滅鼠
derelict	漂流物	漂流物
derivative regulator	微分调节器	微分調整器
derived envelope	导出包络	導出包絡
derrick	吊杆	吊桿
derrick heel	吊杆叉头	吊桿根
derrick rest	吊杆托架	吊桿承座
derrick rigging	吊杆索具	吊桿索具
de salting apparatus	淡化设备	除鹽器
desalting kit	淡化器具	去鹽設備
descending node	降交点	降交點
description of goods	货名	貨名
designed latitude	设计纬度	設計緯度
despatch	速遣	派遣
de-spread	解扩	解擴散
destination code	目的地码	目的地碼
destroyer	驱逐舰	驅逐艦
detachable link	可拆链环	拆合環
detecting system	探测系统	探火系統
detection	探测	探查
detection of iceberg	冰山探测	冰山探測
detention of ship	扣船	扣留船舶
detergent/dispersant additive	清净分散剂	清潔分散劑
determination of range of audibility of sound signal	号笛音响度测定	音響信號聽距測定
determination of range of visibility for navigation light	号灯照距测定(**航行灯照距测定**)	航行燈照距測定
Det Norske Veritas(NV)	挪威船级社	挪威驗船協會
detonation	爆燃	爆震
deviation	①自差 ②绕航	①自差 ②變更航程
deviation compensation device	自差补偿装置	自差補償設施
deviation curve	自差曲线	自差曲線
deviation of the vertical	垂线偏角	垂線偏差
deviation report	绕航变更报告	偏航報告,變更航程報告

英　文　名	祖国大陆名	台湾地区名
deviation table	自差表	自差表
devil's claw	制链爪	止鏈爪,吊鏈鉤
dew-point [temperature]	露点[温度]	露點
DGPS(=differential global positioning system)	差分全球定位系统	差分全球定位系统
dial type governor	表盘式调速器	針盤調速器
diamond formation	菱形编队	菱形編隊
diamond [shape]	菱形号型	菱形號標
diaphone	低音雾笛	霧號器
diaphragm	①横隔板 ②膜片	①隔膜 ②膜片
diesel and/or gas turbine power plant	柴油-燃气联合动力装置	柴油機及(或)燃氣渦輪機動力裝置
diesel cycle	狄塞尔循环	狄賽爾循環
diesel-electric drive	柴油机电力传动	柴油機電力傳動
diesel-electric propulsion plant	柴油机电力推进装置	柴油機電力推進裝置
diesel engine characteristic	柴油机特性	柴油機特性
diesel engine lubricating oil	柴油机机油	柴油機潤滑油
diesel geared drive	柴油机齿轮传动	柴油機齒輪傳動
diesel index	柴油指数	柴油指數
diesel knock	敲缸	柴油爆震
difference crank spread	臂距差	曲柄臂距差
difference of latitude	纬差	緯差
difference of longitude	经差	經差
difference of meridianal parts(DMP)	纬度渐长率差	緯度漸長比數差
differential cylinder	差动油缸	差動氣缸
differential determination	差分测定	差分測定
differential global positioning system	差分全球定位系统	差分式全球定位系統
differential Loran-C	差转罗兰 C	差分羅遠 C
differential measurement	差分测量	差分測量
differential observation	差分观测	較差觀測
differential Omega	差分奥米伽	差分亞米茄
differential positioning	差分定位	差分定位
diffuser	扩压器	擴散器
digital display unit	数字显示装置	數位顯示裝置
digital line system	数字有线系统	數位有線系統
digital radio system	数字无线系统	數位無線電系統
digital selective calling(DSC)	数字选择呼叫	數位選擇呼叫
digital selective calling installation	数字选择呼叫设备	數位選擇呼叫裝置

英　文　名	祖国大陆名	台湾地区名
digital selective calling system	数字选择呼叫系统	數位選擇呼叫系統
digital-to-analog converter	数-模转换器	數位-模擬轉變器
dip	①眼高差 ②俯角 ③降旗礼	①眼高差 ②俯角 ③低旗敬禮
dip circle	磁倾仪	地磁俯角儀
dip correction	眼高差改正	眼高差修正
dip needle	磁倾角针	地磁俯角針
dip net	抄网	抄網
dip of horizon	眼高差	地平俯角
dipper	铲斗	鏟頭
dipstick	测液深标尺	量油尺,量液深尺
dip to	落旗致敬	落旗致敬
direct acting steam pump	蒸汽直接作用泵	直聯蒸汽泵
direct bill of lading	直达提单	直達載貨證券
direct cargo	直达货	直達貨
direct evaporating air cooler	直接蒸发式空气冷却器	直接蒸發空氣冷卻器
direct filling line	直接装注油管	直接注入管路
directional antenna	定向天线	定向天線
directional control valve	方向控制阀	方向控制閥
directional gyroscope	方位[陀螺]仪	定向迴轉儀
directional light	定向导航灯	指導向燈
directional radio beacon	定向无线电信标	定向無線電示標
directional stability	方向稳定性	方向穩定性
direction effect	方向效应	方向效應
direction finder	测向仪	測向儀
direction finder sensitivity	测向灵敏度	方探靈敏度
direction of current	流向	流向
directive error of magnetic compass	磁罗经指向误差	磁羅經指向誤差
directive force	指向力	指向力
direct loading pipe line(= direct filling line)	直接装注油管	直接注入管路
direct port	直达港	直達港
direct-printing telegraphy	直接印字电报	直接印字電報
direct spiral test	正螺旋试验	渦漩試驗
direct tide	上中天潮	順潮
direct transshipment	直接换装	直接換裝
direct transmission	直接传动	直接傳動
dirty ballast	不洁压舱水	不潔壓艙水

英　文　名	祖国大陆名	台湾地区名
dirty cargo	污秽货	污穢貨
dirty oil tank	污油舱	污油櫃
discharge	①放电 ②排泄 ③卸载	①放電 ②排洩 ③卸載
discharge characteristic curve	放电特性曲线	放電特性曲線
discharge current	排出流	排出流
discharge head	排出压头	流出落差
discharge in violation of regulations	违章排放	違規排洩
discharge manifold	排放集管	排洩歧管
discharge valve	排出阀	排出閥
discharging	卸货	卸貨
discharging rate	放电率	放電率
disc ratio	盘面比	盤面比
disc wheel	砂轮	盤輪
dishwater	洗碗水	洗盤水
displacement	排水量	排水量
displacement ship	排水船	排水型船
display mode	显示方式	顯示方式
disposal of wastes	处置废物	廢棄物處理
disposition axis	配置轴	序列軸
dispute	争议	爭端
distance abeam	正横距离	正橫距離
distance between twin trawl	网档间距	網間距
distance by engine's RPH	主机航程	主機(每時轉數計)航程
distance by log	计程仪航程	測程儀航程
distance by vertical angle	垂直角距离	垂直角距離
distance line	距离索	距離索
distance made good	推算航程	實際距離,終結距離
distance run	航程	航程
distance table	里程表	浬程表
distance to closest point of approach, DCPA	最近会遇距离	最接近點距離
distance to the horizon from object	物标能见地平距离	物標水平視距
distant fishery	远洋渔业	遠洋漁業
distillation method	蒸馏法	蒸餾法
distillation plant	蒸馏装置	蒸餾設備
distillation temperature	蒸馏温度	蒸餾溫度
distiller	蒸馏器	蒸餾器

英　文　名	祖国大陆名	台湾地区名
distress	遇险	遇難,遇險
distress acknowledgement	遇险收妥承认	遇險收到確認
distress alerting	遇险报警	遇險警報
distress alert relay	遇险报警转发	遇險報中繼
distress and safety communications	遇险和安全通信	遇險與安全通信
distress call	遇险呼叫	遇難呼叫,求救呼叫
distress call format	遇险呼叫格式	遇險呼叫格式
distress call procedure	遇险呼叫程序	遇險呼叫程序
distress cargo	急运货	急運貨,急裝貨
distress channel	遇险信道	遇險頻道
distress communication	遇险通信	遇險通信
distressed seaman	遇险船员	遇險船員
distress flare	遇险火光信号	遇難照明彈
distress message	遇险报告	遇難信文
distress phase	遇险阶段	遇險階段
distress priority	遇险优先等级	遇險優先順序
distress priority call	遇险优先呼叫	遇險優先呼叫
distress-priority request message	遇险优先申请信息	優先請求遇險信文
distress procedures	遇险求救程序	遇險程序
distress signal	遇险信号	遇險信號,遇難信號
distress telephone call	遇险电话呼叫	遇險電話呼叫
distress telex call	遇险电传呼叫	遇險電傳呼叫
distress traffic	遇险通信业务	遇險業務
distribution	①分配 ②配电	①配送 ②配電
distribution box	分线盒	分配箱
distribution system	配电系统	配電系統
ditch	迫降	迫降
diurnal [apparent] motion	周日视运动	每日視運動(天體)
diurnal inequality	日潮不等	週日差
diurnal phase change	相位日变化	相位日變
diurnal tide	全日潮	週日單潮,週日潮
diver	潜水员	潛水人,潛水員
divergence	辐散	輻散
divergence line	辐散线	輻射線
divergent wave	散波	發散波
diverging shafting	外斜轴系	
diver's boots	潜水靴	潛水靴
diver's descending line	潜水导索	潛水導索

英　文　名	祖国大陆名	台湾地区名
diver's helmet	潜水头盔	潛水頭盔
diver's sign language	潜水手语	潛水手語
divided combustion chamber	分开式燃烧室	分離式燃燒室
diving	潜水	潛水
diving apparatus	潜水装具	潛水器具
diving bell	潜水钟	潛水鐘
diving boat	潜水工作船	潛水工作船
diving equipment	潜水设备	潛水設備
diving suit	潜水服	潛水衣
diving telephone	潜水电话	潛水電話
DMP（＝difference of meridianal parts）	纬度渐长率差	緯度漸長比數差
dock cargo	随船入坞货	船塢貨
docking	进坞	進塢
docking maneuver	进坞操纵	進塢操縱
docking survey	坞内检验	入塢檢驗
docking system	船舶靠泊系统	船舶靠泊系統
dock receipt	场站收据	倉庫收貨單
dock repair	坞修	進塢檢修
dock trial	系泊试验	繫泊試俥
dock warrant	码头收货单	碼頭埠單
document of compliance	合格证书	符合證書
dog down	关闭［水密门窗］	關閉水密［門窗］
·dog watch	二时更	暮更
doldrums	赤道无风带	赤道無風帶
dolphin	①系船桩 ②艇碰垫	①繫船樁 ②小碰墊
domestic waste	生活废弃物	生活廢棄物
domestic water system	生活用水系统	生活用水系統
donkey boiler	辅锅炉	副鍋爐
door to door	门到门	戶到戶
Doppler count	多普勒计数	都卜勒計數
Doppler location	多普勒定位	都卜勒定位
Doppler log	多普勒计程仪	都卜勒計程儀
Doppler position information	多普勒位置信息	都卜勒位置資訊
Doppler shift	多普勒频移	都卜勒頻移
dot pattern	点图	點圖型
double-acting cylinder	双作用汽缸	雙動汽缸
double-acting engine	双作用式发动机	雙動［動力］機
double bearing rudder	双支承舵	雙支承舵

英 文 名	祖 国 大 陆 名	台 湾 地 区 名
double boat purse seine	双船围网	雙船圍網
double bottom	双层底	[二]重底
double bottom tank	双层底舱	[二]重底艙
double cam reversing	双凸轮换向	雙凸輪換向
double check valve	双向止回阀	雙向止回閥
double column	双纵队	雙縱隊
double day tide	隔日潮	隔日潮
double difference	双差	雙差
double drum winch	双卷筒绞车	雙筒絞車
double hook	山字钩	雙鉤
double-hose rigs	双管设备	雙管設備
double hull	双层船壳	雙層船殼
double line abreast	双横队	雙橫隊
double non-return valve(=double check valve)	双向止回阀	雙向止回閥
double-pipe condenser	套管式冷凝器	二管冷凝器
double purchase	1–1 绞辘	雙吊桿聯合作業
double side-band (DSB)	双边带	雙邊帶
double-skin ship	双壳船	雙殼船
double-state compass	双态罗经	雙態羅經
double up	双绑	帶雙[纜]
double up and secure	双绑系牢	各纜打雙
doubling angle on the bow	船首倍角法	艏倍角定位法
doubling plate	覆板	複板,加力板
dovetail groove	燕尾槽	鳩屋槽
down-bound vessel	下行船	下行船
down helm	上风舵	上風舵
down stream	顺水	下游
downstream vessel	顺流船	顺流船
downstroke	下行行程	下行衝程
DR(=dead reckoning)	积算	推算
draconitic month	交点月	交點月
draft	吃水	吃水
draft indicating system	吃水指示系统	吃水指示系統
drag and suction device	耙吸装置	耙吸設施
drag force	阻力	拖曳力
dragging anchor	拖锚,走锚	拖錨
drag net	曳网	地曳網

英　文　名	祖国大陆名	台湾地区名
drain cock	残水旋塞	排泄旋塞
draining method	排水法	排洩法
draining system	疏水系统	排洩系統
draught survey	水尺检量	測量水尺
draw bridge	曳开桥	曳開橋
dredged channel	人工航槽	濬深水道
dredger	挖泥船	挖泥船,控泥機
dredging	①拖锚航行 ②疏浚	①拖錨航行 ②濬深
dredging facility	挖泥工具	挖泥設施
dredging pump	泥[浆]泵	泥泵
drier	①干燥器 ②干燥剂	①乾燥器 ②催乾劑,乾燥劑
drift angle	漂角,流压差	流偏角,偏流角
drift fishing	流网作业	流網捕魚
drift fishing boat	漂流渔船	流網漁船
drift ice	流冰	流冰
drifting	漂航	漂流
drift net	流刺网	流網
drift net shaker	流刺网振网机	流刺網振網機
drilling platform	钻井平台	鑽油平台
drilling rig	钻探设备	鑽探設備
drilling rigs at sea	海上钻井架	海上鑽油台
drilling vessel	钻探船	鑽探船
d-ring	D 型环	D 型環,拉繫環
drinking water ozone disinfector	饮用水臭氧消毒器	飲用水臭氧消毒器
drinking water pump	饮水泵	飲用水泵
drinking water system	饮用水系统	飲用水系統
drip proof type	防滴型	防滴型
drive	压向下风	乘風而駛
drive motor	拖动电动机	拖動馬達
driven gear	从动齿轮	從動齒輪
driving gear	主动齿轮	主動齒輪
drizzle	毛毛雨	毛毛雨
drop lubrication	滴油润滑	滴油潤滑
droppable equipment	空投设备	空投設備
drop point	滴点	落點
drop test	坠落试验	墜落試驗
DRT(=dead-reckoning tracer)	航迹积算仪	推算航迹儀

英　文　名	祖国大陆名	台湾地区名
drum	①卷筒 ②鼓轮	①捲筒(起重機) ②鼓輪
drum controller	鼓形控制器	圓筒控制器,鼓形控制器
drum rotor	鼓形转子	鼓形轉子,鼓形輪子
dry adiabatic lapse rate	干绝热直减率	乾絕熱線直減率
dry bulk container	干散货箱	乾散貨櫃
dry cargo	干货	乾貨
dry cargo ship	干货船	乾貨船
dry certificate	干舱证书	[貨艙]乾燥證明
dry compass	干罗经	乾羅經
dry cylinder liner	干式缸套	乾式缸套
dry dock	干船坞	乾[船]塢
dry ice	干冰	乾冰
drying height	干出高度	出水高度
drying rock	干出礁	出水礁石
drying room	干燥室	乾燥室
dry powder fire extinguishing system	干粉灭火系统	乾粉滅火系統
dry sump lubrication	干底润滑	乾油槽潤滑
dry-type evaporator	干式蒸发器	乾式蒸發器
DSB(=double side-band)	双边带	雙邊帶
DSC(=digital selective calling)	数字选择呼叫	數位選擇呼叫
DSC distress alerts	DSC 遇险报警	數位選擇呼叫遇險警報
dual-duct air conditioning system	双风管空气调节系统	雙風管空調系統
dual-fuel diesel engine	双燃料柴油机	雙燃料柴油機
dual purpose officer	双职高级船员	雙專長甲級船員
dual responsibility	双方责任	雙方責任
dual system	双机系统	雙機系統
Dubhe	天枢(大熊α)	天樞,北斗一(大熊α)
ducted propeller	导管推进器	導罩螺槳
due date	应付日期	到期日
due diligence	谨慎处理	必要的注意
dumb card compass(=pelorus)	哑罗经	啞羅經
dummy piston	平衡活塞	均衡活塞
dumping ground	垃圾倾倒区	垃圾傾倒區
dumping of wastes	倾废	廢[棄]物傾倒
dunnage	衬垫,垫舱物料	墊材,墊艙,襯材
dunnage cargo	垫舱货	填墊貨

英　文　名	祖国大陆名	台湾地区名
duplex	双工	雙工
duplex operation	双工操作	雙工作業
duplex strainer	双联滤器	複式過濾器
duplication of equipment	双套设备	雙套設備
durability	耐久性	耐久性
duration clause	期限条款	期限條款
duration of ebb	落潮持续时间	退潮持續期
duration of flood	涨潮持续时间	漲潮持續期
dust filter	滤尘器	濾塵器
dust-tight	尘密	塵密
dusty cargo	扬尘货	粉狀貨
DW(=dead weight)	[总]载重量	載重量
dye marker	海水染色标志	海水染色標誌
dyke	堤坝	堤
dynamical balance	动平衡	動力平衡
dynamical heeling angle	动横倾角	動橫傾角
dynamical heeling lever	动横倾力臂	動橫傾力
dynamical heeling moment	动横倾力矩	動橫傾力矩
dynamical stability	动稳性	動穩度
dynamical stability lever	动稳性力臂	動穩度力臂
dynamical viscosity	动力黏度	動力黏度
dynamic braking	能耗制动	動力煞車,動力制軔
dynamic course stability	动航向稳定性	動向穩度
dynamic head	动压头	動落差
dynamic positioning(=kinematic positioning)	动态定位	動態定位
dynamic similarity	动力相似	動態相似性
dynamic simulation	动态模拟	動態模擬
dynamic state	动态	動態
dynamometer	测力计,测功器	測功計

E

英　文　名	祖国大陆名	台湾地区名
Earth	地球	地球,地
earth axis	地轴	地軸
earth ellipsoid	地球椭圆体	地球橢圓體
earth-ionospheric waveguide	地球-电离层波导	地球-電離層波導

英　文　名	祖国大陆名	台湾地区名
earth pole	地极	地極
earth shape	地球形状	地球形狀
earth station	地球站	地球台
ease	缓松	回舵,緩鬆
ease her	回舵	回舵,鬆舵
ease off	松开	放鬆,鬆開
easterly wave	东风波	東風波
easting	向东航程	東橫距
east mark	东方标	東方標
easy running fit	轻转配合	輕轉配合
easy slide fit	轻滑配合	輕滑配合
ebb channel	落潮水道	退潮水道
ebb current	落潮流	退潮流
ebb stream(=ebb current)	落潮流	退潮流
ebb strength	落潮流强度	退潮最大速率
EBL(=electronic bearing line)	电子方位线	電子方位線
eccentric anomaly	偏近点角	偏近點角
eccentric gear	偏心传动装置	偏心機構
eccentricity of earth	地球偏心率	地球偏心率
ECDIS(= electronic chart display and in- formation system)	电子海图显示与信息系 统	電子海圖顯示與資訊系 統
echelon formation	梯队	梯隊
echo sounder	回声测深仪	回聲測深儀
echo sounder error	测深仪误差	回聲測深儀誤差
echo sounding	回声测深	回音測深
ecliptic	黄道	黃道
ecliptic latitude	黄纬	黃緯
ecliptic longitude	黄经	黃經
ecliptic pole	黄极	黃道天極
economical power	经济功率	經濟功率
economical speed	经济速度,经济航速	經濟速[率]
economic working condition	经济工况	經濟工作狀況
economizer	经济器	節熱器
ED(= existence doubtful)	疑存	疑有
eddy	漩水	渦流,旋渦,漩渦
eddy making resistance	涡流阻力	興渦阻力
edge of coverage	覆盖区边缘	涵蓋邊緣
effective compression ratio	有效压缩比	有效壓縮比

英　文　名	祖国大陆名	台湾地区名
effective date	生效日	生效日
effective delivery stroke	有效供油行程	有效供油衝程
effective efficiency	有效效率	有效效率
effective head	净压头	有效落差
effective isotropically radiated power	有效全向辐射功率	有效等方向性輻射電力
effective mean pressure	平均有效压力	有效平均壓力
effective power	有效功率	有效功率
effective radiated power	有效辐射功率	有效輻射功率
efficient sound signal	有效声号	有效聲號
effluent	流出物	流出物
effluent concentration	排放浓度	流出物濃度
effluent standard	排放标准	流出物標準
EGC(=enhanced group call)	强化群呼	強化群呼
EGC decoder	增强群呼译码器	強化群呼解碼器
EGC receiver(=enhanced group calling receiver)	增强群呼接收机	強化群呼接收機
EIRP(=equivalent isotropically radiated power)	等效全向辐射功率	等效等向輻射功率
ejection seat mechanism	弹射座机构	彈座機構
elastic coupling	弹性联轴器	彈性聯軸節
elastic hysteresis	弹性滞后	彈性遲滯
elastic limit	弹性极限	彈性限界
elastic modulus	弹性模数	彈性係數
elasto-hydrodynamic lubrication	弹性流体动力润滑	彈性流體動力潤滑
electric actuator	电力执行机构	電力引動器
electrical balancer	电动[力矩]平衡器	電平衡器
electrical engineer	电机员	電機工程師
electrical fishing	电气捕鱼	電氣捕魚
electrical interlocking	电气联锁	電聯鎖
electrical log book	电气日志	電機日誌
electrical propulsion motor room	电力推进电机间	電力推進馬達室
electrical starting	电起动	電力起動
electric cargo winch	电动起货机	電動起貨機
electric defrost	电热融霜	電除霜
electric defrost timer	电热融霜定时器	電除霜定時器
electric drive	电力拖动	電力驅動
electric drive apparatus	电力拖动装置	電力驅動設備
electric-hydraulic servo actuator	电液伺服机构	電動液力伺服致動器

英　文　名	祖国大陆名	台湾地区名
electric-pneumatic remote control system for main engine	电-气式主机遥控系统	電動氣力遙控系統
electric propulsion	电力推进	電力推進
electric propulsion ship	电力推进船	電力推進船
electric safety lamp	安全电灯	安全電燈
electric steering engine	电动舵机	電舵機
electric torch	手电筒	手電筒
electric transformer	变压器	變壓器
electrochemical corrosion	电化学腐蚀	電化腐蝕
electrodialysis method	电渗析法	電透析法
electro-hydraulic directional control valve	电液换向阀	電動液力方向控制閥
electro hydraulic drive	电动液压传动	電動液壓驅動
electro-hydraulic servo valve	电液伺服阀	電動液力伺服閥
electro-hydraulic steering engine	电动液压舵机	電動液力舵機
electrolyte	电解液	電解質,電解液
electromagnetically controlled gyrocompass	电磁控制罗经	電磁控制電羅經
electromagnetic deviation	电磁自差	電磁自差
electromagnetic log(EM log)	电磁计程仪	電磁計程儀
electromagnetic pendulum	电磁摆	電磁擺
electromagnetic wave distance measuring instrument	电磁波测距仪	電磁波測距儀
electronic bearing line(EBL)	电子方位线	電子方位線
electronic bill of lading	电子提单	電子載貨證券
electronic chart	电子海图	電子海圖
electronic chart data base	电子海图数据库	電子海圖資料庫
electronic chart display and information system(ECDIS)	电子海图显示与信息系统	電子海圖顯示與資訊系統
electronic data interchange	电子数据交换	電子數據交換
electronic governor	电子[式]调速器	電子調速器
electronic mail	电子邮件	電子郵件
electronic mail service	电子邮件业务	電子郵件業務
electronic maintenance at sea	电子[设备]海上维修	海上電子維修
electronic navigation	电子航海	電子儀航行術
electronic regulator	电子调节器	電子調整器
electronic remote control system for main engine	电子式主机遥控系统	主機電子遙控系統
electro-pneumatic transducer	电-气变换器	電動氣力轉變器
electro pneumatic type	电动气动式	電動氣動式

英　文　名	祖 国 大 陆 名	台 湾 地 区 名
electrostatic discharge	静电放电	靜電放電
electrostrictive effect	电致伸缩效应	電伸縮效應
element breakdown accident	元件破损事故	元件破損事故
element burnout accident	元件烧毁事故	元件燒毀事故
elements of ship formation pattern	舰艇编队队形要素	船艦編隊隊形要素
elevated pole	仰极	仰極
elevation	高程	標高,仰角
elevation of light	灯高	燈[距水面]高
ellipsoid	椭球体	橢球體
elliptical polarization	椭圆极化	橢圓極化
elliptical stern	椭圆[型]艉	橢圓艉
elongation	天体与太阳角距	天體與太陽之角距
ELT（=emergency locator transmitter）	紧急示位发信机	應急示位發射機
embargo	①封港 ②禁运 ③扣留	①封港 ②禁運 ③扣船
embargo on ship	禁止船舶通航	禁止船舶通航
embarkation charge	搭载费	[搭載]船費
embarkation ladder	登船梯	乘載梯
emergence	出现,露出	露出量
emergency（=urgency）	紧急	緊急
emergency air compressor	应急空气压缩机	應急用空氣壓縮機
emergency antenna	应急天线	應急天線
emergency blower	应急鼓风机	應急鼓風機
emergency brake	紧急刹车	緊急軔,緊急煞俥
emergency breakaway tools	紧急断缆工具	緊急斷纜工具
emergency call（=urgency call）	紧急呼叫	緊急呼叫
emergency cell	应急电池	應急電池
emergency compass	应急罗经	應急羅經
emergency electric equipment	应急电气设备	應急電力設備
emergency fire pump	应急消防泵	應急救火泵
emergency generator	应急发电机	應急發電機
emergency generator diesel engine	应急发电柴油机	應急柴油發電機
emergency instruction	应急说明	應急說明
emergency lighting	应急照明	應急照明
emergency lighting system	应急照明系统	應急照明系統
emergency locator transmitter（ELT）	紧急示位发信机	應急示位發射機
emergency maneuvering	应急操纵	應急操縱
emergency overriding	紧急强制用车	緊急運作
emergency phase	紧急阶段	緊急階段

英 文 名	祖 国 大 陆 名	台 湾 地 区 名
emergency position-indicating radio beacon	紧急无线电示位标	應急指位無線電示標
emergency power source	应急电源	應急電源
emergency power station	应急电站	應急動力站
emergency preparedness	应急准备	應急準備
emergency response team	紧急响应工作队	緊急應對工作隊
emergency service	应急业务	應急業務
emergency shut-down	紧急停堆,紧急关闭	緊急關閉
emergency shut-down device	应急停车装置	緊急關閉設施
emergency source of electrical power (=emergency power source)	应急电源	應急電源
emergency starting	应急起动	應急起動
emergency steering gear	应急操舵装置	應急操舵裝置,應急舵機
emergency stop	应急停车	緊急停俥
emergency stop push button	紧急停车按钮	緊急停俥按鈕
emergency switchboard	应急配电板	應急配電板
emergency turn	紧急回转	緊急迴轉
emergency warning	紧急警报	緊急警報
emission	发射	發射
emission performance	排放性能	排放性能
EM log(=electromagnetic log)	电磁计程仪	電磁測程儀
emulsifying oil	乳化油	乳化油
encoded position data	编码位置数据	編碼之位置數據
encounter	会遇	會遇
encounter rate	会遇率	會遇率
end for end	头尾对换	首尾顛倒換索
endless rope	环索	回頭索,環索
end link	末端环	尾環,鏈端環
end of evening twilight	昏影终	曙昏終了
end on	对遇	正對
endorsement in blank	空白背书	空白背書
endorsement of bill of lading	提单背书	載貨證券之背書
endurance	续航力	續航力,持久
endurance limit	疲劳极限	持久限界
engaging and disengaging gear	离合机构	離合機構
engine block	机体	引擎體
engine control room control	集控室控制	主機控制室控制
engineering officer	轮机员	輪機員

英 文 名	祖 国 大 陆 名	台 湾 地 区 名
engineering ship	工程船	工程船
engine layout diagram	柴油机运转范围图	引擎運輸範圍圖
engine load diagram	柴油机负荷图	引擎負載圖
engine orders	车令	俥令
engine room	机舱	輪機室,機艙
engine room automation	机舱自动化	機艙自動化
engine room emergency bilge suction valve	机舱应急舱底水阀	機艙應急舢水吸入閥
engine room lighting system	机舱照明系统	機艙照明系統
engine room log book	轮机日志	輪機記事簿
engine speed	主机航速	機速
engine telegraph	车钟	俥鐘
engine telegraph alarm	车钟报警	俥鐘警報
engine telegraph order indicator	车令指示器	俥令指示器
engine trial	试车	試俥
Engler viscosity	恩氏黏度	恩氏黏度
enhanced group call(EGC)	强化群呼	強化群呼
enhanced group calling receiver(EGC receiver)	增强群呼接收机	強化群呼接收機
enhanced group call service code	增强群呼业务码	強化群呼業務碼
Enhanced Group Call System	增强群呼系统	強化群呼系統
ensign	国旗	國旗,船籍旗
ensign staff	艉旗杆	艉旗桿
entrance prohibited	不准入境	禁止入境
entry guide	导口	導槽入口
EP(=estimated position)	推算船位	估計船位
ephemeris	星历	天文曆
ephemeris data	星历数据	衛星運行數據表,衛星軌道數據,天文曆數據
EPIRB(=emergency position-indicating radio beacon)	紧急无线电示位标	應急指位無線電示標
EPIRB identification	紧急无线电示位标识别	應急指位元無線電示標識別
equal altitude method	等高法	等高法
equalizing regulation	均功调节	均衡調節
equation of equal altitude	等高差	等高差
equation of time	时差	時差
equator	赤道	赤道

英　文　名	祖国大陆名	台湾地区名
equatorial horizontal parallax	赤道地平视差	赤道地平視差
equatorial mile	赤道里	赤道里
equatorial radius	地球长半轴	赤道半徑
equatorial system of coordinates	赤道坐标系	赤道坐標系統
equatorial telescope	赤道仪	赤道儀
equatorial tide	赤道潮汐	赤道潮
equiangular projection	等角投影	等角投影
equilibrium vapor pressure	饱和水汽压	平衡蒸汽壓
equilibrium vapor pressure of ice surface	冰面饱和水汽压	冰面平衡蒸汽壓
equinoctial spring tides	分点大潮	春秋分大潮
equinoctial tide	分点潮	春秋分點潮,二分潮
equinox	分点	春秋分點
equipment number	舾装数	船具規號,屬具數
equipment receipt	设备交接单	設備接收單
equivalent	①等效 ②等值	①當量 ②同等設備
equivalent isotropically radiated power （EIRP）	等效全向辐射功率	等效等向輻射功率
erosion	侵蚀	浸蝕
error ellipse of position	[船位]误差椭圆	船位誤差橢圓
error in calculation	计算误差	計算誤差
error of observed position	观测船位误差	觀測船位誤差
error parallelogram	[船位]误差平行四边形	誤差平行四邊形
escape	①逃生 ②逸出	①逃出 ②逸出
escape trunk	逃生通道	安全通道,逃生通道
escorting	海上护送	護航
estimated course	推算航迹向	估計航向
estimated latitude	推算纬度	估計緯度
estimated longitude	推算经度	估計經度
estimated position（EP）	推算船位	估計船位
estimated time of arrival （ETA）	预计到达时间	預計到達時間
estimated time of departure （ETD）	预计开航时间	預計離開時間
estuary	河口	河口
ET（ =equation of time）	时差	時差
ETA（ =estimated time of arrival）	预计到达时间	預計到達時間
ETD（ =estimated time of departure）	预计开航时间	預計離開時間
evacuation pump（ =vacuum pump）	真空泵	真空泵
evacuation-slide launching	泄滑下水	洩滑下水

英 文 名	祖 国 大 陆 名	台 湾 地 区 名
evaporating coil	蒸发盘管	蒸發盤管
evaporator pressure regulator(=back pressure regulator)	蒸发压力调节阀(**背压调节阀**)	背壓調整器
evaporator tube bank	蒸发管束	蒸發管排
evasion maneuvre	规避机动	迴避操縱
evasive course	规避航向	迴避航向
evasive steering	规避操舵	迴避操舵法
even keel	平吃水	縱平浮
evolution	发展	隊形變換
examination of leakage and breakage	漏损检测	漏損檢查
except otherwise herein provided	另有相反的规定除外	除另有明文規定外
excess air coefficient	过量空气常数	過量空氣係數
excess of hour angle increment	超差	時角增量超差
exciting frequency	激励[振]频率	激振頻率
exclusive economic zone	专属经济区	專屬經濟區
exclusive fishery zone	专属渔区	專屬漁業區
excursion boat	游览船	遊覽船
exemption	豁免	豁免
exemption certificate	豁免证书	豁免證書
exemption clause	除外条款	豁免條款
exercise area(=practice area)	演习区	演習區
exhaust casing	排气壳体	排氣外殼
exhaust chest	排汽室	排氣室
exhaust gas heat exchanger	废气锅炉	廢氣熱交換器
exhaust hood(=exhaust casing)	排气壳体	排氣外殼
exhaust particulate	排气颗粒	排氣顆粒
exhaust port	排气口	排氣口
exhaust purification	废气净化	排氣淨化
exhaust steam system	排汽系统	排汽系統
exhaust stroke	排气行程	排氣衝程
exhaust temperature	排气温度	排氣溫度
exhaust turbine generating set	废气涡轮发电机组	廢氣渦輪發電機組
exhaust turbo compound system	废气涡轮复合系统	廢氣渦輪複合系統
exhaust unit	排气装置	排氣裝置
exhaust valve	排气阀	排氣閥
existence doubtful(ED)	疑存	疑有
existing ship	现有船	現存船舶,現成船
ex-meridian	近中天	近中天

英　文　名	祖国大陆名	台湾地区名
expanding square search	扩展方形搜寻	擴展方形搜索
expansion joint	伸缩接头	伸縮接頭
expansion ratio	膨胀比	膨脹比
expansion-scavenging stroke	膨胀-换气行程	膨脹-驅氣衝程
expansion stroke	膨胀行程	膨脹衝程
expansion tank	膨胀柜	膨脹櫃
ex pier	目的港码头交货	碼頭交貨
expiry date	期满日,失效日	期滿日,失效日
explosimeter	测爆仪	測爆儀
explosion prevention	防爆	防爆
explosion proof fan	防爆式风机	防爆風扇
explosion-proof type	防爆型	防爆型
explosive	爆炸品	炸藥
explosive cargo	易爆货	易爆炸貨
explosive fog signal	爆炸雾号	音爆霧號
explosive signal	爆炸信号	爆炸信號
export permit	出口许可	出口許可
ex quay(=ex pier)	目的港码头交货	碼頭交貨
ex ship	目的港船上交货	船上交貨
extended protest	延伸海事声明	遭難證明書
extension alarm	延伸报警	延伸警報
extension shaft	延伸轴,中介轴	延伸軸
extension survey	展期检验	延期檢驗
external characteristic of steam turbine	汽轮机外特性	蒸汽渦輪機外特性
external combustion engine	外燃机	外燃機
externally mounted equipment	舱外安装设备	裝於艙外之設備
extra-length charge	超长货附加费	超長貨附加費
extratropical cyclone	温带气旋	溫帶氣旋
ex wharf(=ex pier)	目的港码头交货	碼頭交貨
ex works	工厂交货	工廠交貨
eye splice	眼环[插]接	眼索接琵琶頭

F

英　文　名	祖国大陆名	台湾地区名
face hardening	表面硬化	表面硬化
facsimile(FAX)	传真	傳真
facsimile weather chart	传真天气图	傳真天氣圖

英　文　名	祖 国 大 陆 名	台 湾 地 区 名
factor of subdivision	分舱因数	隔艙因數
factual presumption of fault	事实推定过失	事實推定過失
fading	衰落	衰落,衰落现象
fairing by flame	火工矫形	火焰矯正
fairlead	导缆器	索導
fairway	航道	主航道
falling snow	降雪	降雪
falling tide	落潮	落潮
false distress alerts	误发遇险警报	假遇險警報
false echo	假回波	假回波
familiarization training	熟悉培训	熟悉訓練
fantail stern	扇形艉	扇形艉
FAS	海上加油	海上加油
FAS delivery station	海上加油站	海上加油整補傳送站
fast coupling	刚性联轴器	剛性聯結器
fastening bolt	紧固螺栓	緊固螺栓
fast ice	固定冰	堅冰
fast settling device	快速稳定装置	快速固定設施
fatigue cracking	疲劳裂纹	疲勞裂紋
fault detection and exclusion(FDE)	故障检测与排除	故障檢測與排除
fault detection and identification(FDI)	故障检测与辨识	故障檢測與識別
fault detector	故障探测器	故障探測器
fault diagnosis	故障诊断	故障診斷
fault signal	故障信号	故障信號
fault tree analysis	事故树分析	故障樹分析
favorable current	顺流	顺流
favorable wind (=tail wind)	顺风	顺風
FAX(=facsimile)	传真	傳真
FCL(=full container load)	整箱货	全貨櫃之貨物
FDE(=fault detection and exclusion)	故障检测与排除	故障檢測與排除
FDI(=fault detection and identification)	故障检测与辨识	故障檢測與識別
FEC(=forward error correction mode)	前向纠错方式	前向偵錯方式
Federal Radionavigation Plan(FRP)	联邦无线电导航计划	聯邦無線電導航計劃
feedback	反馈	反饋
feeder	添注漏斗	灌斗,艙口灌斗
feeder link	馈线链路	饋線鏈路
feeder panel	配电屏	饋電屏
feeder service	支线运输	輔助運務

英 文 名	祖 国 大 陆 名	台 湾 地 区 名
feed water ratio	给水倍率	給水率
fender	碰垫,靠把	碰垫
ferro-concrete vessel(=reinforced concrete vessel)	钢筋水泥船	鋼筋混凝土船
ferrous metal	黑色金属	鐵金屬
ferry	渡船	渡船
fetch	风区	受風區
fiberglass reinforced plastic boat	玻璃钢船	玻[璃]纖[維]強化塑膠船
fiber rope	纤维绳	纖維[繩]索
fiber rope highline rig	高架纤维绳传递装置	纜索高線傳遞設備
field test	现场试验	現場試驗
fighting boat	战斗船	戰艦
figure of eight knot	8 字结	8 字結
figure of mark pronunciation	数字拼读法	數位拼音
filing date	交发日期	交發日期
filing time	交发时间	交發時間
filled compartment	满载舱	滿載艙間
filler cargo	填隙货	填隙貨
fillet area	圆角区	内圓角區
fillet welding	填角焊	填角熔焊
filling line	注入管	注入管
filling pipe(=filling line)	注入管	注入管
filtered alert data	报警数据滤除	篩選之警報數據
final act	最后文件	蔵事文件
final course	终航向	終結航向
final diameter	旋回直径	終結直徑
final report	最终报告	最終報告
fine alignment	精对准	精校準
fine thread	细牙螺纹	細牙螺絲,密螺紋
finishing signal	终结信号	終止信號
finite-element calculation	有限元计算	有限元素計算
finned-surface evaporator	肋片式蒸发器	鰭面蒸發器
fin shaft	鳍轴	鰭軸
fin stabilizer	减摇鳍装置	鰭板穩定器
fin-tilting gear	转鳍机构	傾轉鰭機構
FIO(=free in & out)	船方不负装卸费用	裝卸自理
fire alarm	失火警报	火警警報

英　文　名	祖国大陆名	台湾地区名
fire alarm system	火警报警系统	火警報警系統
fire boat	消防船	救火船,消防艇,消防船
fire control plan	防火控制图	火災控制圖
fire control station	消防控制站	火警控制站
fire door	防火门	防火門
fire extinguisher	灭火器	滅火器
fire extinguishing appliance	灭火装置	滅火裝備
fire extinguishing system	消防系统	滅火系統
fire fighting drill	消防演习	消防演習
fire fighting station	消防部署	消防部署
fire hose	水龙带,消防软管	水龍帶
fire insurance	火险	火險
fire main	消防总管	主消防水管,救火主水管
fireman's outfit	消防员装备	消防員裝具
fire patrol system	消防巡逻制度	火警巡邏系統
fire proof bulkhead	防火舱壁	防火艙壁
fire protected lifeboat	耐火救生艇	防火救生艇
fire protecting rules	消防规范	防火規則
fire pump	消防泵	救火泵
fire-tight	耐火	耐火
fire tube boiler	火管锅炉	火管鍋爐
firing order	发火顺序	點火順序
first aid at sea	海上急救	海上急救
first-aid outfit	急救医疗器具	急救醫藥用品
first-class radio electronic certificate	一级无线电电子证书	第一級無線電電子員證書
first quarter	上弦	上弦
fish buying boat	收鱼船	收魚船
fish channel	输鱼槽	輸魚槽
fisheries company	渔业公司	漁業公司
fisherman's bend	锚结	漁人扣
fishery administration	渔政	漁業行政
fishery administration vessel	渔政船	漁政船
fishery agreement	渔业协定	漁業協定
fishery guidance boat	渔业指导船	漁業指導船
fishery patrol boat	渔业巡逻船	漁業巡邏船
fishery research vessel	渔业调查船	漁業研究船

英　文　名	祖国大陆名	台湾地区名
fishery resource	水产资源	水産資源
fishery rules and regulations	渔业法规	漁業法規
fish finder	鱼探仪,鱼群探测器	魚群探尋器,魚群探測器
fish group indicating buoy	渔群指示标	魚群指示浮標
fishing area(=fishing zone)	渔区	漁區
fishing chart	渔场图	漁場圖
[fishing] closed season	禁渔期	禁漁期
fishing gear	渔具	漁具
fishing ground	渔场	漁場
fishing harbor	渔港	漁港
fishing lamp	集鱼灯	集魚燈
fishing light boat	灯光船	集魚燈船
fishing machinery	捕捞机械	漁撈機械
[fishing] mass migration	鱼类回游	魚群洄游
[fishing] migration route	回游路线	洄游路線
fishing operation	渔捞作业	漁撈作業
fishing restriction	渔捞限制	漁撈限制
fishing right	渔业权	漁業權
fishing rod box	钓竿箱	釣竿盒
fishing season(=catching season)	渔汛	漁期
[fishing] season off	休渔期	休漁期
fishing stake	渔栅	漁栅
fishing supervision	渔监	漁業監督
fishing tackle(=fishing gear)	渔具	魚具
fishing technology	捕鱼技术	漁撈學
fishing vessel	渔船	漁船
fishing zone	渔区	漁區
fish meat factory ship	鱼粉加工船	捕魚加工船
fish nursery ground	幼鱼栖息场	幼魚棲息場
fish pump	吸鱼泵	吸魚泵
fish reef	渔礁	漁礁
fish school	鱼群	魚群
fish shelter	人工鱼礁	人工魚礁
fission energy	裂变能	分裂能
fission neutron	裂变中子	分裂中子
fission product	裂变产物	核子分裂产物,分裂产物

英　文　名	祖国大陆名	台湾地区名
fitness for duty	适于当值	適於當值
fix	船位	定位
fixed deck foam system	固定式甲板泡沫系统	固定甲板泡沫系統
fixed-displacement oil motor	定量油马达	固定排量油馬達
fixed forth fire-extinguishing system	固定泡沫灭火系统	固定泡沫滅火系統
fixed gas fire extinguishing system	固定式气体灭火系统	固定氣體滅火系統
fixed high expansion forth fire-extingui-shing system	固定高膨胀泡沫灭火系统	固定高脹力泡沫滅火系統
fixed light	定光	定光
fixed loop antenna	固定环形天线	固定環形天線
fixed net	固定渔网	定骨網
fixed oil production platform	固定式采油平台	固定式産油平台
fixed pitch propeller(FPP)	固定螺距桨	[固]定[螺]距螺槳
fixed pressure water-spraying forth fire-extinguishing system	固定压力喷水灭火系统	固定壓力噴水滅火系統
fixing by bearing and distance	方位距离定位	方位距離定位
fixing by cross bearings	方位定位	交叉方位定位
fixing by distances	距离定位	距離定位
fixing by horizontal angle	水平角定位	水平角定位
fixing by landmark	陆标定位	岸標定位
fixing by vertical angle	垂直角定位	垂直角定位
fixing letter	成交函	成交書
fixture note	订租确认书	成交書(論程傭船需用)
flag at halfmast	下半旗	下半旗
flag chest	旗柜	旗箱
flag discrimination	船旗歧视	船旗歧視
flag of convenience	方便旗	權宜船籍
flag signaling	旗号通信	旗號通信
flag staff	旗杆	旗桿
flag state	船旗国	船旗國
flame and water forming	水火成形	水火成形
flame cutting	火焰切割	火焰截割
flame screen	防火网	防焰網
flame trap	阻火器	防焰網
flammable liquid	易燃液体	易燃液體
flammable solid	易燃固体	易燃固體
flammable solid liable to spontaneous com-	易自燃固体	易自燃固體

英　文　名	祖国大陆名	台湾地区名
bustion		
flammable solid when wet	遇水易燃固体	遇濕易燃固體
flanking rudder	倒车舵	倒俥舵
flank speed	最高速度	飛速[率],強速
flap-type rudder	襟翼舵	襟翼舵
flash chamber	闪发室	急驟蒸發室
flash evaporation	闪发蒸发	急驟蒸發
flashing light	闪光灯	閃光燈
flashing light for signaling	通信闪光灯	通信閃光燈
flashing light signaling	灯光通信	燈號通信
flash point	闪点	閃點
flat seizing	平扎	平纏
flat-surface probe	平面传感器	平面感應器,平面探針
flat-surface sensor(=flat-surface probe)	平面传感器	平面感應器,平面探針
flattening of earth	地球扁率	地球扁率
fleet	船队	艦隊,船隊
fleetNET	船队通信网	船隊通信網
flexibility gyrocompass	挠性罗经	撓性電羅經
flexibility gyroscope	挠性陀螺仪	撓性迴轉儀
flexible coupling	弹[挠]性联轴器	撓性聯結器
flexible rotor	柔性转子	柔性轉子
flexible shafting	挠性轴系	撓性軸系
flexible stay plate	柔性支持板	柔性撐板
flexible waveguide	软波导	撓性導波管
flicker reset	闪光复位	閃爍復位
flight information region	飞行信息区	飛航情報區
Flinders'bar	佛氏铁	校磁鐵棒
float	浮子	浮子
float-free launching	漂浮式下水	自由浮離下水
floating anchor	浮锚	浮錨,海錨
floating crane	起重船	起重船,水上起重機
floating dock	浮船坞	浮塢,浮船塢
floating object	漂浮物	漂浮物
floating oil loading hose	浮式输油软管	浮式輸油軟管
floating oil production platform	浮式采油生产平台(**浮式采油平台**)	浮式産油平台
floating pile driver	打桩船	打樁船
floating production storage and offloading	浮式生产储油及卸载设	浮動型石油生産貯藏及

英 文 名	祖 国 大 陆 名	台 湾 地 区 名
facility（FASO）	施	卸载设施
floating production storage unit（FPSU）	浮式生产储油装置	浮式生产储存装置
floating storage unit（FU）	浮式储油装置	浮動型貯油装置
floating substance（=floating object）	漂浮物	漂浮物
floating trawl	浮拖网	浮曳網
float line	浮子纲	浮繩
float oil layer sampler	浮油层取样器	浮油層取樣器
float on/float off	浮装	浮載
floe ice	浮冰	浮冰
flood	浸水	水浸
floodable length	可浸长度	可浸長度
flood current（=flood stream）	涨潮流	漲潮流
flooded evaporator	浸没式蒸发器	浸沒式蒸發器
flooding angle	进水角	泛水角
flooding method	灌注法	灌注法
flooding stability（=damaged stability）	破舱稳性	破損穩度,受損穩度
flood mark	涨潮标志,泛滥标	洪水位,高潮位標誌,泛水標誌
flood peak	洪峰	洪峰
flood stream	涨潮流	漲潮流
flood［tide］	涨潮	漲潮
floor	肋板	底肋板
flotsam	［难船]漂浮物	遇難船漂浮物
flow-control valve	流量控制阀	流量控制閥
flowmeter	流量表	流量計
flow moisture point	流动水分点	注動水分點
flow regulator	流量调节器	流量調整器
fluid cargo	液货	液狀貨
fluid motor（=hydraulic motor）	液压[油]马达	液壓馬達
fluke	锚爪	錨掌
flush deck vessel	平甲板船	平甲板船
flushing arrangement	冲洗装置	沖洗裝置
flush socket	平插座	平式套座
flux gate compass	磁通门罗经	磁通門羅經
fly fishing	甩竿钓	甩竿釣(蚊鉤釣)
fly wheel	飞轮	飛輪
foam fire extinguishing system	泡沫灭火系统	泡沫滅火系統
FOB（=free on board）	离岸价格	出口地船上交貨

英 文 名	祖 国 大 陆 名	台 湾 地 区 名
fog	雾	霧
fog gong	雾锣	霧鑼
fog horn	雾角	霧角
fog signal	雾号	霧號
fog siren	雾雷	霧雷
fog warning	雾警报	霧警報
fog whistle	雾笛	霧笛
foldable canopy	可折顶篷	可摺頂篷
folding box pallet	折叠式箱型货盘	摺叠式箱形托貨板
folding fin stabilizer	折叠式减摇鳍装置	摺鰭穩定器
folding hatchcover	折叠式舱盖	摺式艙蓋
folio label	海图夹标签	海圖夾標籤
folio list	海图夹目录	海圖夾目錄
folio number	海图夹编号	海圖夾編號
following at a distance	尾随行驶	尾隨行駛
following sea	顺浪	順浪
follow-up control	随动控制	追蹤控制,追隨控制
follow-up ship	后续舰	後續船艦
follow-up speed	随动速度	追蹤速度
follow-up system	随动系统	追蹤系統
Fomalhaut	北落师门(南鱼 α)	北落師門(南魚 α)
food ration	口粮	口糧
food stuff refrigerated storage	伙食冷库	糧食冷凍庫
food wastes	食物垃圾	食物廢棄物
foot line	下纲	沉子綱
forbidden fishing zone	禁渔区	禁漁區
forbidden zone	禁航区	禁航區
forced circulation boiler	强制循环锅炉	強制循環鍋爐
forced lubricating oil system	强力滑油系统	強力潤滑油系統
forced vibration	强制振动	強制振動
force majeure	不可抗力	不可抗力
fore-and-aft distance between two ships	舰间纵距	船艦間縱距
fore and after sail	纵帆	縱帆
fore-and-aft line	艏艉线	艏艉線
fore-and-aft magnet	纵向磁棒	縱向磁棒
fore-and-aft sail (=fore and after sail)	纵帆	縱帆
fore body	艏部	艏部
forecasted wave direction	预报波向	預報波向

英　文　名	祖国大陆名	台湾地区名
forecastle	艏楼	艏樓
forecastle deck	艏楼甲板	艏樓甲板
foreman	装卸长,领班	領班,工頭
fore mast	前桅	前桅
forepeak bulkhead	艏尖舱壁,防撞舱壁	艏艙壁
fore peak tank	艏尖舱	艏尖艙
fore perpendicular	艏垂线	艏垂標
fore sail	前帆	前帆
fore spring	艏倒缆	前倒纜
forged type crankshaft	锻造型曲轴	鍛造曲柄軸
forge weld	锻接	鍛接
fork lift truck	万能装卸机	堆高機
formation	队列,编队	編隊
formation angle	队列角	編隊角
formation axis	队列轴	編隊軸
formation bearing	队列方位	編隊方位
formation coefficient	驳船编队系数	編隊係數
formation line	队列线	編隊線
for scheduled broadcast	定时广播	定時廣播
forward	发出	發出
forward back spring leading aft	艏向后倒缆	艏向後倒纜
forward bow spring	助艏缆	前艏纜
forward error correction	前向纠错	前向偵錯
forward error correction mode	前向纠错方式	前向偵錯方式
forwarding agent	货运代理人	轉運代理人
forward-leading springs	向前倒缆	向前斜出各纜
forward ship	前行舰	正前方船
foul anchor	链缠锚	錨障
foul berth	不良锚地	不良船席
foul bill of lading	不清洁提单	不潔載貨證券
fouled hull propeller curve	污底螺旋桨特性曲线	污船體螺槳特性曲線
foul ground	险恶地	障礙地區
fouling	污底	積垢
fouling hawse	锚链绞缠	纏鏈
fouling resistance	污底阻力	積垢熱阻
foul water	危险水域	航行危險水域
foul wind	逆风	逆風
founder	沉没	沈沒

英　文　名	祖国大陆名	台湾地区名
four point bearing	四点方位	四點方位
four stroke diesel engine	四冲程柴油机	四衝程柴油機
fourth engineer	三管轮	三管輪
FPA(=free from particular average)	货物平安险	單獨海損不賠
FPP(=fixed pitch propeller)	固定螺距桨	[固]定[螺]距螺槳
FPSU(=floating production storage unit)	浮式生产储油装置	浮式生產儲油裝置
fragile cargo	易碎货	易碎貨
frame	①肋骨 ②机架	①肋骨 ②框架
frame frequency	帧频	圓框頻率
frame number	肋骨号数	肋骨編號
frame space	肋距	肋骨間距
franchise	相对免赔额	起賠限額
free alongside ship	船边交货	船邊交貨
freeboard	干舷	乾舷
freeboard deck	干舷甲板	乾舷甲板
freedom of the open seas	公海自由	公海自由
free-fall launching	自由降落下水	自由降落下水
free from particular average(FPA)	货物平安险	單獨海損(不賠)
free gyroscope	自由陀螺仪	自由迴轉儀
free in & out(FIO)	船方不负装卸费用	裝卸自理
free in and out stowed	船方不负责装卸积载费	裝卸堆裝費自理
free of riot and civil commotion	暴动和内乱不保	暴動和內亂不保
free on board(FOB)	离岸价格	出口地船上交貨
free on rail	车上交货	車上交貨
free on truck(=free on rail)	车上交货	車上交貨
free piston gas turbine	自由活塞燃气轮机	自由活塞燃氣渦輪機
free pratique	检疫证书	檢疫完成
free surface	自由液面	自由液面
free surface correction	自由液面修正值	自由液面修正值
free surface effect	自由液面效应	自由液面效應
free vibration	自由振动	自由振動
freezing point	凝点	凝固點
freight	①运费 ②货运	①運費 ②貨運
freight contractor	货运签约人	貨運簽約人
freighter	货船	貨船
freight insurance	运费保险	運費保險
freight manifest	运费清单	運費清單
freight rate	运费率	運費率

英　文　名	祖国大陆名	台湾地区名
freight to collect	到付运费	到付運費
freight tons	运费吨	運費噸,載貨容積噸,貨物噸
freon	氟利昂	氟氯烷冷凍劑
frequency assignment	频率指配	頻率指配
frequency band	频带	頻帶
frequency diversity	频率分集	頻率分集
frequency division system	频率分隔制	頻率分隔制
frequency list	频率表	頻率表
frequency shift keying(FSK)	频移键控	移頻按鍵[制]
frequency shift telegraphy	移频电报	移頻電報
frequency standard	频率标准	頻率標準
frequency tolerance	频率容限	頻率容許差度
fresh air	新风	新鮮空氣
fresh breeze	5 级风	清勁風(5 級)
freshen the nip	防磨损换位	索位掉頭
fresh water circulating pump	淡水循环泵	淡水循環泵
fresh water filling pipe	淡水注入管	淡水注入管
fresh water load line	淡水载重线	淡水載重線
fresh water pump	淡水泵	淡水泵
fresh water system	淡水系统	淡水系統
frictional resistance	摩擦阻力	摩擦阻力
frigate	护卫舰	巡防艦
front	锋	鋒
frontal passage	锋面过境	鋒過境
Froude's number	弗劳德数	佛勞數
frozen cargo	冷冻货	冷凍貨
FRP(=Federal Radionavigation plan)	联邦无线电导航计划	聯邦無線電導航計劃
fruit carrier	水果船	青果船
FSK(=frequency shift keying)	频移键控	移頻按鍵[制]
FTL(=full truck load)	整车货	整車貨物
fuel cladding	燃料包壳	燃料包蓋
fuel consumption	燃油消耗量	燃料消耗量
fuel consumption per ton n mile	每吨海里燃油消耗量	每噸海里燃油消耗量
fuel element	燃料元件	燃料元素
fueling-at-sea station	海上补油站	加油整補站
fuel injection advance angle	喷油提前角	噴油提前角
fuel injection valve cooling pump	喷油器冷却泵	燃料噴射閥冷卻泵

英 文 名	祖国大陆名	台湾地区名
fuel injector	喷油器	燃油噴射器
fuel oil	燃油	燃油
fuel oil drain system	燃油泄放系统	燃油排洩系統
fuel oil filling pipe	燃油注入管	燃油注入管
fuel oil heater	燃油加热器	燃油加熱器
fuel oil homogenizer	燃油均化器	燃油均化器
fuel oil preheating chart	燃油预热图	燃油預熱圖
fuel oil purifying system	燃油净化系统	燃油淨化系統
fuel oil transfer pump	燃油输送泵	燃油輸送泵
fuel oil transport system	燃油驳运系统	燃油轉駁系統
fuel oil viscosity-temperature diagram	燃油黏-温图	燃油黏度溫度圖
fuel supply advance angle	供油提前角	供油提前角
fuel valve opening pressure	喷油器启阀压力	燃油閥開啓壓力
full and bye	满帆通风	滿帆逆戧
full and complete cargo	满载货	滿載貨
full and down	满舱满载	完全滿載
full built-up crankshaft	全组合曲轴	全組合曲軸
full cargo	满舱货	整船貨
full carrier emission	全载波发射	全載波發射
full container load(FCL)	整箱货	全貨櫃之貨物
full container ship	全集装箱船	全貨櫃船
full dress	挂满旗	掛滿旗
full floating gudgeon pin	全浮动式活塞销	全浮動式活塞銷
full landing area mark	全降区标志	全降區標誌
full load displacement	满载排水量	滿載排水量
full moon	满月	滿月,望月
full operational status	完全运行状态	完全運作狀況
full operation capability	完全运行能力	完全運作能力
full truck load(FTL)	整车货	整車貨物
fully-floating sleeve	全浮动式轴承套筒	全浮動式套筒
fumigation	薰舱	煙燻法,燻蒸法
function	功能	功能
function test	功能试验	功能試驗
fundamental fuel lower calorific value	基准燃油低热值	基本燃料低熱值
funnel mark	烟囱标记	煙囪標記
funnel paint	烟囱漆	煙囪漆
furl	折,卷	收,捲
furnace	炉膛	爐膛

G

英　文　名	祖国大陆名	台湾地区名
GA（=general average）	共同海损	共同海損
gaff	斜桁	［縱帆］斜桁
Galaxy	银河系	銀河系
gale	8级风	大風(8級)
gale warning（GW）	大风警报	大風警報
galley	厨房	廚房
galvanic corrosion	电［化腐］蚀	電流侵蝕
galvanization	①电镀 ②镀锌	①電鍍 ②鍍鋅
gang board	跳板	跳板
garbage	垃圾	垃圾
garbage boat	垃圾船	垃圾船
gas analysis	烟气分析	氣體分析
gas carrier	气体运输船	氣體載運船
gas distribution system for diver	潜水供气系统	潛水供氣系統
gas-dust cloud	气尘云	氣塵雲
gaseous fuel	气态燃料	氣體燃料
gases dissolved under pressure	加压溶解气体	加壓溶解氣體
gas-freeing	驱气	消除油氣,清除有害氣體
gas generator	燃气发生器	氣體發生器
gasoline	汽油	汽油
gas turbine ship	燃气轮机船	燃氣渦輪機船
gas welding	气焊	氣焊
gate valve	闸阀	閘閥
gauge pressure	表压力	表壓力
GB（=gyrocompass bearing）	陀罗方位	電羅經方位
GC（=gyrocompass course）	陀罗航向	電羅經航向
GCB（=great circle bearing）	大圆方位	大圓方位
GCC（=great circle course）	大圆航向	大圓航向
gear drive	齿轮传动［装置］	齒輪驅動
gear oil	齿轮油	齒輪油
gear pump	齿轮泵	齒輪泵
Geislinger flexible coupling	盖斯林格弹性联轴器	蓋氏可撓聯結器
general arrangement	总布置图	總佈置圖

英 文 名	祖 国 大 陆 名	台 湾 地 区 名
general atmospheric circulation	大气环流	大氣環流
general average(GA)	共同海损	共同海損
general average act	共同海损行为	共同海損行爲
general average adjustment	共同海损理算	共同海損理算
general average adjustment statement	共同海损理算书	共同海損理算書
general average clause	共同海损条款	共同海損條款
general average contribution	共同海损分摊	共同海損分攤
general average deposit	共同海损分摊保证金	共同海損保證金
general average disbursement insurance	共同海损保险	共同海損支付保險
general average expenditure	共同海损费用	共同海損費用
general average loss or damage	共同海损损失	共同海損損失
general average sacrifice	共同海损牺牲	共同海損犧牲
general average security	共同海损担保	共同海損擔保
general call to all station	普遍呼叫	普通呼叫
general cargo	杂货	雜貨
general cargo ship	杂货船	雜貨船
general certificate	通用证书	通用證書
general chart	总图	總海圖,沿岸通用海圖
general communication	一般通信	一般通信
general communication channels	常规通信信道	正常通信管道
general direction of traffic flow	船舶总流向	船流總向
general lighting	一般照明	一般照明
general operator's certificate	普通操作员证书	普通值機員證書
general radiocommunication	常规无线电通信	一般無線電通信
general service pump	通用泵	通用泵,常用泵
generating set	发电机组	發電機組
generator control panel	发电机[控制]屏	發電機控制屏
generator excited system	发电机励磁系统	發電機激磁系統
Geneva Convention on Territorial Sea and Contiguous Zone	日内瓦领海和毗连区公约	日內瓦領海與鄰接區公約
Geneva Convention on the High Seas	日内瓦公海公约	日內瓦公海公約
gentle breeze	3级风	微風(3級)
geocentric latitude	地心纬度	地心緯度
geodesic	测地线	測地線
geodetic survey	大地测量	大地測量
geographical area group call	地理区域群呼	地理區域群呼
geographic coordinate	地理坐标	地理坐標
geographic information system (GIS)	地理信息系统	地理資訊系統

英　文　名	祖国大陆名	台湾地区名
［geographic］latitude	［地理］纬度	地理緯度
［geographic］longitude	［地理］经度	地理經度
geographic sorting	地理分类	地區劃分
geoid	大地水准面	重力球體
geoidal height map	大地水准面高度图	大地水平面高度圖
geomagnetic equator	地磁赤道	地磁赤道
geomagnetic pole	地磁极	地磁極
geometrical similarity	几何相似	幾何相似
geometric inertial navigation system	几何惯性导航系统	幾何慣性導航系統
geo-navigation	地文航海	地文航海術
geopotential height	位势高度	地勢高度
GEOSAR invalid alert received	接收静止搜救卫星无效报警	接收無效之定置搜救衛星警報
GEOSAR invalid alert transmitted	发射静止搜救卫星无效报警	發射無效之定置搜救衛星警報
GEOSAR traffic ratio	静止搜救卫星通信率	定置搜救衛星通訊比率
geostationary earth orbit	静止地球轨道	定置軌道,地球同步軌道
geostationary meteorological satellite	静止气象卫星	定置氣象衛星
geostationary operational environmental satellite	静止运行环境卫星	定置運作環境衛星
geostationary SAR satellite system	静止搜救卫星系统	定置搜救衛星系統
geostationary satellite（=stationary satellite）	静止卫星	定置衛星
geostationary satellite orbit	静止卫星轨道	定置衛星軌道
geostationary satellite service	静止卫星业务	定置衛星業務
geostrophic wind	地转风	地轉風
German Lloyd's	德国船级社	德國驗船協會
GHA（=Greenwich hour angle）	格林尼治时角	格林［威治］時角
ghost signal	假信号	鬼信號
gill net	刺网	刺網
gimballing error	框架误差	水平環誤差
girder depth	桁材深度	縱桁深度
girding	横拖	橫拖
GIS（=geographic information system）	地理信息系统	地理資訊系統
give-way vessel	让路船	避讓船,讓路船舶
glass-fiber lifeboat	玻璃钢救生艇	玻璃纖維救生艇
global coverage mode	全球覆盖方式	全球覆蓋模式

英　文　名	祖国大陆名	台湾地区名
global maritime distress and safety system （GMDSS）	全球海上遇险安全系统	全球海上遇險及安全系統
global-mode coverage	全球覆盖	全球覆蓋
global navigation satellite system （GLONASS）	全球导航卫星系统	全球導航衛星系統
global positioning system（GPS）	全球定位系统	全球定位系統
globe valve	球阀	球閥
GMDSS（＝global maritime distress and safety system）	全球海上遇险安全系统	全球海上遇險及安全系統
GMDSS area	全球海上遇险安全系统区域	全球海上遇險及安全系統區域
GMT（＝universal time）	世界时	世界時
gnomonic projection	日晷投影	日晷投影
goal post	龙门桅	龍門柱
gong	号锣	鑼
goniometer	测角器	量角器,测角器,测向器
good seamanship	良好船艺	良好船藝
goose neck	吊杆转轴	鵝頸
gooseneck bracket	吊杆座	鵝頸形吊桿座
governing stage	调节级	調節級
government telegram	政务电报	公務電報
GPS（＝global positioning system）	全球定位系统	全球定位系統
GPS fix	GPS 船位	全球定位系統船位
GPS time	GPS 时间	GPS 時間
grab	抓斗	抓斗
grab dredger	斗式挖泥船	抓斗挖泥船
graded fairway	等级航道	分級航道
graded region	内河分级航区	分級航區
gradient of position line	位置线梯度	位置線梯度
gradient wind	梯度风	梯度風
grain	谷物	穀類
grain capacity	散装容积	散裝容積
grain fitting	谷物防动装置	穀類防動裝置
grain-tight	谷密	穀密
grain transverse volumetric upsetting moment	谷物横倾体积矩	穀類體積橫傾力矩
grain upsetting arm	谷物倾侧力臂	穀類傾側力臂
granny knot	错平结,祖母结	假平結

英　文　名	祖国大陆名	台湾地区名
grating hatch	格子舱盖	格子窗口
gravity anomaly chart	重力异常图	重力異常圖
gravity davit	重力式吊艇架	重力吊架(小艇)
gravity disc	比重环	比重盤
gravity feed system	重力式供给系统	重力供給系統
gravity forced-feed oiling system	压力–重力式滑油系统	重力進給潤滑系統
gravity mooring tower	重力系泊塔	重力繫泊塔
gravity oil tank	重力油柜	重力油櫃
gravity platform	重力式平台	重力式鑽油台
gravity tank	重力柜	重力櫃
great circle	大圆	大圈
great circle bearing	大圆方位	大圓方位
great circle chart	大圆海图	大圓海圖
great circle course	大圆航向	大圓航向
great circle distance	大圆距离	大圓距離
great circle sailing	大圆航线算法	大圓航法
great dipper	北斗七星	北斗七星
greatest brilliancy	最亮(金星)	最亮(金星)
greatest eastern elongation	东大距(行星)	東大距(行星)
greatest western elongation	西大距(行星)	西大距(行星)
green flash	绿闪光	綠閃光
green star-signal	绿星信号	綠星信號
green water	舾上浪	覆浪
Greenwich hour angle(GHA)	格林尼治时角	格林[威治]時角
Greenwich mean time	格林尼治平时	格林威治平均時
Greenwich meridian	格林尼治子午线	格林[威治]子午線
Greenwich sidereal time(GST)	格林尼治恒星时	格林[威治]恒星時
greywater	灰水	洗滌水
GRI(=group repetition interval)	组重复周期	組重複週期
gridded chart	有坐标格网海图	有格海圖
grid navigation	格网航法	方格航法
grid variation	格网偏差	方格偏差
gross error	粗差	人爲誤差
gross tonnage(GT)	总吨位	總噸位
groundage	停泊费,滞留费	船舶滯留費,碇泊費,入港稅
ground-based navigational system	陆基导航系统	陸基導航系統
ground detecting lamp	接地检查灯	接地檢查燈

英　文　名	祖国大陆名	台湾地区名
ground rope	沉子纲	沈子網
ground search and rescue processor	地面搜救处理器	地面搜救處理器
ground segment	地面部分	地面部分
ground segment operator	地面部分操作员	地面部分運作者
ground station	地面台	地面台
ground swell	海涌(台风或地震引起)	長浪,激湧
ground tackle	锚设备	錨具
ground wave	地波	地波(無線電)
ground wave to sky wave correction	地天波改正量	地波對天波修正值
group alarm	组合报警(**基群报警**)	群警報
group call broadcast service	群呼广播业务	群呼廣播業務
group coast station call identity	海岸电台群呼识别	海岸電台群呼識別
group flashing light	联闪光	聯閃光
group occulting light	联暗光	聯頓光
group of waves	群波	波群
group repetition interval(GRI)	组重复周期	組重複週期
group ship station call identity	船舶电台群呼识别	船舶電台群呼識別
group velocity	群速	(波)群速度
GST(=Greenwich sidereal time)	格林尼治恒星时	格林[威治]恒星時
GT(=gross tonnage)	总吨位	總噸位
guarantee engineer	保修工程师	保證工程師,保固技師
guard rail	栏杆	欄桿
gudgeon pin	活塞销,轴头销	軸頭銷
guide book	指南	指南
guide plate	导板	導板
guide vane	导向叶片	導葉片
guiding star	导星	導星
gulf	海湾	海灣
gumming	结胶	膠結
gust	阵风	風陣
guy	稳索	牽索
guy eye	稳索眼板	牽索眼板
GW(=gale warning)	大风警报	大風警報
gymnasium	健身房	運動室
gypsy	绞缆滚筒	絞纜滾筒
gypsy wheel(=sprocket)	链轮	鏈輪
gypsy winch	绞缆机	絞索絞機

英　文　名	祖国大陆名	台湾地区名
gyro	陀螺仪	迴轉儀,迴轉機
gyrocompass	陀螺罗经	電羅經
gyrocompass bearing（GB）	陀罗方位	電羅經方位
gyrocompass course（GC）	陀罗航向	電羅經航向
gyrocompass error	陀罗差	電羅經誤差
gyrocompass north	陀罗北	電羅經北
gyro compass room	陀螺罗经室	電羅經室
gyro drift	陀螺漂移	迴轉儀漂移
gyro-magnetic compass	陀螺磁罗经	電磁羅經
gyroscope（=gyro）	陀螺仪	迴轉機,迴轉儀
gyroscopic inertia	定轴性	迴轉慣性
gyroscopic precession	旋进性	迴轉偏移
gyro［scopic］stabilizer	陀螺式减摇装置	迴轉穩定器
gyro sextant	陀螺六分仪	電動水平六分儀
gyro sphere	陀螺球	迴轉球

H

英　文　名	祖国大陆名	台湾地区名
HA（=hour angle）	时角	時角
hack chronometer	次级天文钟	次級天文鐘
hack watch	船表	對時錶
Hague Rules	海牙规则	海牙規則
hair crack	发裂纹	［毛]細裂痕
hair wire	游丝	遊絲
half beam	半梁	半梁
half built-up crankshaft	半组合曲轴	半組合曲柄軸
half-convergency	大圆改正量	半輻合角
half flood	半涨潮	半漲潮
half height container	半高箱	半高櫃
half hitch	半结	半套結
half lethal concentration	半致死浓度	半致死濃度
half lethal dose	半致死剂量	半致死劑量
half moon	半月	半月
half tide	半潮	半潮
half-tide level	平均潮面	中潮海平面
half tide rock	半潮礁	半潮礁
halon fire extinguishing system	卤化物灭火系统	海龍滅火系統

英　文　名	祖国大陆名	台湾地区名
Hamal	娄宿三(白羊 α)	娄宿三(白羊 α)
Hamburg Rules	汉堡规则	漢堡規則
hand expansion valve	手动膨胀阀	手動膨脹閥
hand flare	手持火焰信号	手把火焰信號
hand hole	手孔	手孔
hand line	手钓	手釣
handling	①操作 ②搬运	①操作 ②輸送
hand rail	扶手	扶手
hand starting	手起动	手起動
hand steering gear	手操舵装置	手操舵裝置,笨舵
hanging rudder	悬挂舵	懸舵
harbor	港口	港
harbor area	港区	港區
harbor basin	港池	港池
harbor boat	港作船	港勤船
harbor boundary (=harbor limit)	港界	港界
harbor dues	港务费	港工捐
harbor launch	港作船	港勤船
harbor limit	港界	港界
harbor plan	港泊图	港圖
harbor radar	港口雷达	港口雷達
harbor service kiosk	港口服务处	港務接洽處
harbor speed	港内速度	港內船速
harbor superintendence administration	港务监督	港務監理
harbor survey	港湾测量	港灣測量
harbor watch	停泊值班	碇泊值班
hard down	下风满舵	下風滿舵
hard lay	硬搓[绳]	硬搓索法
hardness number	硬度数	硬度數
hard over	满舵	滿舵
hard-over angle	最大舵角	滿舵角
hard port	左满舵	左滿舵
hard starboard	右满舵	右滿舵
hard up	上风满舵	上風滿舵
harmful cargo	有害货	有害貨物
harmful interference	有害干扰	有害干擾
harmful substance	有害物质	有害物質
harmful substance report （HS）	有害物质报告	有害物質報告

英 文 名	祖 国 大 陆 名	台 湾 地 区 名
harmonic analysis	谐波分析	諧波分析
harmonic emission	谐波发射	諧波發射
harmonic vibration	谐波振动	諧波振動
harpoon gun	扑鲸枪	捕鯨槍
hatch	舱口	艙口
hatch batten	舱口压条	艙口壓條
hatch battening arrangement	封舱设备	艙口壓緊裝置
hatch beam	舱口梁	艙口梁
hatch boom	舱口吊杆	艙口吊桿
hatch coaming	舱口围板	艙口緣圍
hatch cover	舱盖	艙蓋
hatch cover driving device	舱盖曳行装置	艙蓋驅動設施
hatch cover［handling］winch	舱口盖绞车	艙口蓋絞車
hatch end beam	舱口端梁	艙口端梁
hatchet	手斧	手斧
hatch man	舱口装卸指挥人	艙口裝卸指揮人
hatch shift beam	舱口活动梁	艙口活動梁
hatch survey	舱口检验	艙口檢驗
hatch tarpaulin	舱盖布	艙口蓋布
hatchway（＝hatch）	舱口	艙口
haul away	拉开	曳開
hauling net	起网	曳網
hauling winch	绞机	牽索絞機
hawse pipe	锚链筒	錨鏈筒
hawser	缆	纜,大索,大纜
hazardous areas	危险区域	危險區域
hazardous weather message	危险天气通报	危險天氣通報
haze	霾	霾
Hdg（＝heading）	艏向	艏向
head aft spring	助艉缆	艏向後倒纜
head and stern mark	艏艉导标	艏艉標
heading（Hdg）	艏向	艏向
heading marker	艏标志	艏標誌
headland（＝cape）	岬角(岬)	岬,岬角
head light	艏灯	艏燈
head line	艏缆	艏纜
head-on situation	对遇局面	迎艏正遇情況
head reach	纵向冲距	縱向衝距

英　文　名	祖国大陆名	台湾地区名
head sea	顶头浪	頂頭浪,逆浪
head up	艏向上	艏向上
head wind	顶头风	頂頭風,逆風
heat balance	热平衡	熱量平衡表
heat engine	热机	熱機
heat fatigue	热疲劳	熱疲勞
heat fatigue cracking	热疲劳裂纹	熱疲勞裂痕
heat-humidity ratio	热湿比	熱濕比
heating container	加热箱	加熱櫃
heating steam	加热蒸汽	加熱蒸汽
heating water ratio	加热水倍率	加熱水率
heat insulating material	绝热材料	絕熱材料
heat output of reactor	堆热功率	反應器熱功率
heat rate	耗热率	耗熱率
heat-resistant paint	防热漆	防熱漆
heat resisting material	耐热材料	耐熱材料
heave	起伏	起伏
heave the lead	打水砣	打水砣
heave to	滞航	頂風緩航
heaving	垂荡	[船身]起伏
heaving line	撇缆	撇纜
heaving line slip knot	撇缆活结	撇纜索活結
heavy and lengthy cargo carrier	重大件运输船	重大件貨運輸船
heavy derrick(=jumbo boom)	重吊杆	重吊桿
heavy fuel oil	重质燃料油	重燃油
heavy-lift cargo	超重货	笨重貨
heavy-lift charge	超重货装卸费	逾重貨附加費
heavy lift derrick cargo winch	重吊起货机	重貨起重機
heavy rainstorms	暴风雨	暴風雨
heavy weather	恶劣天气	惡劣氣候
heavy weather navigation operating mode management	大风浪航行工况管理	惡劣氣候航行作業形式管理
hectopascal	百帕	百巴斯噶
heel block	[吊杆]跟部滑车,艉龙骨墩	[吊桿]跟部滑車
heeling adjustor	倾差仪	傾差儀
heeling deviation	倾斜自差	傾側自差
heeling error(=heeling deviation)	倾斜自差	傾側自差

英 文 名	祖 国 大 陆 名	台 湾 地 区 名
heeling error instrument(=heeling adjustor)	倾差仪	傾差儀
heeling moment	横倾力矩	傾側力矩
height above the hull	船体以上高度	［號燈］距船身高度
height clearance(=air draft)	净空高度	淨空高度
height difference	潮高差	潮高差
height of center of buoyancy	浮心高度	浮心高
height of center of gravity	重心高度	重心高
height of eye	眼高	眼高
height of tide	潮高	潮高
height of wave and swell combined	合成波高	合成波高
height rate	潮高比	潮高比
helical flow pump	旋涡泵	旋流泵
helicopter operation	直升机作业	直升機作業
helicopter rescue strop	直升机救生套	直升機救生環索
heliometer	日光仪	量日儀
heliostat	定日镜	定日鏡
helium-oxygen diving	氦氧潜水	氦氧潛水
helm angle	舵角	舵角
helm order	舵令	舵令
hermetically sealed refrigerating compressor unit	全封闭式制冷压缩机	全封閉冷凍壓縮機
hermetic deck valve	密封甲板阀	封密甲板閥
HF communication	高频通信	高頻通信
HHW(=higher high water)	高高潮	較高高潮(一日中)
hierarchical analysis	层次分析	層次分析
high and low pressure relay	高低压继电器	高低壓繼電器
higher high water(HHW）	高高潮	較高高潮(一日中)
higher low water	高低潮	較高低潮(一日中)
high focal plane buoy	高灯芯浮标,大型浮标	
high holding power anchor	大抓力锚	高抓著力錨
highline rig	高架索补给装置	高架索補給裝置
highline winch	高架索绞车	高線絞機
high precision positioning system	高精度定位系统	高精準定位系統
high ［pressure]	高压	高氣壓
high sea	狂浪(7 级)	高浪
high seas	公海	公海
high seas maritime safety information	公海海上安全信息	公海海上安全資訊

英　文　名	祖国大陆名	台湾地区名
high speed craft	高速艇筏	高速艇筏
high speed diesel engine	高速柴油机	高速柴油機
high swell	强涌	強湧
high temperature corrosion	高温腐蚀	高溫腐蝕
high velocity induction air conditioning system	高速诱导空气调节系统	高速誘導空調系統
high water(HW)	高潮	高潮
high water time	高潮时	高潮時
Himalaya clause	喜玛拉雅条款	喜瑪拉雅條款
hinged door	铰链门	鉸鏈[式]門
hiring of ship	船舶租赁	船舶租賃
history of marine navigation(=nautical history)	航海史	航海史
HLW(=higher low water)	高低潮	較高低潮(一日中)
h-mode vibration	H 波型振动	H 型振動
hogging	舯拱	舯拱
hogshead	大桶(合63加仑)	液量單位(合63加侖)
hoisting signal	吊升信号	吊升信號
hoisting sling	吊货索环	起吊索具
hold	①货舱 ②握住	①貨艙 ②挽住
holder of bill of lading	提单持有人	載貨證券持有人
holding down bolt	地脚螺栓	壓緊螺栓
holding power of anchor	锚抓力	錨抓著力
holding power of chain	链抓力	鏈抓著力
holding tank	集污舱	貯留艙
hold what you've got	抓到就打住	打住,拉平各纜
home port(=port of registration)	船籍港	船籍港
homeward voyage	返航	返航
homing	①追踪 ②引导	①追蹤 ②導向
homing behavior	回归习惯	回歸習性
homing signal	引航信号	引領信號
homogeneous cargo	均质货	勻質貨
homogenizer	均质器	均質機
hook	钩	鉤
hook cycle	钩吊周期	吊鉤週期
hooper barge	开底泥驳	開底泥駁
hooper dredger	开底挖泥船	斗式挖泥船
horizon	地平	地平

英 文 名	祖 国 大 陆 名	台 湾 地 区 名
horizontal beam width	水平波束宽度	水平波束寬度
horizontal coordinate system	地平坐标系	水平坐標制
horizontal parallax	地平视差	地平視差
horizontal positioning of lights	号灯水平位置	號燈水平位置
horizontal rudder	水平舵	橫舵
horizontal sector	水平光弧	水平弧區
horizontal working range	水平工作范围	水平工作範圍
horse latitude	副热带无风带	馬緯度
horsepower limiter	马力限制器	馬力限制器
hose	软管	軟管
hose nozzle	软管喷嘴	軟管噴嘴
hose test	冲水试验	沖水試驗,射水試驗
hose tie rack	软管固定架	油管固定柵架
hospital ship	医院船	醫院船
hot brine defrost	热盐水融霜	熱鹽水除霜
hot gas defrost	热气融霜	熱氣除霜
hot shut-down	热停堆	熱關閉
hot starting	热态起动	熱起動
hot water circulating pump	热水循环泵	熱水循環泵
hot water heating system	热水供暖系统	熱水加熱系統
hot water tank	热水柜	熱水箱,熱水櫃
hot well	热水井	熱阱,熱井
hour angle(HA)	时角	時角
hour circle	时圈	時圈
hours of service	工作时间	工作時間
housefall rig	联动补给装置	聯動補給裝置
house flag	公司旗	公司旗
housing anchor	收锚	收錨
hovercraft	气垫船	氣墊船
HS(=harmful substance report)	有害物质报告	有害物質報告
hull construction	船体结构	船身構造
hull insurance	船壳险	船身保險
hull maintenance	船体保养	船體保養
hull resistance characteristic	船舶阻力特性	船體阻力特性
hull war risk	船舶战争险	船舶戰爭險
humidifier	加湿器	給濕器
humidity	湿度	濕度
hummock	冰丘	冰丘

英 文 名	祖 国 大 陆 名	台 湾 地 区 名
hummocked ice	堆积冰	圆丘冰
hurricane	12级风	颱風(12级)
HW(=high water)	高潮	高潮
hybrid navigation system	混合导航系统	混合導航系統
hydrant	消防栓	消防栓,水龍頭
hydraulic accumulator	液压蓄能器	液壓儲蓄器
hydraulic actuating gear	液压执行机构	液力致動機構
hydraulically actuated exhaust valve mech-anism	液压式排气阀传动机构	液力致動排氣閥機構
hydraulic amplifier	液压放大器	液壓放大器
hydraulic blocking device	液压锁闭装置	液力鎖閉設施
hydraulic booster	液压升压器	液力增壓器
hydraulic brake	液压制动器	液壓靭,液壓煞俥,液力剎
hydraulic buffer	液压缓冲器	液力緩衝器
hydraulic cargo winch	液压起货机	液壓起貨機
hydraulic control non-return valve	液控单向阀	液控止回閥
hydraulic control valve	液压控制阀	液壓控制閥
hydraulic directional control valve	液压换向阀	液力方向控制閥
hydraulic efficiency	水力效率	液力效率
hydraulic exhaust valve rotation system	排气阀液压旋转系统	排氣閥液壓轉動系統
hydraulic [friction] clutch	液压离合器	液壓離合器
hydraulic governor	液压式调速器	液力調速器
hydraulic hatch cover	液压舱盖	液壓艙口蓋
hydraulic joint	液压接头	液壓接頭
hydraulic lock	液压锁	液壓鎖
hydraulic moment converter(=hydraulic moment variator)	液压变矩器	液壓變矩器
hydraulic moment variator	液压变矩器	液壓變矩器
hydraulic motor	液压[油]马达	液壓馬達
hydraulic oil	液压油	液壓油
hydraulic oil tank	液压油柜	液壓油櫃
hydraulic operated cargo valve	液压操纵货油阀	液力操作液貨閥
hydraulic operated valve	液压操纵阀	液力操縱閥
hydraulic pressure test	液压试验	液壓試驗
hydraulic pump	液压泵	液力泵
hydraulic reduction gear	液压减速[传动]装置	液壓減速裝置
hydraulic servo-motor	液压伺服马达	液力伺服馬達

英 文 名	祖 国 大 陆 名	台 湾 地 区 名
hydraulic servo valve	液压伺服阀	液力伺服閥
hydraulic steering engine	液压舵机	液壓操舵裝置
hydraulic system	液压系统	液壓系統
hydraulic telemotor	液压遥控传动装置	液壓遙控裝置
hydraulic top bracing	液压上支撑	液壓上支撐
hydraulic transmission [drive]	液压传动	液力傳動
hydraulic [transmission] gear	液压传动装置	液力傳動機構
hydraulic transmitter	液压发送器	液壓傳動器
hydraulic variable speed driver	液压变速[传动]装置	液力變速驅動裝置
hydro-blasting	高压水冲洗	高壓水冲洗
hydro cylinder	液压缸	液壓缸
hydrodynamic force	水动力	流體動力
hydrodynamic force coefficient	水动力系数	流體動力係數
hydrodynamic lubrication	液体动力润滑	流體動力潤滑
hydrofoil craft	水翼艇	水翼船
hydrographic survey	海道测量	海道測量
hydrography	①海道测量学 ②水文学	①海道測量學 ②水文學,水理學
hydro jet boat	喷水推进船	噴水推進船
hydrophone intercommunicator	水声对讲机	水聲對講機
hydrostatic curves plan	静水力曲线图	静水[性能]曲線圖
hydrostatic lubrication	液体静力润滑	流體静力潤滑
hydrostatic release unit	静水压力释放器	静力釋放裝置
hygrograph	湿度计	濕度計
hygrometer	湿度表	濕度表
hyperbolic navigation system	双曲线导航系统	雙曲線導航系統
hyperbolic position line	双曲线位置线	雙曲線位置線
hypothecation	[船舶或船货]抵押	船舶或船貨抵押
hypothetical outflow of oil	假想出油量	假想油流出量

I

英 文 名	祖 国 大 陆 名	台 湾 地 区 名
ice anchor	冰锚	繋冰錨,冰錨
ice atlas	冰况图集	冰況地圖
ice avalanche	冰崩	冰崩
ice barrier	冰封地带	冰障
icebound	冰困	冰封

英　文　名	祖 国 大 陆 名	台 湾 地 区 名
ice boundary	冰区界限线	冰區界線
icebreaker	破冰船	破冰船
icebreaker bow	破冰[型]艏	破冰型艏
icebreaker stem (=icebreaker bow）	破冰[型]艏	破冰型艏
ice chart	冰图	冰圖
ice clause	冰冻条款	冰凍條款
ice cover	冰盖	覆冰[量]
ice covered area	冰封区域	冰封區域
ice crusher	碎冰机	碎冰機
ice edge	冰缘线	冰緣線
ice fathometer	回声测冰仪	測冰儀
ice field	冰原	冰原,冰野
ice free	无冰区	無冰區
ice free port	不冻港	不凍港
ice fringe	沿岸冰带	冰繐
ice hold	冰舱	冰艙
ice navigation	冰区航行	冰區航行
ice patrol service	冰情巡逻服务	冰區巡邏服務
ice period	冰冻期	冰凍期
ice report	冰况报告	冰況報告
ice rind	冰壳	脆冰殼
ice shelf	冰架	連岸冰,冰灘
ice strengthening	抗冰加强	抗冰加強
ice thickness	冰厚	冰層厚度
ice warning	冰情警报	冰情警報
icing	结冰	結冰
ID(=identify）	识别	識別
ideal cycle	理想循环	理想循環
identification	辨识	辨識
identification data	识别数据	識別數據
identification of Loran ground and sky waves	罗兰天地波识别	羅遠天地波識別
identifier	识别器	識別器
identify	识别	識別
identity signal	识别信号	識別信號
idle speed	空转转速	惰速[率]
idle time	惰转时间	空轉時間
idling gear	惰轮,中间轮	惰輪

英 文 名	祖 国 大 陆 名	台 湾 地 区 名
IG(=inert gas)	惰性气体	惰性氣體
ignition point	燃点	點火點
ignition quality	发火性能	點火性
IGS(=inert gas system)	惰性气体系统	惰[性]氣[體]系統
illegitimate last voyage	最后不合法航次	最後違法航程
immediate cause	直接原因	直接原因
immediate danger	紧急危险(**紧迫危险**)	立即危險
immersion pump(=submersible pump)	潜水泵	潛水泵
immersion ratio of propeller	螺旋桨浸深比	螺槳浸深比
immersion suit	救生服	浸水衣
immunity of state-owned vessel	国有船舶豁免权	國有船舶豁免權
IMO(=International Maritime Organization)	国际海事组织	國際海事組織
IMO hazard class	国际海事组织危险类别	國際海事組織危險類別
impact	冲击	衝擊
impact load	冲击负载	衝擊負載
impeller(=blade wheel)	叶轮	葉輪
imperfect combustion	不完全燃烧	不完全燃燒
impingement corrosion	冲击腐蚀	衝擊腐蝕
important load	重要负载	重要負載
import permit	进口许可	輸入許可
impulse-reaction turbine	冲动–反动式涡轮机	衝動反動渦輪機
impulse stage	冲动级	衝動輪級
impulse steam turbine	冲动式汽轮机	衝動式汽輪機
impulse turbine	冲动式涡轮机	衝動式渦輪機
inboard	舷内	船内
inboard end chain	末端链节	船内端錨鏈節
inboard plain bearings	内置式平面轴承	内置平面軸承
incident	事故	事故
incident data	事故资料	事故數據
incineration at sea	海上焚烧	海上焚化
incinerator	焚烧炉	焚化爐
inclining test	倾斜试验	傾側試驗
incomplete combustion	燃烧不完全	不完全燃燒
index arm [of sextant]	[六分仪]指标杆	[六分儀]指標桿
index chart	索引图	索引圖,目錄示圖
index correction	指标改正量	指標改正
index error	指标差	器差,指標誤差

英　文　名	祖国大陆名	台湾地区名
index-mirror	指标镜	[六分儀]器鏡
index of adjoining chart	邻图索引	鄰接海圖索引
index of cooperation	合作指数	協同指數
Indian Ocean Region(IOR)	印度洋区	印度洋區域
indicated mean effective pressure	平均指示压力	指示平均有效壓力
indicated power	指示功率	指示功率
indicated specific fuel oil consumption	指示燃油消耗率	指示燃油消耗率
indicated thermal efficiency	指示热效率	指示熱效率
indicator	指示器	指示器
indicator valve	示功阀	指示閥
indirect air cooler	间接冷却式空气冷却器	間接空氣冷卻器
indirect echo	间接回波	間接回波
indirect transshipment	间接换装	間接換裝
indirect transmission	间接传动	間接傳動
induction ratio	诱导比	感應比
induction unit	诱导器	感應器
inert gas(ICT)	惰性气体	惰性氣體
inert gas blower	惰性气体风机	惰氣鼓風機
inert gas generator	惰性气体发生器	惰氣產生器
inert gas smothering system	惰性气体窒息灭火系统	惰氣窒火系統
inert gas system(IGS)	惰性气体系统	惰[性]氣[體]系統
inertia force	惯性力	慣性力
inertial navigation system(INS)	惯性导航系统	慣性導航系統
inertial stopping distance	停车冲程	慣性停俥距離
inertial trial	惯性试验	慣性試驗
inevitable accident	不可避免的事故	不可避免的事故
infectious substance	感染性物质	傳染性物質
inference engine	推理机	推理機
inferior fuel	劣质燃料	劣質燃料
inflammable limit	易燃限度	易燃限度
inflatable appliance	气胀设备	充氣設備
inflatable lifejacket	气胀救生衣	充氣救生衣
inflatable liferaft	气胀[救生]筏	充氣救生筏
inflated appliance	已充气设备	已充氣設備
information flow	信息流	資訊流
information processing	信息处理	資訊處理
information processing machine	信息处理机	資訊處理機
information receiving station	信息接收台	資訊接收台

英 文 名	祖国大陆名	台湾地区名
information sending station	信息发送台	資訊發送台
inhaul [line]	牵索	內牽
inhaul winch	拉进绞车	拉進絞機
inherent frequency	固有频率	固有頻率
inherent vice	固有缺陷	固有缺陷
in iron	顶风停船	頂風停船
initial alignment	初始对准	初始校準
initial classification	初次入级	初次入級
initial course	始航向	起程航向
initial metacenter	初稳心	初定傾中心
initial metacentric height	初稳性高度	初定傾中心高
initial metacentric height above baseline	初稳心高度	初定傾中心在基線以上高度
initial metacentric radius	初稳心半径	初定傾半徑
initial operational capability(IOC)	初步运行能力	初始運轉能力
initial report	初始报告	初步報告
initial stability	初稳性	初穩度
initial stability height	初稳性高度	初定傾[中心]高度
initial steam parameter	初始蒸汽参数	初始蒸汽參變數
initial survey	初次检验	初次檢驗
injection start pressure	启阀压力	噴射啓動壓力
injection station	注入站	注入站
injection timing	喷油正时(喷油定时)	噴油定時
injector testing equipment	喷油器试验台	噴油器試驗設備
inland depot	内陆站	內陸倉庫
inland navigation	内河航行	內河航行
inland rules	内河航行规则	內河航行規則
inland vessel(=river boat)	内河船	內河船
inland waterway	内陆水道	內陸水道
inland waterway navigation aids	内河航标	內陸水道助航標誌
inland waterway navigation and pilotage	内河引航	內水領航
inlet advance crank angle	进气提前角	進氣提前角
inlet lag crank angle	进气滞后角	進氣滯後角
inlet valve	进气阀	進氣閥
in-line diesel engine	直列式柴油机	直列式柴油機
INMARSAT(=international maritime satellite system)	国际海事卫星系统	國際海事衛星系統
INMARSAT A ship earth station	国际海事卫星 A 船舶	國際海事衛星 A 船舶

英 文 名	祖国大陆名	台湾地区名
	地球站	電台
INMARSAT B ship earth station	国际海事卫星 B 船舶地球站	國際海事衛星 B 船舶電台
INMARSAT CES(=INMARSAT coast earth station)	国际海事卫星海岸地球站	國際海事衛星海岸電台
INMARSAT coast earth station(INMARSAT CES)	国际海事卫星海岸地球站	國際海事衛星海岸電台
INMARSAT C ship earth station	国际海事卫星 C 船舶地球站	國際海事衛星 C 船舶電台
INMARSAT land earth station (INMARSAT LES)	国际海事卫星陆地地球站	國際海事衛星陸地電台
INMARSAT LES(=INMARSAT land earth station)	国际海事卫星陆地地球站	國際海事衛星陸地電台
INMARSAT mobile number	国际海事卫星移动号码	國際海事衛星行動碼
INMARSAT M ship earth station	国际海事卫星 M 船舶地球站	國際海事衛星 M 船舶電台
INMARSAT network coordination station	国际海事卫星网络协调站	國際海事衛星網路協調站
INMARSAT SES(=international maritime satellite ship earth station)	国际海事卫星船舶地球站	國際海事衛星船舶電台
inner bottom	内底	內底
inner bottom construction plan	内底结构图	內底結構圖
inner bottom longitudinal	内底纵骨	內底縱材
inner bottom plate	内底板	內底板
inner sea	内海	內海
input axis	输入轴	輸入軸
INS(=inertial navigation system)	惯性导航系统	慣性導航系統
inscrutable fault	不明过失	不明過失
in service training	在职培训	在職訓練
inshore current	近岸流	近岸流
inshore traffic zone	沿岸通航带	沿岸通航區
inside admission	内进汽	內側進汽蒸汽機
inside micrometer	内径千分尺	內分厘卡,內測微計
inside passage	内线航道	內線航道
in sight of one another	互见中	互見中
inspection by boarding	登轮检查	登輪檢查
inspection by notary public	公证鉴定	公證檢定
inspection of chamber	舱室鉴定	艙室檢查

英　文　名	祖　国　大　陆　名	台　湾　地　区　名
inspection of hold	货舱鉴定	貨艙檢查
inspection of package	包装鉴定	包裝檢查
inspection of port state control	港口国管理检查	港口國管制檢查
inspection of rat evidence	鼠患检查	鼠患檢查
inspection of tank	液舱鉴定	液艙櫃檢查
inspection of weight	重量鉴定	重量檢定
inspection on hatch and/or cargo	载损鉴定	貨損檢查
instantaneous rate of discharge of oil content	油分瞬时排放率	油份瞬間排洩率
instantaneous state	瞬态	瞬時狀態
in stays	顶风停住	頂風停住
in step	同步	起伏一致
instructional chart	教学海图	教學海圖
instruction decoder	指令译码器	指令譯碼器
instructive mark	指示标志	指示標誌
instrument error	[六分仪]器差	儀器差,儀器誤差
insubmersibility	不沉性	不沈性
insufficiently packed cargo	包装不牢货	包裝不固貨
insulating container	隔热箱	絕熱櫃
insulating material	绝缘材料	絕緣材料
insurance claim	保险索赔	保險索賠
insured	被保险人	被保險人
insured value	保险价值	保險價值
intact stability	完整稳性	完整穩度
intake silencer	进气消音器	進氣消音器
integral regulator	积分调节器	積分調整器
integral shaft generator	同轴发电机	共軸發電機
integrated barge	分节驳船	組合駁船
integrated barge train	分节驳船队	組合駁船隊
integrated mode	组合模式	組合模式
integrated navigation system	组合导航系统	組合導航系統
integrating gyroscope	积分陀螺仪	積分迴轉儀
integration and test	组合与测试	整合及測試
integrity monitor	完善性监测台	整合監視器
intelligent collision avoidance system	智能避碰系统	智慧型避碰系統
intensity(=strength)	强度	強度
intensity of lights	号灯发光强度	號燈照明強度
interaction between ships	船吸效应	船間相互作用

英　文　名	祖国大陆名	台湾地区名
intercardinal point	隅点	象限點
interchangeable parts	可互换部件	可互換配件
INTERCO(=international code symbol)	国际信号码组符号	國際信號代碼符號
intercommunication telephone set	对讲电话机	電話對講機
interface	界面	界面,介面
interference between longer and shorter circle path signals	长短大圆信号干扰	長短大圓信號干擾
interlock protection of shore power connection	岸电联锁保护	岸電聯鎖保護
intermediate bearing	中间轴承	中間軸承
intermediate casing	中间壳体	中間殼體
intermediate fuel oil	中间燃料油	中間燃油
intermediate point	三字点	中間點
intermediate point of great circle	大圆分点	大圓中分點
intermediate port	中途港	中途港
intermediate shaft	中间轴	中間軸
intermediate survey	中期检验	中期檢驗
intermodulation products	互调产物	互調產物
internal efficiency	内效率	內效率
internal mounted equipment	舱内安装设备	裝於艙內設備
internal power	内功率	內功率
internal safety audits	内部安全检查	內部安全稽核
internal sea(=inner sea)	内海	內海
internal transfer	船内驳运	船內轉駁
internal waters	内水	內水
internal wave	内波	內波,潛波
International Agreement on the Use of IN-MARSAT ship Earth Station within the Territorial Sea and Ports	关于在领海和港内使用国际海事卫星船舶地面站的国际协议	在領海及港内使用國際海事衛星船舶地球台之國際協約
International Association of Light house Authorities	国际航标协会	國際燈塔協會
International Bulk Chemical Code	国际散装化学品规则	國際散裝化學品章程
international code symbol(INTERCO)	国际信号码组符号	國際信號代碼符號
International Convention for Maritime Search and Rescue	国际海上搜寻救助公约	國際海上搜索與救助公約
International Convention for Preventing Collisions at Sea(SOLAS)	国际海上避碰公约	海上避碰國際公約
International Convention for Safe Contain-	国际集装箱安全公约	安全貨櫃國際公約

英　文　名	祖国大陆名	台湾地区名
ers		
International Convention for Safety of Life at Sea	国际海上人命安全公约	海上人命安全國際公約
International Convention for the Prevention of Pollution from Ships	国际防止船舶造成污染公约	防止船舶污染國際公約
International Convention for the Prevention of Pollution of the Sea by Oil	国际防止海洋油污染公约	防止海水油污染國際公約
International Convention for the Safety of Life at Sea, 1974	1974 年国际海上人命安全公约	一九七四年海上人命安全國際公約
International Convention for the Unification of Certain Rules Concerning the Immunity of State-owned Ships	统一国有船舶豁免某些规定的国际公约	國有船舶豁免權統一規定國際公約
International Convention for the Unification of Certain Rules of Law with Respect to Collision between Vessels	统一船舶碰撞某些法律规定的国际公约	船舶碰撞法律統一規定國際公約
International Convention for the Unification of Certain Rules Relating to Assistance and Salvage of Aircraft or by Aircraft at Sea	统一对水上飞机的海难援助和救助及由水上飞机施救的某些规定的国际公约	救助與撈救海上航空器及由航空器施救之統一規定國際公約
International Convention for the Unification of Certain Rules Relating to Penal Jurisdiction in Matters of Collision or Other Incidents of Navigation	统一船舶碰撞或其他航行事故中刑事管辖权某些规定的国际公约	關於碰撞或其他航行事件之統一刑事管轄國際公約
International Convention for the Unification of Certain Rules Relating to the Arrest of Seagoing Ships	统一海船扣押某些规定的国际公约	統一海船假扣押規定國際公約
International Convention on Certain Rules Concerning Civil Jurisdiction in Matters of Collision	船舶碰撞民事管辖权某些规定的国际公约	關於船舶碰撞事件民事管轄國際公約
International Convention on Civil Liability for Oil Pollution Damage	国际油污损害民事责任公约	油污損害民事責任國際公約
International Convention on Load Lines (LL)	国际载重线公约	國際載重線公約
International Convention on Maritime Liens and Mortgage	船舶优先权和抵押权国际公约	船舶優先權及抵押權國際公約
International Convention on Oil Pollution Preparedness, Response and Cooperation	国际油污防备、响应和合作公约	油污防備、因應與合作國際公約

英 文 名	祖国大陆名	台湾地区名
International Convention on Salvage	国际救助公约	國際救助公約
International Convention on Standard of Training, Certification and Watching-keeping for seafarers, 1978	1978 年海员培训、发证和值班国际公约	一九七八年航海人員訓練、發證及當值標準國際公約
International Convention on Standard of Training, Certification and Watch Keeping for Seafarers	国际海员培训、发证和值班标准公约	航海人員訓練、發證及當值標準國際公約
International Convention on the Establishment of an International Fund for Compensation for Oil Pollution Damage	设立国际油污损害赔偿基金国际公约	設立油污損害國際賠償基金國際公約
International Convention on Tonnage Measurement of Ships	国际船舶吨位丈量公约	船舶噸位丈量國際公約
International Convention Relating to Intervention on the High Seas in Case of Oil Pollution Casualties	国际干预公海油污事故公约	油污染事故在公海行使干涉國際公約
International Convention Relating to Stowaways	国际偷渡公约	處理偷渡人國際公約
International Convention Relating to the Limitation of the Liability of Owners of Seagoing Ships	海船所有人责任限制国际公约	海船所有人責任限制國際公約
International COSPAs-SARSAT Programme Agreement	国际低极轨道搜救卫星计划协议	國際衛星輔助搜救計劃協議
International COSPAs-SARSAT Programme Agreement	国际低极轨道搜救卫星系统计划协定	國際搜救衛星系統計劃協約
international custom and usage	国际惯例	國際慣例
international direct dialing	国际直拨	國際直撥
international DSC frequencies	国际数字选择呼叫频率,国际 DSC 频率	國際數位選擇呼叫頻率
International Hydrographic Organization	国际海道测量组织	國際海道測量組織
international ice patrol bulletin	国际冰况巡视报告	國際冰況巡邏布告
international information service	国际信息业务	國際資訊業務
International Maritime Organization (IMO)	国际海事组织	國際海事組織
International Maritime Organization Class (IMO class)	国际海事组织类号	國際海事組織類號
international maritime satellite	国际海事卫星	國際海事衛星
International Maritime Satellite Organization (INMARSAT)	国际海事卫星组织	國際海事衛星組織

英　文　名	祖 国 大 陆 名	台 湾 地 区 名
international maritime satellite ship earth station (INMARSAT SES)	国际海事卫星船舶地球站	國際海事衛星船舶電台
international maritime satellite system (INMARSAT)	国际海事卫星系统	國際海事衛星系統
International Meteorological Organization	国际气象组织	國際氣象組織
International Mobil Satellite Organization	国际移动卫星组织	國際海事衛星組織
international NAVTEX service	国际航行警告业务	國際航行警告電傳業務
International Oil Pollution Prevention Certificate (IOPP Certificate)	国际防止油污证书	國際防止油污證書
International Organization for Standardization (ISO)	国际标准化组织	國際標準組織
International Pollution Prevention For The Carriage of Noxious Liquid Substances in Bulk	国际防止散装运输有毒液体物质污染证书	國際載運有毒液體物質防止污染證書
International Safety Management Code (ISM Code)	国际安全管理规则	國際安全管理章程
international safety NET	国际安全通信网	國際安全通信網
international sea area	国际海域	國際海域
International Sewage Pollution Prevention Certification (ISPP Certification)	国际防止生活污水污染证书	國際防止污水污染證書
international shore connection	国际通岸接头	國際岸上接頭
international signal code	国际信号码	國際信號代碼碼組
international signal flag	国际信号旗	國際信號旗
interpellation clause	质询条款	解釋條款
interrupt service routine	中断服务程序	中斷服務常規
interrupt system	中断系统	中斷系統
intertropical convergence zone (ITCZ)	热带辐合带	熱帶輻合地帶
intervening obstruction	居间障碍物	居間障礙物
in the wind	在上风,向上风	居上風,頂風
inward turning	内旋	內向旋轉,內旋
IOC(= initial operational capability)	初步运行能力	初始運轉能力
ionosphere	电离层	游離層
ionospheric refraction correction	电离层折射改正	游離層折射修正
IOR(= Indian Ocean Region)	印度洋	印度洋區域
irradiation inspection tube	辐照监督管	照射檢測管
irregular wave	非规则波	不規則波浪
island shelf	岛架	島架
ISO ambient reference condition	国际标准环境状态	國際標準組織環境狀況

英　文　名	祖国大陆名	台湾地区名
isobar	等压线	等壓線
isobaric surface	等压面	等壓面
isogonic chart	等磁差图	等偏差線,等角線圖
isolated danger mark	孤立危险物标志	孤立危險物標誌
isolating method	隔离法	隔離法
isolating switch	隔离开关	隔離開關
isolating valve	隔离阀	隔離閥
isotherm	等温线	等溫,等溫線
ISO viscosity classification	国际标准化组织黏度分级	國際標準組織黏度分級
ITCZ(=intertropical convergence zone)	热带辐合带	熱帶輻合地帶

J

英　文　名	祖国大陆名	台湾地区名
jacket cooling water pump	缸套冷却水泵	缸套冷卻水泵
jacking system	升降系统	頂舉系統
jack-knife	水手刀	水手刀
jack staff	艏旗杆	艏旗桿
jackstay	主钢缆(**撑杆**)	撐桿
jack-up rigs	升降式平台	升降式鑽油台
Japanese Maritime Corporation (NK)	日本船级社	日本海事協會
Jason clause	杰森条款	詹森條款(超額條款)
jet propulsion	喷水推进	噴射推進
jet pump	喷射泵	噴射泵
jetting pipeline	冲桩管线	水沖管路
jettison	抛弃	[海難]投棄,海難投棄貨物
jetty	①突码头 ②栈桥	①突堤,碼頭 ②棧橋
jew's harp	锚环	錨環
joining link	连接链环	連接鏈環
joining shackle	连接卸扣	連接接環
joint and several liability	连带责任	連帶責任
joint inspection	联检	聯檢
jumbo boom	重吊杆	重吊桿
junk	中国帆船	中國帆船
Jupiter	木星	木星
jurisdiction clause	管辖权条款	管轄權條款

英 文 名	祖 国 大 陆 名	台 湾 地 区 名
jurisdiction of ship collision	船舶碰撞管辖权	船舶碰撞管轄權
jurisdiction over ship	船舶管辖权	船舶管轄權
jury rudder	应急舵	應急舵

K

英 文 名	祖 国 大 陆 名	台 湾 地 区 名
katabatic	下降风	頹風
kedge anchor	小锚	小錨,艉錨,小移船錨
keelson	中桁材	內龍骨
keep plate	压紧板	押板
Kelvin scale	绝对温标	熱力學溫度標
kerosene engine	煤油发动机	煤油引擎
kerosene test	涂煤油试验	煤油試驗
key lock switch	锁定开关	鎖定開關
keyway	键槽	鍵槽
kinematic positioning	动态定位	動態定位
kinematic simulation	运动模拟	動力模擬
kinematic viscosity	运动黏度	動力黏度
kinetic energy	动能	動能
king bolt	主螺栓	大螺栓
king post	吊杆柱	主柱
kingpost outrigger	吊杆柱平台	主柱突出枭
king-size	特大的	特大
kn. (=knot)	节	節
knock	爆震	爆震
knot	节	節
knowledge base	知识库	知識庫
knowledge presentation technique	知识表达技术	知識表達技術
Kochab	帝(小熊β)	帝,北極二(小熊β)
Kuroshio	黑潮	黑潮
kW power	有功功率	瓩功率

L

英　文　名	祖国大陆名	台湾地区名
label	标签	標籤
labeling	加标签	標籤
laboratory analysis report	实验室化验报告	實驗室分析報告
labyrinth gland	曲径式密封	曲折填函蓋
lagan	系浮标的投海货物	附有浮標之投海貨物
lamp attracting	灯光诱鱼	燈光誘魚
lamp room	灯具间	燈具室
Lanchester balancer	兰氏平衡器(往复惯性 力矩平衡器)	蘭氏均衡器
land breeze	陆风	陸風
land earth station（LES）	陆地地球站	衛星陸上電台
land engine	陆用发动机	陸用引擎
landfall	初见陆地	初見陸地
landing permit(=shore pass)	登岸证	登岸證
landing ship	登陆舰	登陸艦艇
landing through surf	浅浪登陆	淺浪登陸
land-line charge	陆线费	陸線費
landmark	陆标	岸標,岸上目標
land mobile-satellite service	陆上移动卫星业务	陸上行動衛星業務
land station	陆地电台	陸上[電]台
land type	陆用式	陸用式
lane	巷	公定航道,巷道
lane identification（LI）	巷识别	巷識別
lane identification meter	巷识别计	巷識別計
lane letter	巷号	巷號
lane set	巷设定	巷設定
lane slip	滑巷	滑巷
lane width	巷宽	巷寬
lap	①余面 ②搭边	①餘面 ②搭邊
lap joint	搭接	搭接
large correction chart	改版图	大改正海圖
large location error	大定位误差	定位顯著誤差
laser sounder	激光测探仪	雷射測深儀
LASH(=lighter aboard ship)	载驳船	子母船,浮貨櫃船

英　文　名	祖国大陆名	台湾地区名
lashing	绑扎	捆縛,縛固
lashing bar	绑扎棒	拉繫桿
lashing cable	绑扎索	拉繫索
lashing chain	绑扎链	拉繫鏈
lashing eye	绑扎环	拉繫環,D 型環
lashing hook	绑扎钩	拉繫鉤
lashing plate	绑扎板	拉繫板
lashing pot	绑扎套筒	拉繫缸
lashing rod	绑扎杆	拉繫桿
last opportunity rule	最后机会原则	最後機會原則
last quarter	下弦	下弦
latent defect	潜在缺陷	潛在缺陷
lateral clearance	横向间隙	側向間隙
lateral mark	侧面标志	側面標誌
lateral vibration	横向振动	側向振動
lateral view	侧视图	側視圖
latitude by account	计算纬度	推算緯度
latitude by pole star	北极星求纬度	極星求緯法
latitude correction	纬度改正量	緯度修正值
latitude effect	纬度效应	緯度效應
latitude error	纬度误差	緯度誤差
latitude error corrector	纬度误差校正器	緯度誤差校正器
latitude factor	纬度因数	緯度因數
latitude of vertex	顶点纬度	頂點緯度
lattice mast	桁架桅	格式桅
launching	下水	下水
launching appliance	下水设备	下水設備
launching arrangement	下水装置	下水裝置
launching station	下水站	下水站
laundry	洗衣间	洗衣間
lavatory(=toilet)	盥洗室	盥洗室
law of mariner	船员法	海員法
law of the flag	船旗国法	船旗國法
laydays	①受载期 ②装卸货天数	①到港期限 ②許可裝卸日數
layer corrosion	片蚀	層蝕
laytime	装卸期限	約定裝卸時間
lay up	闲置船	停航

英　文　名	祖国大陆名	台湾地区名
lazaret	隔离病房	隔離病院
l-band EPIRB	l-波段紧急无线电示位标	L频带應急指位無線電示標
l-band EPIRB system	l-波段紧急无线电示位标系统	频带應急指位無線電示標系統
LBP(=length between perpendiculars)	垂线间长	垂標間距
L/C(=letter of credit)	信用证	信用狀
LCL(= less than container load)	拼箱货	不足一個貨櫃之貨物
lead	①导程 ②铅锤	①導程 ②測錘,測深錘
lead-base bearing alloy	铅基轴承合金	鉛基軸承合金
lead base white metal bearing	铅基白合金轴承	鉛基白合金軸承
leading beacon	导标	引導示標
leading block	引导滑车	導滑車
leading edge	导边	導緣(螺槳)
leading line	导航线	導航線
leading ship	前导舰	前導船艦
lead lane	冰间水道	冰間巷道
lead of slide valve	滑阀导程	滑閥導程
leak	泄漏	洩漏
leakage and breakage	漏损和破损	漏損和破損
leakage test	渗漏试验	漏洩試驗
leakprofness	密封性	密封性
leak stopper	堵漏器材	堵漏器材
leak stopping	堵漏	堵漏
leap day	闰日	閏日
leap month	闰月	閏月
leap second	闰秒	閏秒
leap year	闰年	閏年
least significant bit	最低有效位	最低有效位元
leaving bodily	平离	平離
leaving bow first	艏离	艏先離
leaving stern first	艉离	艉先離
leaving wharf	离码头	離碼頭
lee anchor	惰锚	下風錨,不著力錨
lee helmsman	预备舵工	預備舵手
lee shore	下风岸	下風岸
lee side	下风舷	下風側,下風舷
leeway angle	风压差	風壓差角

英　文　名	祖国大陆名	台湾地区名
leeway coefficient	风压差系数	風壓差係數
left flank ship	左翼舰	左翼船艦
left-hand circular polarization	左旋圆极化	左旋圓形極化
left-hand rotation	左向旋转,逆时针旋转	左向旋轉,逆時針旋餞
left-hand rotation diesel engine	左旋柴油机	左旋柴油機
left-hand turning	左旋	左轉
legal time	法定时	法定時
legitimate last voyage	最后合法航次	最後合法航程
length between perpendiculars(LBP)	垂线间长	垂標間距
length-breadth ratio	长宽比	長寬比
length of formation	队形长度	隊形長度
length of tow	拖带长度	拖帶長度
length overall(LOA)	总长	全長
lengthy cargo	长件货	超長貨
LES(=land earth station)	陆地地球站	衛星陸上電台
lesser ebb	较弱落潮(一日中)	較弱落潮(一日中)
lesser flood	较强涨潮(一日中)	較強漲潮(一日中)
less than container load	拼箱货	拼櫃貨
less than truck load(LTL)	拼车货	拼車貨
LES TDM channel	陆地地球站时分多路复 　用信道	衛星陸上電台劃時多制 　頻道
let go anchor	抛锚	抛錨
letter of credit(L/C)	信用证	信用狀
letter of indemnity	保函	賠償責任保證書,賠償 　保證書
letter of subrogation	索赔代理让书	代位求償書
letter pronunciation	字母拼读法	字母拼音
level gauge	水准仪	液位計
level ice	平整冰	平整冰
leveling charge	平舱费	平艙費
leveling of engine bed	机座找平	定機座水平
leveling survey	水准测量	水平測量
level of responsibility	责任级	責任層級
lever of form stability	形状稳性力臂	形狀穩度力臂
lever of weight stability	重量稳性力臂	重量穩度力臂
lever-type hydraulic governor	杆式液压调速器	槓桿式液壓調速器
lex fori	法院地法	審判地法,法院地法 　[庭]

英　文　名	祖国大陆名	台湾地区名
lex loci contractus	从签约地法	從簽約地法
LHA(=local hour angle)	地方时角	當地時角
LHW(=lower high water)	低高潮	較低高潮(一日中)
LI(=lane identification)	巷识别	巷識別
liability insurance	对第三者负责的保险	責任保險
liability of towage	拖带责任	拖帶責任
lien	留置权	留置權
lifeboat	救生艇	救生艇
lifeboat compass	救生艇罗经	救生艇用羅經
lifeboat deck	救生艇甲板	救生艇甲板
lifeboat with self-contained air support system	自供空气救生艇	自供空氣系統救生艇
lifebuoy	救生圈	救生圈
life equipment	救生属具	救生設備
life float	救生浮	救生浮具
lifejacket	救生衣	救生衣
life line	救生索	扶手索,救生索
life raft	救生筏	救生筏
life salvage	人命救助	(船)生命救助費
life saving appliance	救生设备	救生器具
life saving signal	救生信号	救生信號
life support system	生命支持系统	生命支援系統
lift force	升力	升力
lift hatch cover	升降式舱盖	吊式艙蓋
lifting beam	吊梁	吊樑
lifting by floating crane	起重船打捞	起重船起吊
lift on/lift off	吊装	吊上吊下船
lift out piston	吊缸	吊缸
light air	1级风	軟風(1级)
light alloy	轻合金	輕合金
light alloy ship	轻合金船	輕合金船
light beacon	灯桩	燈標,燈浮標
light breeze	2级风	輕風(2级)
light buoy crane	航标起重机	吊航標機
light cargo(=bulky cargo)	轻泡货	輕笨貨(泡貨)
light condition	空载状态	輕載船況
light displacement	空船排水量(**空载排水量**)	輕載排水量

英　文　名	祖国大陆名	台湾地区名
lighted buoy	灯浮标	燈浮標
lighted mark	灯标	燈標
lighter aboard ship(LASH)	载驳船	子母船,浮貨櫃船
lighterage	过驳	轉駁
light-float	船形灯标	船形燈標
light fuel oil	轻质燃料油	輕燃料油
lighthouse	灯塔	燈塔
lighthouse tender	灯塔供应船	燈塔補給船
light press fit	轻压配合	輕壓配合
light-purse seine	光诱围网	燈誘圍網
light range	灯光射程	燈光射程
light refined products	轻质炼制品	輕質精煉油
lights	号灯	號燈
light signal	灯光信号	燈光信號
light-vessel	灯船	燈船
lignum vitae bearing	铁梨木轴承	鐵梨木軸承
limber	污水道	通水小孔
limber board	污水道盖板	通水道蓋板
limitation of liability	责任限制	(船東)責任限制
limited characteristic	限制特性	限制特性
limit error	极限误差	極限誤差
limiting latitude	限制纬度	限制緯度
limiting speed	限速	限制速度
limit mark	界限标	界限標
limit of liability for loss of or damage to luggage	行李损坏赔款限额	行李損失責任限制
limit of liability for personal injury	人身伤亡赔款限额	人身傷亡責任限制
limit of sector	光弧界限	光弧界限
line abreast	横队	橫隊
line arrangement machine	干线理线机	理繩機
line casting machine	干线放线机	放繩機
line fishing boat	钓船	釣船
line formation	行列编队	行列編隊
line hauler	干线起线机	捲繩機
linen locker	卧具储藏室	寢具儲存室
line of equal bearing	恒位线	等方位線
line of position(LOP)	位置线	位置線
liner	班轮	定期船

英　文　名	祖国大陆名	台湾地区名
liner bill of lading	班轮提单	定期船載貨證券
liner conference	班轮公会	定期船公會
liner service	班轮运输	定期船業務
liner term	班轮条款	定期船條件
[liner] wear rate	[缸套]磨损率	缸套磨損率
line throwing appliance	抛绳设备	抛繩器
line throwing gun	撒缆枪	射繩槍,撒纜槍,抛繩槍
line winder	盘线装置	盤繩機
lining	衬料	襯料
link joint	铰链连接	鉸鏈連接
linoleum	油毡	油氈
liquefied chemical gas(LNG)	液化化学气体	液化化學氣體
liquefied gas	液化气体	液化氣體
liquefied gas carrier	液化气船	液化氣體船
liquefied natural gas(LPG)	液化天然气	液化天然氣
liquefied natural gas carrier	液化天然气船	液化天然氣船
liquefied petroleum gas	液化石油气	液化石油氣
liquefied petroleum gas carrier	液化石油气船	液化石油氣船
liquid bulk cargo	液体散货	液體散貨
liquid cargo(=fluid cargo)	液货	液狀貨
liquid cargo ship(=tanker)	液货船	液貨船
liquid chemical tanker	液体化学品船	化學液體船
liquid compass	液体罗经	濕羅經
liquid container	储液缸	貯液櫃
liquid damping vessel	液体阻尼器	液體阻尼器
liquid floated gyroscope	液浮陀螺仪	液浮迴轉儀
liquid lubrication	液体润滑	液體潤滑
liquid pump	液货泵	液貨泵
liquid substance	液体物质	液體物質
liquid-tight	液密	液密
list	表	表
list of coast station	海岸电台表	海岸電台表
list of lights	航标表	燈標表
list of radio signals	无线电信号表	無線電信號表
list of ship station	船舶电台表	船舶電台表
lithium bromide water absorption refrigerating plant	溴化锂吸收式制冷装置	溴化鋰水吸收冷凍裝置
litter	担架	擔架

英　文　名	祖国大陆名	台湾地区名
littoral current (=coastal current)	沿岸流	沿岸流
livestock cargo	活动物货	牲口貨
livestock carrier	牲畜运输船	牲口運輸船
living quarter	住舱	住艙
living resources	生物资源	生物資源
living resources of the sea	海洋生物资源	海洋生物資源
lizard	有眼环的短索	末端附眼環短索
LL (=International Convention on Load Lines)	国际载重线公约	國際載重線公約
Lloyd's Weekly Casualty Report	劳氏海事周报	勞氏海事週報
LLW (=lower low water)	低低潮	較低低潮(一日中)
LMT (=local mean time)	地方[平]时	地方平均時,地方平時
LNG (=liquefied natural gas)	液化天然气	液化天然氣
LOA (=length overall)	总长	全長
load and fouling hull operating mode management	装载和污底工况管理	裝載與污底作業形式管理
load carrying properties	[滑油]承载特性	負載特性
load characteristic	负荷特性	負載特性
load-down program	减负荷程序	減載方案
loaded performance	加载性能	加載性能
load governor	负载调节器	負載調節器
load indicator	负荷指示器	負荷指示器
loading	装载	裝載
loading list (=cargo list)	装货清单	裝貨清單
loading manual	载装手册	裝載手冊
load limit knob	负荷限制旋钮	負荷限制旋鈕
load line	载重线	載重線,載重水線
load line area	载重线区域	載重線區域
load line assignment	载重线勘划	載重線勘劃
load line certificate	载重线证书	載重線證書
load line disc	载重线圆圈	載重線圈
load line mark	载重线标志	載重線標誌
load line survey	载重线检验	載重線檢驗
load on top (LOT)	顶装法	上層積載
load-sharing	负荷分配	負載分配
load-up program	加负荷程序	加載方案
load water line	满载水线	滿載水線
local area augmentation system	本地增强系统	當地增強系統

英 文 名	祖国大陆名	台湾地区名
local area differential GPS	本地差分 GPS	當地差分全球定位系統
local area network	局域网络	局部區域網路
local cargo	区间货	區間貨
local clause	地区条款	地區條款
local control	机旁控制	現場控制
local hour angle(LHA)	地方时角	當地時角
local mean time(LMT)	地方[平]时	地方平均時,地方平時
local-mode coverage	区域覆盖	區域覆蓋
local mode of operation	区域性作业模式	區域性作業模式
local scale	局部比例尺	局部比例尺
local sidereal time(LST)	地方恒星时	當地恒星時
local strength	局部强度	局部強度
local stress	局部应力	局部應力
local time	地方时	地方時
local user terminal(LUT)	本地用户终端	地面終端台
local warning	本地警告	當地警告
local water level	当地水位	當地水位
located alert	已定位报警	經定位之警報
locating	寻位	定位
locating and homing signals	寻位与归航信号	定位與導向信號
locating signal	寻位信号	定位信號
location	位置	部位,位置
location acquisition probability	获得位置概率	獲得位置機率
location grid	定位格架	定位格
log	①计程仪 ②圆木	①計程儀 ②圓形木材
log book	航海日志	航海日誌,航泊日誌
logical valve	逻辑阀	邏輯閥
log in	入网	上網
log out	脱网	下網
log raft	排筏(**木排筏**)	木排
log reading	计程仪读数	計程儀讀數
long flash	长闪光	長閃光
long form bill of lading	长式提单	長式載貨證券
longitude correction	经度改正量	經度修正值
longitudinal bulkhead	纵舱壁	縱艙壁
longitudinal distance of center of buoyancy from midship	浮心距中距离	縱向浮心與舯距離
longitudinal distance of center of floatation	漂心距中距离	縱向浮面中心與舯距離

英　文　名	祖国大陆名	台湾地区名
from midship		
longitudinal distance of center of gravity from midship	重心距中距离	縱向重心與舯距離
longitudinal frame system	纵骨架式	縱肋系統
longitudinal metacenter	纵稳心	縱定傾中心
longitudinal metacentric height	纵稳性高度	縱定傾中心高
longitudinal metacentric height above baseline	纵稳心高度	縱定傾中心在基線以上高度
longitudinal metacentric radius	纵稳心半径	縱定傾半徑
longitudinal section plan	纵剖面图	縱剖面圖
longitudinal shifting board	纵向止移板	縱防動板
longitudinal stability	纵稳性	縱穩度
longitudinal stability height	纵稳性高度	縱定傾[中心]高度
longitudinal stability lever	纵稳性力臂	縱穩度力臂
longitudinal strength	纵[向]强度	縱向強度
longitudinal vibration damper	轴向减振器	縱向振動阻尼器
long line	延绳钓	延繩釣,長繩釣
long-range scanning	远程扫描	長距程掃描
long splice	长[插]接	長接
long-stroke diesel engine	长行程柴油机	長衝程柴油機
long time delay	长延时	長延時
long ton	长吨	長噸
look-out	瞭望	瞭望
look-out on forecastle	瞭头	在艏瞭望
loop alignment error	环形天线装调误差	環形天線對準誤差
loop antenna	环形天线	環形天線
loop scavenging	回流扫气	環狀驅氣
loose fit	松配合	鬆配合
loose hardware	可卸硬件	可卸艤品
loose propeller blade	可卸螺旋桨叶	可卸螺槳片
LOP(=line of position)	位置线	位置線
Loran	罗兰	羅遠儀
Loran-A	罗兰 A	羅遠 A
Loran-A receiver	罗兰 A 接收机	羅遠 A 接收機
Loran-C	罗兰 C	羅遠 C
Loran-C alarm	罗兰 C 告警	羅遠 C 警報
Loran chart	罗兰海图	羅遠海圖
Loran-C receiver	罗兰 C 接收机	羅遠 C 接收機

英　文　名	祖国大陆名	台湾地区名
Loran fix	罗兰船位	羅遠定位
Loran position line	罗兰位置线	羅遠位置線
Loran table	罗兰表	羅遠表
loss of coolant accident	失水事故	冷卻水損失事故
LOT(=load on top)	顶装法	上層積載
lot cargo	成批货	整批貨,大宗貨
loud speaker signaling	扬声器通信	揚聲器通信
lounge	休息室	休息室
low-duty	轻型	輕型
low earth orbit	低地球轨道	低空地球軌道(低空軌道)
low Earth orbit SAR satellite system	低轨道搜救卫星系统	低軌道搜救衛星系統
lower branch of meridian	子圈	子午線下半部
lower calorific value	低热值	低熱值
lower culmination	下中天	[天體]下中天
lower deck	下甲板	下甲板
lower high water(LHW)	低高潮	較低高潮(一日中)
lower hold	底舱	底[貨]艙
lower low water(LLW)	低低潮	較低低潮(一日中)
lower reach	下游	下游
lower transit (=lower culmination)	下中天	[天體]下中天
lower tween deck	下二层甲板	下方中甲板
low flame spread	低播焰性	低度火焰蔓延
low-frequency stress	低频压力	低頻應力
low-lubricating oil pressure trip device	低滑油压力保护装置	低潤滑油壓跳脫設施
low [pressure]	低压	低壓,低氣壓
low pressure steam generator	低压蒸汽发生器	低壓蒸汽發生器
low speed diesel engine	低速柴油机	低速柴油機
low temperature corrosion	低温腐蚀	低溫腐蝕
low-vacuum protective device	低真空保护装置	低真空保護設施
low-voltage protection	低电压保护	低電壓保護
low-voltage release	低电压释放	低電壓釋放
low water(LW)	低潮	低潮
low water time	低潮时	低潮時
LPG(=liquefied petroleum gas)	液化石油气	液化石油氣
LST(=local sidereal time)	地方恒星时	當地恒星時
LTL(=less than truck load)	拼车货	拼車貨
lubber line error	基线误差	艏線誤差

英 文 名	祖 国 大 陆 名	台 湾 地 区 名
lub-oil dilution	机油稀释	潤滑油稀釋
lubricating grease	润滑脂	滑脂
lubricating oil batch purification	滑油间歇净化	潤滑油分批淨化
lubricating oil filling pipe	滑油注入管	[潤]滑油注入管
lubricating oil pass purification	滑油连续净化	潤滑油連續淨化
lubricating oil pump	滑油泵	[潤]滑油泵
lubricating oil transfer pump	滑油输送泵	[潤]滑油輸送泵
lubrication system	润滑系统	潤滑系統
luff tackle	2-1 绞辘	2-1 轆轤
luggage(=baggage)	行李	行李
luggage room(=baggage room)	行李间	行李間
lumber cargo ship	木材船	木材[運載]船
luminescence of the sea	海发光	海發光
luminous range	光力射程,照距	光照距,光強度視程
lumpsum freight	总付运费	[總額]包載運貨
lunar diurnal tide	太阴全日潮	太陰日週潮
lunar eclipse	月蚀	月食
lunar phases	月相	月相
lunar tide	月潮	太陰潮
lunitidal interval	月潮间隙	月中天潮汐間歇
lurch	突然倾斜	[船]突傾側
LUT(=local user terminal)	本地用户终端	地面終端台
LUT in a LEOSAR system	低轨道搜救卫星系统本地用户终端	低軌道衛星搜救系統地面終端台
LW(=low water)	低潮	低潮
lying on the keel block	落墩	坐墩

M

英 文 名	祖 国 大 陆 名	台 湾 地 区 名
machine finishing	机械精加工	機械加工
machinery space	机器处所,机炉舱	機艙[空間],機器空間,機械室
machinery space bilge	机舱污水井	機艙舭水
machinery space of category A	A 类机器处所,A 类机舱	甲種機艙空間
macroscopic test	肉眼检查	肉眼檢查
made-to-order	定制的	訂製者

英　文　名	祖国大陆名	台湾地区名
magazine	弹药舱	彈藥艙
magnetic annual change	年差	年磁差
magnetic anomaly	异常磁区	磁異常區
magnetic bearing(MB)	磁方位	磁方位
magnetic compass	磁罗经	磁羅經,磁羅盤
magnetic compass adjustment	磁罗经校正	磁羅經校正
magnetic course(MC)	磁航向	磁航向
magnetic deflection	磁偏角	磁偏轉
magnetic dip	磁倾角	地磁傾角,磁傾角
magnetic equator	磁赤道	地磁赤道,磁赤道
magnetic meridian	磁子午线	磁子午線
magnetic needle	磁针	磁針
magnetic north	磁北	磁北
magnetic storm	磁暴	磁暴
[magnetic] variation(=Var)	磁差	地磁差
magnetization of transducer	换能器充磁	轉變器充磁
magnetostrictive effect	磁致伸缩效应	磁效伸縮效應
maiden voyage	处女航,首航	處女航
mail and baggage room	邮件及行李间	郵件及行李間
mailbox service	邮箱业务	郵箱業務
mail room	邮件舱	郵件室
mail ship	邮船	郵船
main antenna	主用天线	主天線
main bearing	主轴承	主軸承
main boiler	主锅炉	主鍋爐
main cargo oil line	货油总管	貨油主管
main channel	主航道	主航道
main circulating pump	主循环泵	主循環泵
main condenser circulating pump	主冷凝器循环泵	主冷凝器循環泵
main coolant system	主冷却剂系统	主冷卻劑系統
main deck	主甲板	主甲板
main engine fault emergency maneuver	主机故障应急处理	主機故障應急操縱
main engine remote control panel	主机遥控屏	主機遙控屏
main engine revolution speedometer	主机转速表	主機轉速表
main engine telegraph	主车钟	主俥鐘
main gas turbine	主燃气轮机	主燃氣渦輪機
main generator diesel engine	主发电柴油机	主柴油發電機
main line	干线	幹繩

英 文 名	祖 国 大 陆 名	台 湾 地 区 名
main line guide pipe	干线导管	幹繩導管
main mark	主标志	主標誌
main propelling machinery room	主推进机舱	主推進機器
main propulsion unit	主推进装置	主推進裝置
main receiver	主用收信机	主接收機
main rotor	主转子	主轉子
main source of electrical power	主电源	主電源
main starting valve	主起动阀	主起動閥
main steam turbine	主汽轮机	主蒸汽渦輪機
main stream	主流	主流
main switchboard	主配电板	主配電盤
maintenance	维护,保养	維護,保養
maintenance waste	维护保养废弃物	維護保養所生廢棄物
main towing line	主拖缆	主拖纜
main transmitter	主用发信机	主發射機
main vertical zone	主竖区	主要垂直區域
major axis	长轴	長軸
major cause	主要原因	主要原因
major conversion	重大改装	重大改裝
make dead	切断	切斷
making way through water	对水移动	對水移動
malodorous cargo	恶臭货	惡臭貨
management for the safe operation of ships	船舶安全营运管理	船舶安全營運管理
management level	管理级	管理級
mandatory pilotage(=compulsory pilotage)	强制引航	強制引水
maneuver	机动操纵	操縱
maneuverability	操纵性	操縱性能,運轉能力
maneuverability identification	操纵性识别	操縱性識別
maneuverability indices	船舶操纵性指数	操縱性能指數
maneuvering board	船舶运动图	運動圖
maneuvering in canal	运河操纵	運河中操縱
maneuvering in narrow channel	狭水道操纵	狹水道中操縱
maneuvering light signal	操纵灯号	操縱號燈
maneuvering period	转舵阶段	轉舵階段
maneuvering ship	机动舰[船]	運轉船
maneuvering signal	操纵信号	運轉信號,操縱信號
maneuvering tank	操纵性试验水池	操縱性試驗池

英　文　名	祖国大陆名	台湾地区名
manhole	人孔	人孔
manifest	载货清单	載貨單,艙單
Manila rope	白棕绳	馬尼拉繩,白棕繩
man-machine communication system	人–机通信系统	人–機通信系統
manned	人工操纵的	人工操縱者
manning of lifecraft	艇筏配员	救生艇筏人員配額
man overboard	人员落水	人員落水
manrope knot	握索结	扶手索結
manual pump	手摇泵	手搖泵
manual setting	手动调整	人工調整
manual telex service	人工用户电报业务	人工電報交換業務
margin line	限界线	邊際線
margin plate	内底边板	舭緣板
marine accident analysis	海事分析	海事分析
marine accident report	海事报告	海事報告
marine air conditioning	船舶空气调节	船舶空調
marine and air navigation light	海空两用灯标	海空兩用航行燈
marine auxiliary machinery	船舶辅机	船舶輔機
marine clutch	船用离合器	船用離合器
marine communication	水上通信	水上通信
marine corrosion	海洋腐蚀	海洋腐蝕
marine coupling	船用联轴器	船用聯軸節
marine court	海事法庭	海事法庭
marine diesel engine	船用柴油机	船用柴油機
marine diesel oil	船用柴油	船用柴油
marine ecological investigation	海洋生态调查	海洋生態調查
marine electric installation	船舶电气设备	船舶電力裝置
marine engineering	轮机工程	輪機工程
marine engineering management	轮机管理	輪機管理
marine engineering survey	海洋工程测量	海洋工程測量
marine environment	海洋环境	海洋環境
marine environmental protection	海洋环境保护	海洋環境保護
marine environment investigation	海洋环境调查	海洋環境調查
marine fishery	海洋渔业	海洋漁業
marine forecast	海上预报	海上預報
marine gas oil	船用轻柴油	船用輕柴油
marine gas turbine	船用燃气轮机	船用燃氣渦輪機
marine gear box	船用齿轮箱	船用齒輪箱

英 文 名	祖 国 大 陆 名	台 湾 地 区 名
marine gravimetric survey	海洋重力测量	海洋重力測量
marine hydrology	海洋水文	海洋水文
marine insurance	海上保险	海上保險
marine investigation	海洋调查	海洋調查
marine magnetic survey	海洋磁力测量	海洋磁力測量
marine main engine	船舶主机	船舶主機
marine medicine	航海医学	航海醫學
marine monitoring	海洋监测	海洋監測
marine multifunction incinerator	船用全功能焚烧炉	船用多功能焚化爐
marine natural reserves	海上自然保护区	海洋自然保護區
marine navigation	航海学	航海學
marine navigation expert system	航海专家系统	航海專家系統
marine nuclear power plant	船舶核动力装置	船用核子動力裝置
marine pollutant	海洋污染物	海洋污染物
marine pollutants report	海洋污染物报告	海水污染物報告
marine pollution	海洋污染	海水污染
marine power plant	船舶动力装置	船舶動力裝置
marine power plant economy	船舶动力装置经济性	船舶動力裝置經濟性
marine power plant maintainability	船舶动力装置可维修性	船舶動力裝置可維修性
marine power plant maneuverability	船舶动力装置操纵性	船舶動力裝置操縱性
marine power plant service reliability	船舶动力装置可靠性	船舶動力裝置可靠性
marine propulsion shafting	船舶推进轴系	船舶推進軸系
marine psychology	航海心理学	航海心理學
marine pump	船用泵	船用泵
marine radar	船用雷达	船用雷達
marine radio navigation	船舶无线电导航	船舶無線電導航
marine radiotelephone	海上无线电话	海上無線電話
marine refrigerating plant	船舶制冷装置	船舶冷凍裝置
marine resources investigation	海洋资源调查	海洋資源調查
marine safety supervision	海上安全监督	海上安全監督
marine salvage	海难救助[打捞]	海上救助
marine science and technology	海洋科学技术	海洋科技
marine search and rescue	海上搜救	海上搜[索與]救[助]
marine shafting	船舶轴系	船舶軸系
marine steam boiler	船舶蒸汽锅炉	船用蒸汽鍋爐
marine steam engine	船舶蒸汽机	船舶蒸汽機
marine steam turbine	船用汽轮机	船用蒸汽渦輪機
marine store	船用物料	船用物料

英　文　名	祖国大陆名	台湾地区名
marine surveillance	海洋监视	海洋監視
marine system	船舶系统	船舶系統
marine transportation	水上运输	水上運輸
marine weather data	海上气象数据	海上氣象數據
maritime arbitration	海事仲裁	海事仲裁
maritime assistance	海事援助	海事援助
maritime cargo insurance	海运货物保险	海運貨物保險
maritime case	海事判例	海事判例
maritime claim	海事请求	海事求償
maritime code	海商法	海商法
maritime court(=marine court)	海事法庭	海事法庭
maritime declaration of health	航海健康申报书	航海健康申報書
maritime distress channel	海上遇险信道	海上遇險頻道
maritime enquiry	海上询问	海上詢問
maritime identification digits(MID)	海上识别数字	水上識別碼
maritime investigation	海事调查	海事調查
maritime jurisdiction	海事管辖	海事管轄權
maritime law(=maritime code)	海商法	海商法
maritime lien	船舶优先权	海事優先擔保權,海事留置權
maritime litigation	海事诉讼	海事訴訟
maritime mediation	海事调解	海事調解
maritime mobile satellite service	卫星海上移动业务	衛星水上行動業務,水上行動衛星通信業務
maritime mobile selective-call identify code	海上移动选择呼叫识别码	水上行動選擇呼叫識別碼
maritime mobile service	海上移动业务	海上行動業務
maritime mobile service identity	海上移动业务识别	水上行動業務識別
maritime radionavigation-satellite service	海上卫星无线电导航业务	水上無線電衛星導航業務
maritime radionavigation service	海上无线电导航业务	海上無線電助航業務
maritime reconciliation	海事和解	海事和解
maritime rescue co-ordination center	海上救助协调中心	海上搜救協調中心
maritime rescue sub-center	海上救助中心	海上救助站
maritime rules and regulations	航海法规	海事法規
maritime safety information	海上安全信息	海事安全資訊
maritime safety information via satellite	卫星转发海上安全信息	衛星轉發之海上安全資訊

英　文　名	祖国大陆名	台湾地区名
maritime SAR plan	海上搜救计划	海上搜救計劃
maritime sovereignty	海洋主权	海洋主權
mark	标志	標誌
Markab	室宿一(飞马α)	室宿一(飛馬α)
marking	标记,记号	標記,記號
marking vessel	标志船	標誌船
marline spike hitch	绳锥结	椎套結
Mars	火星	火星
mass moment of inertia	质量惯性矩	質量慣性矩
mast	桅	桅
master control circuit	主控线路	主控線路
master controller	主令控制器	主控制器
master control station	主控站	主控站
master pedestal	主台座	主托架
master signal	主台信号	主台信號
master station	主台	羅遠主台
masthead light	桅灯	桅頂燈
matching	匹配	配合,相配
matching member	配合件	配合件
materials hazardous in bulk	散装时危险物质	散裝時危險物質
material which may liquefy	易流态化物质	易液化物質
mate's receipt	收货单	收貨單
matrix signal	矩阵信号	矩陣信號
maximum allowable pressure	最大容许压力	最大容許壓力
maximum angle of dynamic inclination	极限动倾角	最大動傾角
maximum breadth	最大宽度	最大寬度
maximum continuous rating	最大持续功率	額定最大連續出力
maximum explosion pressure gauge	最高爆发压力表	最高爆發壓力錶
maximum explosive pressure	最高爆发压力	最高爆發壓力
maximum fuel limit screw	最大燃油量限制螺钉	最大燃油量限制螺釘
maximum heeling moment	极限横倾力矩	最大傾側力矩
maximum height	最大高度	最大高度
maximum height of lift	最大起升高度	最大提升高度
maximum measuring depth	最大测量深度	最大測量深度
maximum no-load speed	最高空载转速	最大無負載速度
maximum permissible stable operation power	最大容许稳定运行功率	容許最大穩定運轉功率
maximum radar range	雷达最大作用距离	雷達最大效程

英　文　名	祖国大陆名	台湾地区名
maximum righting moment	最大复原力矩	最大扶正力矩
maximum stability lever	最大稳性力臂	最大穩度力臂
mayday relay	转发无线电话遇险信号	轉發遇險通報(無線電話用語)
MB(=magnetic bearing)	磁方位	磁方位
MC(=magnetic course)	磁航向	磁航向
MCC(=mission control center)	任务控制中心	任務管制中心
MCC service area	任务控制中心服务区	任務管制中心服務區
M combustion chamber	M 燃烧室	M 燃燒室
mean anomaly	平近点角	平近點角
mean draft	平均吃水	平均吃水
mean ecliptic	平黄道	平黃道
mean effective pressure limit	等平均有效压力限	平均有效壓力限制
mean equator	平赤道	平赤道
mean error	平均误差	平均誤差
mean high water interval(MHWI)	平均高潮间隙	平均高潮間隙
mean latitude	平均纬度	平均緯度
mean longitude	平均经度	平均經度
mean low water interval(MLWI)	平均低潮间隙	平均低潮間隙
mean piston speed	活塞平均速度	活塞平均速度
mean pressure meter	平均压力计	平均壓力計
mean sea level(MSL)	平均海面	平均海平面
means of escape	逃生方法	逃生方法
means of going astern	倒车装置	倒俥裝置
mean sun	平太阳	平均太陽,平太陽
mean time	平时	平均時
mean time between failures	故障间隔平均时间	平均故障間隔[時間]
mean-time-to-repair	维修前平均使用时间	平均修復間隔時間
measured mile trial	标柱测速试验	標柱速率試驗
measured value	测定值	測定值
measurement cargo	容积货物	呎碼貨,容積貨
measurement ton	容积吨	呎碼噸
measurement tonnage	丈量吨位	丈量噸位
measuring means	测量方法	測計方法
measuring the explosive limit	测爆	爆發限度測計
measuring unit	测量单元	測定單位
meat chopper	碎肉机	碎肉機
meat mill	磨肉机	磨肉機(捏和機)

英　文　名	祖国大陆名	台湾地区名
meat separator	鱼肉分选机	魚肉採取機
mechanical efficiency	机械效率	機械效率
mechanical governor	机械式调速器	機械調速器
mechanical impurities	机械杂质	機械雜質
mechanically actuated valve mechanism	机械气阀传动机构	機械致動閥機構
mechanically propelled lifeboat	机械推进救生艇	機械推進救生艇
mechanical purchase	机械滑车	機械複滑車
mechanical rust removal	机械除锈	機械除銹
mechanical strength	机械强度	機械強度
mechanical supercharging	机械增压	機械增壓
mechanical top bracing	机械式上支撑	機械式上支撐
mechanical transmission	机械传动	機械傳動
medical advice	医疗指导	醫療指導
medical assistance	医疗援助	醫療協助
medical care	医疗	醫療
medical examination	健康检查	健康檢查
medical first aid	急救	急救
medical messages	医疗电文	醫療信文
medical premises	医务室	醫務室
medical transports	医疗运输	醫療運送,醫療傳送
Mediterranean mooring	地中海系泊法	地中海繫泊法
medium carbon steel	中碳钢	中碳鋼
medium force fit	中级压紧配合	中級壓緊配合
medium-range forecast	中期天气预报	中期預報
medium speed diesel engine	中速柴油机	中速柴油機
Meiyu	梅雨	梅雨
melting point	熔点	熔點
Menkar	天囷一(鲸鱼 α)	天囷一(鯨魚 α)
Merak	天璇(大熊 β)	天璇,北斗二(大熊 β)
Mercator chart	墨卡托海图	麥卡托海圖
Mercator projection	墨卡托投影	麥卡托投影法
Mercator's sailing	墨卡托算法	麥氏航海
merchant ship	商船	商船
merchant ship flag	商船旗	商船旗
merchant ship search and rescue manual	商船搜救手册	商船搜救手冊
Mercury	水星	水星
meridian	子午线	子午線,經線
meridian altitude	中天高度	中天高度

英 文 名	祖国大陆名	台湾地区名
meridian angle	子午角	子午線角
meridian circle	子午圈	子午圈
meridian gyro	主陀螺	子午迴轉儀
meridian instrument	子午仪	子午儀,中星儀
meridian observation	中天观测	中天高度觀測
meridian passage	中天	中天
meridian sailing	在子午线上航行	經線航法
meridian-seeking moment	指向力矩	尋子午線力矩
meridian-seeking torque(=meridian see-king moment)	指向力矩	尋子午線力矩
meridian zenith distance	中天顶距	[過]子午圈天頂距
meridional parts(MP)	纬度渐长率	緯度漸長比
MES message channel	移动地球站信息信道	行動地球台資訊頻道
message	报文	信文
message field	信文范围	信文欄
message filtering factor	信文筛选因素	信文篩檢因素
message format	电文格式	電文格式
message marker	信文标志	信文標誌
message transfer time	信文转换时间	信文轉換時間
messenger	引缆	傳遞索
messenger return line	引缆回收索	回收索
metacenter	稳心	定傾中心
metacentric height	稳心高度	定傾[中心]高度
metacentric radius	稳心半径	定傾半徑
metal-edge type strainer	金属片状粗滤器	金屬層片過濾器
metal lifeboat	金属救生艇	金屬救生艇
metallographic inspection	金相检查	金相檢查
meteorological aids service	气象辅助业务	氣象輔助業務
meteorological bulletins	气象报告	氣象報告
meteorological chart	气象要素图	氣象圖
meteorological element	气象要素	氣象要素
meteorological forecast	气象预报	氣象預報
meteorological information	气象信息	氣象資訊
meteorological messages	气象电文	氣象信文
meteorological radar	气象雷达	氣象雷達
meteorological-satellite service	卫星气象业务	衛星氣象業務
meteorological service	气象服务	氣象服務
meteorological visibility	气象能见度	氣象能見距

英　文　名	祖国大陆名	台湾地区名
meteorological warning	气象警告	氣象警告
meteorology echo	气象回波	氣象回波
metering pump	计量泵	計量泵
metering rod	量杆	量桿
method of altering course by two half-angles	两半角[转向]法	兩半角轉向法
method of internal compensation	内补偿法	内補償法
method of intersection	交会法	交會法
method of intersection by sextant	六分仪交会法	六分儀交會法
method of outer compensation	外补偿法	外補償法
method of random sampling	随机取样法	隨機取樣法
metric coarse thread	公制粗牙螺纹	公制粗螺紋
metric ton	公吨	公噸
MET warning(=meteorological warning)	气象警告	氣象警告
MF communication	中频通信	中頻通信
MF/HF radio installation	中/高频无线电设备	中/高頻無線電裝置
MFNT(=most favored nation treatment)	最惠国待遇	最惠國待遇
MF radio installation	中频无线电设备	中頻無線電裝置
MHWI(=mean high water interval)	平均高潮间隙	平均高潮間隙
microcomputer	微机	微電腦
microcomputer control system	微机控制系统	微電腦控制系統
microcomputer remote control system for main engine	微机控制主机遥控系统	微電腦遙控主機系統
micrometer depth gauge	深度千分尺	深度測微計
micro processor	微处理器	微處理機
microwave ranging system	微波测距系统	微波測距系統
MID(=maritime identification digits)	海上识别数字	水上識別碼
mid-channel buoy	中央浮标	航道中流浮標
middle latitude	中分纬度	中間緯度
middle latitude sailing	中分纬度算法	中緯航法
middle reach	中游	中游
midship	舯	舯
midship section coefficient	舯剖面系数	舯剖面係數
mid-water trawl	中层拖网	中層拖網
military navigation	军事航海	軍事航海
Milky Way	银河	銀河
mineclearance operation	清除水雷作业	清除水雷作業
mine hunter	猎雷舰	獵雷艦

英　文　名	祖国大陆名	台湾地区名
mine-laying navigation service	布雷航海勤务	佈雷航海勤務
mineralizing equipment of drinking	饮用水矿化器	飲用水礦化器
mineral oil	矿物油	礦物油
mineral resources of the sea	海洋矿物资源	海洋礦產資源
mine-sweeping formation	扫雷队形	掃雷隊形
mine-sweeping navigation service	扫雷航海勤务	掃雷航海勤務
minimum capsizing moment	最小倾覆力矩	最小翻覆力矩
minimum distance	最近距离	最小距離
minimum freight	最低运费	起碼運費
minimum freight bill of lading	最低运费提单	起碼運費載貨證券
minimum freight ton	最低运费吨	起碼運費噸
minimum luminous intensity	最低发光强度	最低照明強度
minimum measuring depth	最小测量深度	最小測量深度
minimum radar range	雷达最小作用距离	雷達最小效程
minimum safe manning	最低安全配员	船員最低安全配額
minimum stable engine speed	最低稳定转速	最低穩定轉速
minimum starting pressure	最低起动压力	最低起動壓力
minor axis	短轴	短軸
minor overhaul	小修	小修
Mirfak	天船三(英仙 α)	天船三(英仙 α)
misalignment value	偏中值	偏中值
miscellaneous dangerous substance	杂类危险物质	雜項危險物質
mislanded	误卸	誤卸
mission control center	任务控制中心	任務管制中心
mist	霭	靄
mixed cargo	混合货	混合貨物,混載貨
mixed current	混合流	混合潮流
mixed cycle	混合循环	混合循環
mixed flow pump	混流泵	混流泵
mixed frame system	混合骨架式	混合肋骨系統
mixed tide	混合潮	混合潮
mixing governing	混合调节	混合調節
MLWI(=mean low water interval)	平均低潮间隙	平均低潮間隙
mobile earth station	移动地球站	行動地球台,行動衛星電台
mobile earth station status	移动地球站状态	行動地球台狀態
mobile offshore drilling unit (MODU)	近岸移动式钻井装置	可動式離岸鑽探平台
mobile-satellite service	卫星移动业务	衛星行動業務

英　文　名	祖国大陆名	台湾地区名
mobile service	移动业务	行動業務
mobile station	移动电台	機動電台
mobile unit	移动单元	行動單元
moderate breeze	4 级风	和風(4 級)
moderate fog	轻雾	中霧
moderate sea	中浪	和浪
moderate swell	中涌	中湧
modified zigzag maneuver test	变 Z 形试验	修正式蛇航試驗,修正式之字航行試驗
MODU（＝mobile offshore drilling unit）	近岸移动式钻井装置	可動式離岸鑽探平台
modular pulse converter supercharging	模件式脉冲转换增压	模組脈衝轉換增壓
modulus of midship section	舯剖面模数	舯剖面模數
molasses tanker	糖浆船	糖蜜船
molded breadth	型宽	模寬
molded depth	型深	模深
mole	突堤	海堤
molecular structure	分子结构	分子結構
momentary overload	瞬时过载	瞬時過載
moment compensator	力矩平衡器	力矩調整器
moment of hydrodynamic force	水动力力矩	流體動力力矩
moment of turning ship	转船力矩	船舶迴轉力矩
moment of wind resistance	风阻力矩	風阻力矩
moment to change trim per centimeter （TPC）	每厘米纵倾力矩	每公分俯仰差力矩
momentum theory	动量理论	動量理論
monitoring	监测	監測
monitoring panel	监视屏	監測屏
monitor station	监测站	監視站
mono-hull ship	单体船	單[胴]體船
monsoon	季风	季節風,季風
monthly inspection	每月检查	每月檢查
moon rise	月出	月出
moon set	月没	月沒
moon's age	月龄	月齡
moon's apparent motion	月球视运动	月視運動
moon's path	白道	白道
moor	一字锚泊	一字雙錨泊
mooring anchor	系泊锚	繫留錨,碇泊錨

英　文　名	祖国大陆名	台湾地区名
mooring bitt	系缆桩	繫船椿
mooring buoy	系船浮[筒]	繫泊浮筒,繫船浮筒
mooring buoy with telephonic or telegraph-ic communications	装有电报或电话通信的系泊浮筒	裝有電報或電話通信之繫泊浮筒
mooring capstan	系泊绞盘	繫泊絞盤
mooring head and stern	艏艉锚泊	艏艉碇泊
mooring line	系缆	繫泊纜索,繫纜
mooring operating mode management	系泊工况管理	繫泊作業管理
mooring orders	带缆口令	帶纜口令
mooring pipe	导缆孔	[繫船]道索管
mooring to two anchors	一字双锚泊	雙錨繫泊
mooring trial(=dock trial)	系泊试验	繫泊試俥
mooring winch	系泊绞车	繫船絞車
morning fog	晨雾	晨霧
Morse code	莫尔斯码	莫斯電碼
Morse code fog signal	莫尔斯码雾号	莫爾斯碼霧號
mortgage registration	抵押登记	抵押登記
most favored nation treatment(MFNT)	最惠国待遇	最惠國待遇
most probable position(MPP)	最概率船位	最可能船位
most significant bit	最有效位	最高有效位元
motor lifeboat	机动救生艇	馬達救生艇
motor man	机工	機工
motor revolution error	电机转速误差	馬達轉速誤差
motor sailer	机帆船	機帆船
motor vessel(MV)	内燃机船	內燃機船
mountain river	山区河流	
mounting fittings	装配附件	裝配件
mousing	止脱结	止脫繩,止脫結
movable fit	动配合	動配合
movable relieving hook	活动解拖钩	脫鉤
moving blade	动叶片	轉動葉片
moving-weight stabilizer	移动重量式减摇装置	多重式穩定器
MP(=meridional parts)	纬度渐长率	緯度漸長比
MP mode chain(=multi-pulse mode chain)	MP 型链	多脈搏型鏈
MPP(=most probable position)	最概率船位	最可能船位
MSI(=maritime safety information)	海上安全信息	海上安全資訊
MSL(=mean sea level)	平均海面	平均海平面

英 文 名	祖 国 大 陆 名	台 湾 地 区 名
multi-beam sounding system	多波束测深系统	多波束測深系統
multi-buoy mooring	多浮筒系泊	多浮筒繫泊
multi-entry pulse converter	多进口脉冲转换器	多進口脈波變換器
multi-entry pulse converter system	多进口脉冲转换器系统	多進口脈衝轉換器系統
multi-hull ship	多体船	多[胴]體船
multimodal transportation	多式联运	多式聯運
multimodal transport bill of lading	多式联运提单	多式聯運載貨證券
multimodal transport operator	多式联运经营人	多式聯運經營人
multipath	多径	多[途]徑
multipath error	多径误差	多徑誤差
multipath propagation	多径传播	多徑傳播
multi-pintle rudder	多支承舵	多舵針舵
multiple-effect evaporation	多效蒸发	多效蒸發
multiple pile driven tower	多能打桩架	多功能打樁架
multiple reflection echo	多次反射回波	多次反射回波
multiple-stage flash evaporation	多级闪发	多級閃蒸發
multi-point mooring system	多点系泊系统	多點繫泊系統
multi-pulse mode chain(MP mode chain)	MP 型链	多脈搏型鏈
multipurpose additive	多效添加剂	多功能添加劑
multipurpose cargo vessel	多用途货船	多用途貨船
multipurpose fishing boat	多用途渔船	多用途漁船
multipurpose towing ship	多用途拖船	多用途拖船
multi-satellite link	多卫星链路	多衛星鏈路
multi-ship collision avoidance	多船避碰	多船避碰
multi-stage compressor	多级压缩机	多級壓縮機
mushroom anchor	菌形锚	菌形錨
mushroom ventilator	菌形通风筒	菌型通風筒
muster list(=station bill)	应变部署表	部署表
muting	哑控	靜音
MV(=motor vessel)	内燃机船	內燃機船

N

英 文 名	祖 国 大 陆 名	台 湾 地 区 名
nadir	天底	天底點
named endorsement	记名背书	記名背書
name plate	名牌	銘牌
narrow band direct print	窄带直接印字	狹頻帶直接印字

英　文　名	祖 国 大 陆 名	台 湾 地 区 名
narrow-band direct-printing telegraph equipment(NBDP)	窄带直接印字电报设备	狹頻帶直接印字電報設備
narrow-band direct-printing telegraphy	窄带直接印字电报	狹頻帶直接電報
narrow channel	狭水道	狹窄水道
national coordinator	国内协调人	國家協調人
national DSC frequencies	国内数字选择呼叫频率	國內數位選擇呼叫頻率
national enquiry	国家询问	國家詢問
nationality of vessel	船舶国籍,船籍	船籍國
national NAVTEX service	国内奈伏泰斯业务	國內航行警告電傳業務
national safety NET	国家安全通信网	國家安全通信網
national safety NET service	国内安全通信网	國家安全通信網業務
natural circulation boiler	自然循环锅炉	自然循環鍋爐
natural feature	自然地貌	自然地貌
natural fiber rope	植物纤维绳	天然纖維索
natural forum	自然行诉	自然源繪地,自然管轄地
nautical almanac	航海天文历	航海曆
nautical chart(=chart)	海图	海圖
nautical charts and publications	航海图书资料	航海圖書刊物
nautical fault	航海过失	航海過失
nautical history	航海史	航海史
nautical instrument	航海仪器	航海儀器
nautical meteorology	航海气象	航海氣象
nautical mile (=n mile)	海里	浬,海里
nautical science	航海科学	航海科學
nautical service	航海保证	航海勤務
nautical table	航海表	航海表
nautical twilight	航海晨昏朦影	航海曚光
naval pipe	锚链管	錨鏈管
naval ship	军船	軍艦
NAVAREA	航行警告区	航行警告區
NAVAREA coordinator	航行警告区域协调国	航行警告區域協調人
NAVAREA warning	航行警告区警告	航行警告區之警告
NAVAREA warning bulletin	航行警告区公告	航行區警告通告
NAVAREA warning service	航行警告区业务	航行警告區業務
navigable bridge-opening	通航桥孔	可航橋孔
navigableness	可航性	可航性
navigable semicircle	可航半圆	可航半圓

英　文　名	祖 国 大 陆 名	台 湾 地 区 名
navigable waters	通航水域	適航水域
navigating in fog	雾中航行	霧航
navigating in heavy weather	风暴中航行	惡劣氣候航行
navigating in narrow channel(=channel navigation)	狭水道航行	狹水道航行
navigating in rocky water	岛礁区航行	多礁水域航行
navigation aids	导航设备	助航設施
navigation aids on canal	运河航标	運河助航標
navigational chart	航用海图	航海用海圖
navigational plan(=sailing plan)	航行计划	航行計劃
navigational planets	航用行星	導航行星
navigational satellite	导航卫星	導航衛星
navigational telex (NAVTEX)	航行警告[电传]系统, 奈伏泰斯	航行警告電傳
navigational warning	航行警告	航行警告
navigational warning signal	航行警告信号	航行警告信號
navigational watch	航行值班	航行當值
navigation consultant	航海顾问	航海諮詢
navigation management	航行管理	航管
navigation mark	航[行]标[志]	航路標誌
navigation parameter	导航参数	導航參變數
navigation radar	导航雷达	導航雷達
navigation safety communication	航行安全通信	航行安全通信
navigation service for landing	登陆航海勤务	登陸海勤
navigation sonar	导航声呐	導航聲納
navigation table(=nautical table)	航海表	航海表
navigation wind(=ship wind)	航行风	航行風
NAVTEX(=navigational telex)	航行警告[电传]系统, 奈伏泰斯	航行警告電傳
NAVTEX important warning	奈伏泰斯重要警告	航行警告電傳之重要警告
NAVTEX message numbering	奈伏泰斯电文编号	航行警告電傳信文編號
NAVTEX priority message	奈伏泰斯优先电文	航行警告電傳優先信文
NAVTEX routine warnings	奈伏泰斯日常警告	航行警告電傳例行警告
NAVTEX vital warnings	奈伏泰斯紧急警告	航行警告電傳緊急警告
Navy Navigation Satellite System(NNSS)	海军导航卫星系统	海軍衛星導航系統
NBDP(=narrow-band direct-printing tele-graph equipment)	窄带直接印字电报设备	狹頻帶直接印字電報設備

英　文　名	祖国大陆名	台湾地区名
NCS(= network coordination station）	网络协调站	網路協調電台
NCS common TDM channel	网络协调站共用时分多路复用信道	網路協調電台共用劃時多工制頻道
NCS/LES signaling channel	网络协调站到陆地地球站信令信道	網路協調電台與陸上衛星電台之信號頻道
NCS/NCS signaling channel	网络协调站到网络协调站信令信道	網路協調電台間之信號頻道
NDB(= non-directional beacon）	无方向性信标	無方向性示標
neap rise(NR）	小潮升	小潮升
nearest land	最近陆地	最近陸地
near gale	7 级风	疾風(7 級)
necessary bandwidth	必要带宽	必須頻帶寬度
needle valve	针阀	針閥
negative feedback	负反馈	負反饋
negative feed back control system	负反馈控制系统	負反饋控制系統
net boat	网船	網船
net carrying pipe	送网管	輸網管
net drum	卷网机	捲網機
net monitor	网位仪	網位儀
net positive suction head	净正吸入压头	淨正吸入水頭
net positive suction height	净正吸高	淨正吸入高
net shifter	理网机	理網機
netting	网衣	網片
net tonnage(NT）	净吨位	淨噸位
net winch	起网机	起網機
network control center	网络控制中心	網路控制中心
network coordination station（NCS）	网络协调站	網路協調電台
network liability system	网状责任制	網狀責任制
neutral current	中性流	中線電流
neutral point	中性点	中性點
neutron power meter	中子功率表	中子功率表
new danger mark	新危险物标志	新危險物標誌
new edition chart	新版图	新版海圖
New Jason clause	新杰森条款	紐哲遜條款
new moon	新月	新月,朔
next ship on the left	左邻舰	左側船艦
n-heptane insoluble	正庚烷不溶物	正庚烷不溶物
nichrome	镍铬耐热合金	鎳鉻合金

英　文　名	祖国大陆名	台湾地区名
night effect	夜间效应	夜間效應
night order book	夜航命令簿	夜航命令簿
night vision sextant	夜视六分仪	夜視六分儀
nimbo-stratus	雨层云	雨層雲
nimonic alloy	镍铬钛[耐热]合金	鎳鉻立克[耐熱合金]
NK(＝Japanese Maritime corporation)	日本船级社	日本海事協會
n mail	海里	浬,海里
NNSS(＝Navy Navigation Satellite System)	海军导航卫星系统	海軍衛星導航系統
no beacon emission	不发射信标	示標無發送
no cure-no pay	无效果-无报酬	無效無償
nodal MCC	结点任务控制中心	結點任務管制中心
node	结[节]点	波節
noise pollution	噪声污染	噪音污染
no load	空载	無載
no-load test	空载试验	空載試驗,無負載試驗
nominal range	额定光力射程	公稱光程
nominal speed loss	自然减速	標稱速降
non-combustible material	不燃材料	不燃材料
non-directional beacon(NDB)	无方向性信标	無方向性示標
non-displacement craft	非排水船舶	非排水型船
non-ferrous metal	非铁金属,有色金属	非鐵金屬
non-hazardous areas	非危险区域	非危險區域
non-line connection	无缆系结	無纜索連結
non-living resources	非生物资源	非生物資源
non-navigational chart	航用参考图	非航用海圖
non-negotiable bill of lading	不可转让提单	不可轉讓載貨證券
non-packed cargo	未包装货	未包裝貨
non-removable	不可拆卸的	不可拆卸者
non-retractable fin stabilizer	非收放型减摇鳍装置	非收放型鰭板穩定器
non-return valve	止回阀	止回閥
non-reversible diesel engine	不可倒转柴油机	不可逆轉柴油機
normal loss	正常损耗	正常損耗
normal service rating	正常使用功率	常用[定額]出力
normal starting sequence	正常起动程序	正常起動程序
north celestial pole	北天极	北天極
northern hemisphere	北半球	北半球
northern light	北极光	北極光

英　文　名	祖国大陆名	台湾地区名
north gyro	北向陀螺	北向迴轉儀
north mark	北方标	北方標
north pole	北极	北極
north star(= Polaris)	北极星	北極星
north up	北向上	北向上
N. O. S. cargo	未列名货	未列名貨
notarial survey	公证检验	公證檢驗
notice mark	注意标志	注意標誌
notice of readiness	准备就绪通知书	(装卸)準備完成通知書
notice to mariners	航海通告	航行通告
notification of country of beacon registration	信标登记国通告	示標登記國通告
notify party	通知方	通知方
noxious cargo(= harmful cargo)	有害货	有害貨物
noxious liquid substance	有害液体物质	有毒液體物質
nozzle	喷嘴	噴嘴
nozzle block	喷嘴组	噴嘴塊
nozzle chamber	喷嘴室	噴嘴室
nozzle dribbling	喷油嘴滴漏	噴油嘴滴漏
nozzle governing	喷嘴调节	噴嘴調節
nozzle ring	喷嘴环	噴嘴環
nozzle valve	喷嘴阀	噴嘴閥
NR(= neap rise)	小潮升	小潮升
NT(= net tonnage)	净吨位	淨噸位
nuclear cargo ship safety certificate	核动力货船安全证书	核子貨船安全證書
nuclear driven	核动力驱动	核能驅動
nuclear fuel	核燃料	核燃料
nuclear measurement system	核测量系统	核子測量系統
nuclear passenger ship safety certificate	核动力客船安全证书	核子客船安全證書
nuclear [powered] ship	核动力船	核子動力船
nuclear propulsion plant	核动力推进装置	核能推進裝置
nuclear reactor	核反应堆	核子反應器
nuclear reactor poisoning	核反应堆中毒	核子反應器中毒
nuclear submarine	核潜艇	核子潛艇
nude cargo	裸装货	裸貨
null point	哑点	電點,無效點,消盡點
numeral flag	数字旗	數字旗

英　文　名	祖 国 大 陆 名	台 湾 地 区 名
nursery ground	繁殖场	繁殖場
nutation	章动	章動(地軸之章動)

O

英　文　名	祖 国 大 陆 名	台 湾 地 区 名
oar	桨	槳
oarlock	桨门	槳架
obligation to render salvage service	救助义务	救助義務
oblique distance between ships	舰间斜距	船艦間斜距
obliquity of the ecliptic	黄赤交角	黃道斜度
obliquity of the moon path	黄白交角	白道斜度
obscured sector	遮蔽光弧	遮光弧
observation	观测	觀測,觀察
observation check	外观检查	外觀檢查
observation tank	观测柜	窺測油櫃
observed altitude	观测高度	觀測高度
observed altitude correction	观测高度改正	觀測高度修正
observed density	视密度	觀測密度
observed latitude	观测纬度	觀測緯度
observed longitude	观测经度	觀測經度
observed position(OP)	观测船位	觀測船位
obstruction	碍航物	障礙
obstruction sounding	障碍物探测	障礙物探測
occasional survey	临时检验	臨時檢查,臨時檢驗
occluded front	锢囚锋	包圍鋒
occulting light	明暗光	頓光
occupied bandwidth	占用带宽	佔用頻帶寬
ocean bill of lading	远洋提单	海運載貨證券
ocean current	海流(**洋流**)	洋流
ocean current chart	洋流图	洋流圖
oceaneering	海洋工程	海洋工程
oceangoing vessel(=ocean trader)	远洋船	遠洋船舶
ocean monitoring ship	海洋监测船	海洋監測船
ocean navigation	大洋航行	大洋航行
oceanographic research vessel	海洋调查船	海洋調查船
ocean passage	大洋航路	大洋航路
ocean region code	洋区码	洋區碼

英　文　名	祖国大陆名	台湾地区名
ocean regions	洋区	洋區
ocean sounding chart	大洋水深图	大洋水深圖
ocean station vessel	海洋定点船	海洋測候船
ocean trader	远洋船	遠洋船
ocean weather report	海洋气象报告	海洋氣象報告
ocean weather vessel	海洋大气船	大洋測候船
octans	南极星座	南極星座
O/D(=office of destination)	收报局	收報局
odorous cargo	气味货	氣味貨
off-centered display	偏心显示	偏心顯示
off-centering	中心偏移	偏心
off course(=off way)	偏航	偏航
off-hire	停租	停租
office of destination(O/D)	收报局	收報局
office of origin(O/O)	发报局	發報局
off-normal lower	下限越界	下限越界
off-normal upper	上限越界	上限越界
off-sea fishery	外海渔业	外海漁業
offset	偏移	偏位
offshore anchor	开锚	離岸錨
offshore area	近海区	離島地區
offshore drilling operation	近海钻井作业	近海鑽探作業
offshore fishery	近海渔业	近海漁業
offshore islands	近海岛屿	外圍島嶼,離島
offshore navigation	近海航行	近海航行
offshore platform	海上平台	海上平台
offshore survey	近海测量	海域測量
offshore tracks	外航路	外航路
offshore wind	吹开风	離岸風
off the wind	离风	離風
off way	偏航	偏航
oil	油类	油類
oil-air separator	油气分离器	油氣分離器
oil boom	围油栏	攔油索
oil burning unit	油燃烧器	燃油器
oil content meter	油分计	油含量計
oil discharge monitoring and control system	排油监控装置	洩油偵控系統
oil fence(=oil boom)	围油栏	攔油索

英 文 名	祖 国 大 陆 名	台 湾 地 区 名
oil-filled electrical transformer	充油变压器	充油變壓器
oil filtering equipment	滤油设备	濾油設備
oil free air compressor	无油润滑空气压缩机	無油潤滑空氣壓縮機
oil fuel	油类燃料	燃油
oil fuel filling station	加油站	加燃油站
oil fuel unit	燃油装置	燃油裝備組
oil loading terminal	装油港站	裝油終端站
oil lubricating	油润滑	油潤滑
oil mist detector	油雾浓度探测器	油霧偵測器
oil pollution	油污染	油污
oil pollution risk	船舶油污险	油污險
oil pressure differential controller	油压压差控制器	油壓差控制器
oil separator	油分离器	油水分離器
oil skimmer	浮油回收船	去油沫器,撇油器(捞油船)
oil storage platform	储油平台	儲油平台
oil storage tanker	储油船	貯油船
oil sump	油底壳	油槽
oil tank coating	油舱涂料	油艙塗料
oil tanker	油船	運油船
[oil] tanker anchorage	油船锚地	油輪錨地
oil terminal	油码头	油終端站
oil-tight	油密	油密
oil/water interface	油水界面	油水分界面
oil water interface detector	油水界面探测仪	油水分界面探測儀
oily bilge water	含油舱底水	含油舺水
oily mixture	含油混合物	含油混合物
oily rags	油破布	含油破布
oily residues	油类沉积物	油[類]殘留物
oily water	油污水	[含]油污水
oily water disposal boat	油污水处理船	油污水處理船
oily-water separating equipment	油水分离设备	油水分離設備
oily water separator	油水分离器	油水分離器
Omega	奥米伽	亞米茄
Omega chart	奥米伽海图	亞米茄海圖
Omega fix	奥米伽船位	亞米茄船位
Omega navigator	奥米伽导航仪	亞米茄航儀
Omega propagation correction	奥米伽传播改正量	亞米茄傳播修正值

英　文　名	祖　国　大　陆　名	台　湾　地　区　名
Omega signal format	奥米伽信号格式	亞米茄信號格式
Omega table	奥米伽表	亞米茄表
omnibus bill of lading	混合提单	合運載貨證券
omnidirectional fairleader	全向导缆器	全向導索器
omnidirectional radio range	全向无线电测距	萬向無線電測距
omnidirection radio beacon	全向无线电信标	萬向無線電示標
omnirange navigation	全向导航	全向導航
on board bill of lading	已装船提单	已裝船載貨證券
on-board communications	船上通信	船上通信
on-board communication station	船上通信电台	船上通信電台
on board training	船上训练	船上訓練
on board training record book	船上训练记录簿	船上訓練紀錄簿
on-condition maintenance	视情维修	視情況維修
on deck bill of lading	舱面货提单	艙面載貨證券
one fuel ship	使用一种燃油船舶	使用一種燃油船
one-way route	单向航路	單向航路
on-off two position regulator	双位式调节器	雙位開關調整器
on-scene commander	现场指挥	現場指揮
on-scene communication	现场通信	現場通信
on shore wind	吹拢风	向岸風
on the bow	前八字	前八字方向
on the fly	①飞行中 ②运动中	①飛行中 ②運動中
on the quarter	后八字	後八字方向
onus of proof	举证责任	舉證責任
O/O(=office of origin)	发报局	發報局
OP(=observed position)	观测船位	觀測船位
OPC(=Omega propagation correction)	奥米伽传播改正量	亞米茄傳播修正值
open bill of lading	空白提单	空白載貨證券
open charter	货港未定租船合同	任務待定備船契約
open combustion chamber	开式燃烧室	開式燃燒室
open cooling water system	开式冷却水系统	開式冷卻水系統
open cup test	开杯试验	開杯法試驗
open cycle gas turbine	开式循环燃气轮机	開口循環式燃氣輪機
open deck	露天甲板	露天甲板
open deck space	露天甲板空间	露天甲板空間
open end container	端开门箱	端開貨櫃
open-loop system	开环系统	開環系統
open network	开放网络	開放網絡

英 文 名	祖 国 大 陆 名	台 湾 地 区 名
open side container	侧开门箱	侧開貨櫃
open top container	敞顶箱	敞頂貨櫃
open type fuel valve	开式喷油器	開式燃油閥
open type hydraulic system	开式液压系统	開式液壓系統
operating draft	营运吃水	航務吃水
operating instruction	操作说明	操作說明
operating latitude	适用纬度	適用緯度
operating ship speed	适用航速	適用船速
operational command	操作指令	作戰指揮
operational level	操作级	操作級
operational pollution	操作污染	操作污染
operational readiness	操作准备	操作準備
operational wastes	作业废弃物	操作所生廢棄物
operation at sea	水上作业	海上作業
operation manual	操作说明书	操作說明書
operations and equipment manual	操作与设备手册	操作與設備手冊
operations control center	作业控制中心	作業控制中心
operation sequence	操作程序	操作程序
opposed piston diesel engine	对置活塞式柴油机	對衝活塞柴油機
optical theodolite	光学经纬仪	光學經緯儀
optimal control	最优控制	最佳控制
optimum decision	最佳决策	最佳決定
optimum load sharing	最佳负荷分配	最佳負載分配
optimum match point	最佳匹配点	最佳吻合點,最佳匹配點
optimum performance	最佳性能	最佳性能
optimum route	最佳航线	最佳航路
optimum routing	最佳航线拟定	最佳航線擬定
optimum speed	最佳航速	最佳航速
optimum track ship routing	最佳航迹定线	最佳船舶航路
optional cargo	选港货,选卸货	卸地待定貨
orange smoke signal	橙色烟号	橙色煙號
orbit prediction	轨道预报	軌道預報
order bill of lading	指示提单	指示載貨證券
order of ship formation	舰艇编队序列	艦艇編隊序列
ordinary flash	一般闪光	一般閃光
ordinary practice of seaman	海员通常做法	海員常規
ordinary seaman(OS)	二级水手	普通水手

英　文　名	祖国大陆名	台湾地区名
ordinary wear and tear	自然磨损	自然耗损
ore carrier	矿砂船	礦砂船
ore-oil carrier	矿砂–石油船	礦砂與油兼用船
organic acid	有机酸	有機酸
organic peroxide	有机过氧化物	有機過氧化物
orion	猎户座	獵戶座
OS(=ordinary seaman)	二级水手	普通水手
OSC(=on-scene commander)	现场指挥	現場指揮
otter board	网板	網板
otter trawler	单拖网船	單桅網漁船
otter trawling	单拖	單拖
outage	停用	停播,停用,停電
outboard	外舷	外檔
outboard bearing	外置轴承	船外軸承
outboard boom	舷外吊杆	舷外吊桿
outboard roller bearings	外置式圆柱滚子轴承	外置滾子軸承
outboard shot	锚端链节	[錨鏈]外短節
outboard work	舷外作业	舷外作業
out-foot	追过	追越
outhaul line	外牵索	外牽索
outhaul winch	拉出绞车	拉出絞機
out-of-band emission	带外发射	頻帶外發射
out of command	操纵失灵	操縱失靈
out of command light	操纵失灵号灯	操縱失靈號燈
out-of-phase diagram	p-V 转角示功图	失相圖
out of trim	前后吃水不当	前後吃水不當
output axis	输出轴	輸出軸
outside admission	外进汽	外邊進氣
outside micrometer	外径千分尺	外分厘卡,外測微計
outward turning	外旋	往外轉
over	通话完毕(无线电话用语)	通話完畢(無線電通話用語)
overboard discharge outlet	舷旁排出口	舷外排出口
overboard discharge valve	舷外排出阀	舷外排洩閥
overboard scupper	舷外排水孔	舷外排水孔
over current	过电流	過量電流
over-delivery	溢卸	溢卸
overflow pipe	溢流管	溢流管

英　文　名	祖国大陆名	台湾地区名
overflow tank	溢油柜	溢流櫃
overflow valve	溢流阀	溢流閥
over haul	大修	翻修,大修
overhaul	检修,拆卸检修	檢修,大修,翻修
overhead power cable	架空电缆	架空電纜
over-landed(=over-delivery)	溢卸	溢卸
overload capacity	过载能力	過載能力
overload limit	超负荷功率限	超載限制
overload of a sling	超重吊货	吊貨超重
overload test	过载试验	過負荷試驗,超載試驗
overload trip	过载脱扣	過載跳脫
overriding operational condition	超载工况	意外之操作狀況
over shoot	超越角	超越角
overshoot	惯性转头角	慣性轉頭角
overspeed governor	超速调节器	超速調速器
overspeed protection device	超速保护装置	超速保護設施
overspeed trip mechanism	超速跳闸机构	超速跳掣機構
overstowing arrangement	倒装	順序不當裝載
overstress	过度应力	超應力
overtaken vessel	被追越船	被超越船,被追趕船
overtaking	追越	追越
overtaking sound signal	追越声号	追越信號
overtaking vessel	追越船	超越船,追趕船
overturn(=capsize)	倾覆	傾覆
over voltage	过电压	過電壓
overwear	过度磨损	過度磨損
oxidation inhibitor	抗氧化剂	抗氧化劑
oxidation stability	氧化安定性	氧化穩定性
oxidizing substance	氧化剂	氧化物質
oxy-acetylene welding	氧乙炔焊	氧乙炔熔接
oxy-arc cutting	氧气电弧切割	氧氣電弧截割
oxygen analyser	测氧仪	氧氣分析器
oxy-hydrogen cutting	氢氧切割	氫氧截割
oxy-hydrogen welding	氢氧气焊	氫氧熔接
ozone generator	臭氧发生器	臭氧產生器

P

英　文　名	祖国大陆名	台湾地区名
PA(=position approximate）	概位	概略位置
P/A(=particular average）	单独海损	單獨海損
Pacific Ocean Region(POR）	太平洋区	太平洋區域
package	包装	包裝
packaged boiler	总装式锅炉	組合鍋爐
packaging code number	包装标号	包裝號碼
packaging group	包装类	包裝分類
packed cargo	包装货	包裝貨
packet switching	分组交换	分組交換
packet switching data network	数据分组交换网	數據分組交換網
packet switching telephone network	电话分组交换网	電話分組交換網
pack ice	浮冰群	塊冰
packing	包装	包裝
packing list	装箱单	裝箱單
packing materials	包装材料	包裝材料
PAD(=predicted area of danger）	预测危险区	預測危險區
paddle wheel	明轮	明輪
paddle wheel vessel	明轮推进器船	明輪船
pad eye	眼板	眼板
paint	①涂料 ②涂层 ③油漆	①塗料 ②塗層 ③油漆,漆
painter	艇首缆	艇首索
painting	涂漆	塗刷
paint room	油漆间	油漆間
paired frequencies	成对频率	成對頻率,頻率對
pair trawling(=twin trawling）	对拖	雙拖
palletized cargo	托盘货	托板貨
Panama Canal Tonnage	巴拿马运河吨位	巴拿馬運河噸位
panting beam	强胸横梁	抗拍梁
pantry	配餐间	配膳室
paper products	纸制品	紙製品
parachute signal	降落伞信号	降落傘信號彈
parallactic angle	星位角,视差角	星位角
parallactic displacement	视差移位	視差位移

英　文　名	祖国大陆名	台湾地区名
parallactic triangle	视差三角形	視差三角形
parallax	视差	視差
parallax in altitude	高度视差	高位視差
parallel body	平行舯体	平行舯段
parallel course	平纬航向	平緯航向
paralleling panel	并车屏	併俥屏
parallel of declination	赤纬圈	赤緯平行圈
parallel of latitude	纬线,纬[度]圈	緯度平行圈
parallel operation reactor	并车电抗器	並聯運轉反應器
parallel-running test	并联运行试验	並聯運轉試驗
parallel track search	平行航线搜寻	平行航線搜索
parameter non-uniform rate	参数不均匀率	參數不匀率
parameter setting	参数设定	參數設定
paramount clause	首要条款	首要條款
parcel freight	包裹运费	包裹運費
parking meter	靠泊表	靠泊表
part cargo	部分货载	部分貨載
partial eclipse	偏蚀	偏蝕,偏食
partially enclosed lifeboat	部分封闭救生艇	部分圍蔽救生艇
particular average	单独海损	單獨海損
particulate emission	颗料排放物	排放顆粒
particulate trap	颗粒收集器	顆粒收集器
partly filled compartment	部分装载舱室	部分裝載艙間
parts list	零件明细表	零件明細表
parts per million（ppm）	百万分之几	百萬分之幾
passage（=route）	航路	航路,航線
passage planning	航线设计	航線設計
passenger	旅客	旅客,乘客
passenger accommodation	旅客起居设备	旅客起居設備
passenger-cargo ship	客货船	客貨船
passenger ship	客船	客船
passenger ship safety certificate	客船安全证书	客船安全證書
passenger ticket	客票	客票
passing through the rapids	过[湍]滩	通過湍流
passing underneath	潜越	潛越
patching	补板	貼補
patrol boat	巡逻船	巡邏艇
patrol boat signal	巡逻艇信号	巡邏艇信號

英　文　名	祖国大陆名	台湾地区名
payload	有效载荷	有效負載
payment	支付	給付
payment of hire	租金支付	租金支付
pay off	偏离风向	轉向下風
pay out	松出	鬆出
PC(=point of collision)	碰撞点	碰撞點
PCA(=polar cap absorption)	极冠吸收	極冠吸收
PCC(=pure car carrier)	汽车运输船	車輛運輸船
P code(=precision code)	P 码	P 碼
PD(=position doubtful)	疑位	可疑位置
Peacock	孔雀(孔雀 α)	孔雀十一(孔雀 α)
peak envelope power	峰包功率	峰包功率
peak value	峰值	尖峰值
pedestal socket	吊艇柱座	凸式套座
peeler	剥皮机	剝皮機
pelagic survey	远海测量	遠洋測量
pelican hook	滑钩	滑鉤
pellet	托盘	托板
pelorus	哑罗经	啞羅經
penalty cargo	高费率货	工資加成貨
pendulous gyrocompass	摆式罗经	擺式電盤經
percentage of log correction	计程仪改正率	計程儀修正率
perfect gas	理想气体	理想氣體
performance monitor	性能监视器	性能監測器
performance of competence	适任	稱職
performance standard	性能标准	性能標準
performance verification test	性能验证测试	性能審認試驗
perigee	近地点	月近點
perihelion	近日点	近日點
perils of the sea	海上风险	海上危險
periodical inspection	周期性检查	定期檢查
periodical survey	定期检验	定期檢驗
periodical survey of cargo gear	起货设备定期检验	貨物裝卸設備定期檢驗
periodic wind	周期风	週期風
period measurement system	周期测量系统	定期測量系統
period of encounter	遭遇周期	遭遇週期
period of grace	宽限期	寬限期
period of hire	租期	租期

英 文 名	祖 国 大 陆 名	台 湾 地 区 名
period of pitching(=pitching period)	纵摇周期	縱搖週期
period of responsibility of carrier	承运人责任期间	運送人責任期
peripheral equipment	外围设备	週邊設備
periscope sextant	潜望六分仪	潛望六分儀
perishable cargo	易腐货	易腐貨
permanent repair	永久性修理	徹底檢查
permanent wind	恒定风	恒定風
permeability	渗透率	浸水率,浸透性
permissible length of compartment	许可舱长	艙區許可長度
permissible stress(=allowable stress)	许用应力	容許應力
permissible wear	容许磨损	容許磨損
perpendicular error	动镜差	垂直誤差
perpendicular replenishment at sea	航行垂直补给	海上垂直補給
personal locator beacon(PLB)	个人示位标	個人示位標
personnel locator beacon	人员定位标	人員定位示標
person in distress	遇险者	遇險人員
petrol engine	汽油机	汽油機
petty average	[港务]杂费	[港務]雜費
Phact	丈人一(天鸽 α)	丈人一(天鴿 α)
phase coding	相位编码	相位編碼
phase comparison	比相	比相
phase difference	相位差	相[位]差
phase modulation	调相,相位调制	相位調變
phase sequence	相序	相序
phase shift keying	相移键控	相移鍵移
phasing	定相	定相(傳真、真蹟用)
phenomena	天象纪要	天文現象
phone telex	话传用户电报	話傳交換電報
PHONETEX(=phone telex)	话传用户电报	話傳交換電報
photoeffect	光电效应	光電效應
photo electric flue gas detector	光电燃气探测器	光電煙道氣探測器
physical start-up	物理起动	物理起動
PI(=protection and indemnity)	保赔	防護賠償
PI club(=protection and indemnity club)	保赔协会	保護及賠償協會
pier	桩基码头	碼頭
piezoelectric effect	压电效应	壓電效應
pigtail	挠性接头,辫子	豬尾式接頭

英　文　名	祖国大陆名	台湾地区名
PI insurance	保赔保险	船舶營運人責任保險
piled mooring tower	桩基系泊塔	樁基繫泊塔
pillar	支柱	支柱
pillar buoy	柱形浮标	柱形浮標
pilot	引航员	引水,引水人
pilotage	①引航费 ②引航学	①引水費 ②引水術
pilot anchorage	引航锚地	引水錨地
pilot chart of inland waterway	内河引航图	内水引水圖
pilot hoist	引航员升降设备	引水人升降機
piloting	引航	地文導航法
piloting team	引航班	導航組
pilot ladder	引航员软梯	引水人梯
pilot station	引航站	引水站
pilot valve	导阀,控制阀	[嚮]導閥
pilot vessel	引航船	引水船
pipeline characteristic curve	管路特性曲线	管路特性曲線
pipeline fittings	管路附件	管路配件
pipeline layer	布管船	佈管船
pipeline mark	管线标	管線標
pipe tunnel	管隧	管道
piracy	海盗行为	海盜行爲
pirate	海盗	海盜
PI risk (=protection and indemnity risk)	保赔责任险	防護及賠償責任險
piston	活塞	活塞
piston assembly	活塞组件	活塞組件
piston-connecting-rod arrangement mis-alignment	活塞运动装置失中	活塞連桿裝置欠對準
piston cooling water pump	活塞冷却水泵	活塞冷卻水泵
piston crown	活塞头	活塞頂
piston crown ablation	活塞顶烧蚀	活塞頂燒蝕
piston pump	活塞泵	活塞式泵
piston ring axial clearance	活塞环平面间隙	活塞環軸向間隙
piston ring breakage	活塞环断裂	活塞環斷裂
piston ring gap clearance	活塞环搭口间隙	活塞環接口間隙
piston ring joint clearance (=piston ring gap clearance)	活塞环搭口间隙	活塞環接口間隙
piston ring sticking	活塞环黏着	活塞環膠著
piston ring wear monitoring system	活塞环磨损监测系统	活塞環磨損監測系統

英　文　名	祖国大陆名	台湾地区名
piston rod stuffing box	活塞杆填料函	活塞桿填料函
piston seizure	咬缸	咬缸，氣缸膠著
piston skirt	活塞裙	活塞裙
piston stroke	活塞行程	活塞衝程
piston underside pumping effect	活塞下部泵气功能	活塞下部泵效應
pitch	①螺距 ②沥青	①螺距 ②瀝青炭
pitch angle	螺距角	螺距角，週節角
pitch angle indicator	螺距角指示器	螺距角指示器
pitch damping control	纵摇阻尼控制器	縱搖阻尼控制
pitching	纵摇	縱搖
pitching period	纵摇周期	縱搖週期
pitch ratio	螺距比	節圓直徑比
pitometer log	水压计程仪	水壓計程儀
pitting	点蚀	斑蝕
pivoting point	枢心	樞心
PL(=proof load)	试验负荷	安全載重，安全負荷
placard	标牌	標貼
plain language	明语	明語
plane chart	平面图	平面圖
plane position indicator	平面位置显示器	平面位置指示器
planet apparent motion	行星视运动	行星視運動
planetary orbit	行星轨道	行星軌道
planing boat	滑行艇	滑航艇
plank stage	［作业］跳板	跳板
plank stage hitch	架板结	跳板結
plastic boat	塑料艇	塑膠艇
plastic garbage bag	塑料垃圾袋	塑膠垃圾袋
plate keel	平板龙骨	平板龍骨
plate-type evaporator	板式蒸发器	板式蒸發器
platform container	平台箱	平台貨櫃
platform deck	平台甲板	台甲板
PLB(=personal locator beacon)	个人示位标	個人示位標
pleasure yacht(=yacht)	游艇	遊艇
plot a course	标绘航线	標繪航線
plot a distance	标绘距离	標繪距離
plotting chart	空白定位图	作業圖
plum rain(=Meiyu)	梅雨	梅雨
plunger and sleeve assembly	柱塞套筒组件	柱塞與套筒組件

英　文　名	祖国大陆名	台湾地区名
plunger matching parts	柱塞偶件	柱塞偶合件
plunger pump	柱塞泵	唧子泵,柱塞泵
pneumatic amplifier	气动放大器	氣力放大器
pneumatic power amplifier	气动功率放大器	氣動功率放大器
pneumatic regulator	气动调节器	氣力調整器
pneumatic remote control system for main 　　diesel engine	柴油主机气动遥控系统	柴油主機氣力遙控系統
point of collision(PC)	碰撞点	碰撞點
point positioning	单点定位	單點定位
poisonous cargo	有毒货	毒性貨
poisonous exhaust composition	排气有害成分	毒性排氣成分
poisonous substance	有毒物质	有毒物質
polar altitude	仰极高度	仰極高度
polar cap absorption(PCA)	极冠吸收	極冠吸收
polar coordinate method	极坐标法	極坐標法
polar distance	极距	極距
polar ice	极地冰	極冰
Polaris	北极星	北極星
Polaris correction	北极星高度改正量	北極星修正
polarization	极化	極化
polarization error	极化误差	極化誤差
polar navigation	极区航行	極區航行
polar orbiting satellite service	极轨道卫星业务	繞極軌道衛星業務
polar slide valve diagram	极坐标滑阀图	滑閥極坐標圖
pole mast	柱状桅	柱狀桅
political officer	政委	政工官
polling	查询	詢訊
pollutants	污染物	污染物
pollution category	污染类别	污染類別
polyvalent officer	多专长高级船员	多專長甲級船員
pontoon	①浮码头 ②趸船	①浮箱 ②駁船
pontoon bridge	浮桥	浮橋
pontoon cover	箱形舱盖	箱形艙蓋
pontoon hatch cover(=pontoon cover)	箱形舱盖	箱形艙蓋
poop	艉楼	艉艛
poop anchor	艉锚	艉錨
poop deck	艉楼甲板	艉艛甲板
population size	资源量	資源量

英 文 名	祖国大陆名	台湾地区名
porpoising	跃水现象	躍水現象
port	驳门	駁門
POR(=Pacific Ocean Region)	太平洋区	太平洋區域
portable fan	可移式风机	可攜式風扇
portable radio apparatus for survival craft	救生艇筏手提无线电设备	艇用輕便無線電設備
portable tank	移动罐柜	可攜式櫃
port capacity	港口通过能力	港口能量
port captain	指导船长	駐埠船長
port charter party	港口租船合同	港口租船契約
port clearance	出口许可证	結關出口
port engine	左舷发动机	左舷引擎
port hand buoy	左侧浮标	左舷通過浮標
port hole	舷窗	舷窗
port management	港口管理	港口管理
port of arrival	到达港	到達港
port of call	挂靠港	寄泊港
port of departure	出发港	出發港
port of destination	目的港	目的港
port of discharge	卸货港	卸貨港
port of loading	装货港	裝貨港
port of origin port of sailing	始发港	始航港
port of refuge	避难港	避難港
port of refuge expenses	避难港费用	避難港費用
port of registration	船籍港	船籍港
port of sailing	始发港	始航港
port of transshipment	转口港	轉口港
port operation service	港口营运业务	港埠營運業務
port regulations	港章	港口規章
port side	左舷	左舷
port state	港口国	港口國
port state control(PSC)	港口国管理	港口國管制
port station	港口电台	港埠電台
port's cargo throughput	港口吞吐量	港口貨物吞吐量
positional stability	位置稳定性	位置穩定性
position approximate(PA)	概位	概略位置
position difference	船位差	船位差
position doubtful(PD)	疑位	可疑位置

英　文　名	祖国大陆名	台湾地区名
position fix	定位	定位
position indicating mark	示位标	示位標
positioning anchor	定位锚	定位錨
positioning of engine frame	机架定位	機架定位
position line by bearing	方位位置线	方位位置線
position line by distance	距离位置线	距離位置線
position line by distance difference	距离差位置线	距離差位置線
position line by horizontal angle	水平夹角位置线	水平角位置線
position line by vertical angle	垂直角位置线	垂直角位置線
position line standard error	位置线标准差	位置線標準誤差
position line transferred	转移位置线	轉移位置線
position report	船位报告	船位報告
position signal code	位置信号码	位置信號碼
positive displacement pump	容积泵	排量式泵
positive feedback	正反馈	正反饋
positive steam distribution	正蒸汽分配	正蒸汽分配
possible point of collision(PPC)	可能碰撞点	可能碰撞點
potential energy	位能,势能	位能,勢能
potential fisheries resources	潜在渔业资源	潛在的漁業資源
pounding	拍击	顫抖
pouring water test	泼水试验	澆水試驗
pour point	倾点	流動
pour point depressant	降凝剂	流動點下降劑
power conversion	功率换算	功率換算
power density	功率密度	輸出密度(原子爐)
power driven vessel	机动船	動力船舶
power limit for continuous running	持续运转功率	連續運轉功率限
power load	有功负荷	動力負載
power margin	功率储备	功率餘裕
power piston	动力活塞	動力活塞
power plant effective specific fuel oil con-sumption	动力装置燃油消耗率	動力裝置燃油消耗率
power plant [effective] thermal efficiency	动力装置[有效]热效率	動力裝置[有效]熱效率
power plant specific mass	动力装置单位质量	動力裝置單位質量
power plant viability	动力装置生命力	動力裝置壽命
power reserve(=power margin)	功率储备	功率餘裕
power stroke	动力行程	動力衝程

英 文 名	祖 国 大 陆 名	台 湾 地 区 名
power take-in drive	功率输入传动装置	功率輸入傳動裝置
power-to-volume ratio	功率容积比	功率容積比
power turbine	动力涡轮	動力輪機
PPC(=possible point of collision)	可能碰撞点	可能碰撞點
PPI(=plane position indicator)	平面位置显示器	平面位置指示器
ppm(=parts per million)	百万分之几	百萬分之幾
PPS(=precise positioning service)	精密定位业务	精密定位業務
practice	实践	實務
practice area	演习区	演習區
practice training	实作训练	實作訓練
preamble	报头	報頭
pre-amplifier	前置放大器	前置放大器
precautionary area	警戒区	警戒區
precession	岁差	歲差
precipitous sea	怒涛(9 级)	怒濤(9 級)
precise ephemeris	精密星历	精密天文曆
precise positioning service	精密定位业务	精密定位業務
precision accuracy	精度	精度
precision code(P code)	P 码	P 碼
precombustion chamber	预燃室	預燃室
predicted area of danger(PAD)	预测危险区	預測危險區
pre-exciting switch	充磁开关	激磁開關
preliminary notice	预告	預告
preliminary voyage	预备航次	預備航程
pressure and vacuum breaker	压力真空切断阀	壓力真空斷路器
pressure-compounded turbine	复式压力级涡轮机	複壓渦輪機
pressure-control fuel valve	启线压力可调式喷油器	壓力控制式燃油閥
pressure-control valve	压力控制阀	壓力控制閥
pressure ratio	压比	壓力比
pressure reducing valve	减压阀	減壓閥
pressure regulator	压力调节器	制壓器,調壓器,壓力調 整裝置
pressure stage	压力级	壓力級
pressure system	气压系统	壓力系統
pressure vacuum relief valve	真空安全阀	呼吸閥
pressure vessel	压力壳	壓力容器
prevailing wind	盛行风	盛行風
prevention of pollution by garbage	防止垃圾污染	防止垃圾污染

英 文 名	祖 国 大 陆 名	台 湾 地 区 名
prevention of pollution by oil	防止油污染	防止油污染
prevention of pollution by sewage	防止生活污水污染	防止污水污染
preventive maintenance	定期预防维修	预防保養
prewash	预洗	预洗
prima facie evidence	初步证据	表面證據
primary distribution system	一次配电系统	一次配電系統
primary field	一次场	一次場
primary loop	一回路	一次迴路
primary-secondary clocks	子母钟	子母鐘
primary shield water system	一次屏蔽水系统	一次屏蔽水系統
primary stress	主应力	主應力
prime mover automatic starter	原动机自动起动装置	原動機自動起動器
prime vertical(PV)	卯酉圈	卯酉圈
priming	起动注油注水	起動注給
priming pump	初给泵	起動泵
printing finished signal	打印结束信号	打印結束信號
prismatic coefficient	棱形系数	[縱向]稜塊係數,稜形係數
private aid to navigation	私用助航标	私用助航標
probable error	概率误差	或然差
probable track area	概率航迹区	或然航跡區
procedures and arrangement manual	程序和布置手册	程序和佈置手冊
procedure signal	程序信号	程序信號
process control	程序控制	程序控制
processing time	处理时间	處理時間
processor mode	处理器模式	處理器模式
process unit	处理设备	處理設備
Procyon	南河三(小犬 α)	南河三(小犬 α)
product carrier	成品油船	成品油[運載]船
professional competence	称职	稱職
proficiency	熟练	熟練
programmable read only memory	可编程只读存储器	可程式唯讀記憶體
programmed control(=process control)	程序控制	程式控制
progressive wave	前进波	前進波
prohibited articles	违禁物品(违禁品)	違禁品
prolonged blast	长声	長聲
promissory note	期票	期票
prompt critical accident	瞬发临界事故	瞬發臨界事故

英　文　名	祖国大陆名	台湾地区名
prompt loading	即期装船	即時裝載
prompt ship	即期船	即期船
proof load(PL)	试验负荷	安全載重,安全負荷
proof test for accommodation ladder	舷梯强度试验	舷梯安全限試驗
proof test for ship cargo handling gear	起货设备吊重试验	貨物裝卸設備安全限試驗
propagation error	传播误差	傳播誤差
propagation path	传播路径	傳播路徑
propeller	推进器	螺槳,推進器
propeller characteristic	螺旋桨特性	螺槳特性
propeller impeller	带罩叶轮	帶罩葉輪
propeller racing	飞车	螺槳空轉
propeller statical equilibrium	螺旋桨静平衡	螺槳靜平衡
propeller submergence	螺旋桨沉深	螺槳深沈
property	财物	財物
proportional band	比例带	比例帶
proportional regulator	比例调节器	比例調整器
proportioner(=proportional regulator)	比例调节器	比例調整器
propulsion characteristic	推进特性	推進特性
propulsion device	推进装置	推進設施
protecting cargo	防护货	防護貨
protection against single-phasing	单相运行保护	單相保護
protection and indemnity(PI)	保赔	防護賠償
protection and indemnity club(P&I club)	保赔协会	保護及賠償協會
protection and indemnity risk(PI risk)	保赔责任险	防護及賠償危險
protective clothing	防护衣	防護衣
protective location	保护位置	保護位置
protective location of segregated tank ballast	专用压载舱保护位置	隔離壓艙水艙保護位置
protective mark	保护标志	保護標誌
Protocol Relating to Intervention on the High Seas in Case of Pollution by Substances other than Oil	干预公海非油类物质污染议定书	油以外物質污染事故在公海行使干涉議定書
provisional release	临时许可	臨時放行
provision room	粮食库	糧食庫
proximate cause	近因	近因
proximity limit switch	极限开关	極限開關
PSC(=port state control)	港口国管理	港口國管制

英　文　名	祖国大陆名	台湾地区名
pseudolite	伪卫星	僞訊號
pseudo range	伪距	僞距
PSK(=phase shift keying)	相移键控	相移鍵移
psychrograph	干湿计	乾濕計
psychrometer	干湿表	空氣濕度計,乾濕表
public correspondence	公众通信	公衆通信
public correspondence service	公众通信业务	公衆通信業務
public space	公用处所	公用空間
public switched data network	公用数据交换网络	公衆交換數據網路
public switched telephone network	公用电话交换网络	公用交換電話網路
pulsating magnetic field	脉动磁场	脈動磁場
pulse converter supercharging	脉冲转换器增压	脈衝轉換器增器
pulse 8 positioning system	脉8定位系统	脈八定位系統
pulse turbocharging	脉冲式涡轮增压	脈衝渦輪增壓
pump auto-change over device	泵自动切换装置	泵自動切換設施
pump capacity	泵流量	泵能量
pump characteristic curve	泵特性曲线	泵特性曲線
pump dredger	吸扬式挖泥船	泵吸挖泥船
pump head	泵压头	泵壓頭
pumping	泵吸	抽出
pump room sea valve	泵舱通海阀	泵室海水閥
pump timing mark	油泵定时标记	泵定時記號
pure car carrier(PCC)	汽车运输船	車輛運輸船
purge	扫气	驅氣
purification system	净化系统	淨化系統
purifier	分水机	淨油機
purse line	括纲	收縮網
purser	事务员	事務員
purse seine	围网	圍網,巾著網
purse seiner	围网渔船	圍網漁船
pushboat	推船	推船
push button "down"	"降速"按钮	"降速"按鈕
push button "up"	"增速"按钮	"增速"按鈕
pushed beam	承推梁	承推梁
pusher(=pushboat)	推船	推船
pusher train	顶推船队	推駁船隊
pushing	顶推	推頂
pushing frame	顶推架	推頂架

英 文 名	祖 国 大 陆 名	台 湾 地 区 名
pushing gear	顶推装置	推頂裝置
pushing post	顶推柱	推頂柱
pushing steering line	顶推操纵缆	推頂操舵纜
push type	推进式	推進式
put away	驶离	駛離
put off	出发,离岸	駛離岸
put out	驶出	駛出
put to sea	出海	出海
PV(=prime vertical)	卯酉圈	卯酉圈
p-V indicated diagram	p-V 示功图	p-V 示功圖
pyrometer	高温计	高溫計
pyrometer probe	测温传感器	高溫計傳感器
pyrotechnic signal	烟火信号	煙火信號
p-ϕ indicated diagram	p-ϕ 示功图	p-ϕ 示功圖

Q

英 文 名	祖 国 大 陆 名	台 湾 地 区 名
q-ship	伪装商船	偽裝商船
quadrantal deviation	象限自差	象限自差
quality assurance	质量保证	品質保證
quality control	质量管理	品[質]管[制]
quality of the bottom	底质	底質
quantity production	大量生产	大量生產
quarantine	检疫	檢疫
quarantine anchorage	检疫锚地	檢疫錨地
quarantine vessel	检疫船	檢疫船
quarter	艉舷	艉部
quartering sea	艉舷浪	艉側浪
quarter ramp	艉斜跳板	艉斜跳板
quartz pressure sensor	石英压力传感器	石英壓力測感器
quenching	淬火	淬火
quick action closing device	速闭装置	快速關閉設施
quick-closing emergency valve	应急速闭阀	應急速閉閥
quick closing valve	速闭阀	快閉閥
quick flashing light	急闪光	快速閃光
quick release buckle	速脱扣	速脱扣
quick release coupling	快速接头	快釋接頭

R

英　文　名	祖国大陆名	台湾地区名
RA（=right ascension）	赤经	赤經
racing	螺旋桨空转	空轉
racing boat	赛艇	競賽艇
rack and pinion	齿条–齿轮	齒條與小齒輪
racon	雷达信标	雷達信標
radar beacon（=racon）	雷达信标	雷達信標
radar chart	雷达海图	雷達海圖
radar echo-box	雷达回波箱	雷達回波箱
radar identification	雷达识别	雷達識別
radar mast	雷达桅	雷達桅
radar navigation	雷达导航	雷達導航
radar navigation chart	雷达引航图	雷達導航圖
radar performance monitor	雷达性能监视器	雷達性能監測器
radar plotting	雷达标绘	雷達測繪
radar reflector	雷达反射器	雷達反射器
radar simulator	雷达模拟器	雷達模擬機
radar transponder	雷达应答器	雷達詢答器
radial ball bearing	径向球轴承	徑向滾珠軸承
radial engine	星型发动机	星型發動機
radial-flow compressor	径流式压缩机	徑向流壓縮機
radial-flow turbine	径流式涡轮	徑向流渦輪機
radial flow turbocharger	径流式涡轮增压器	徑向流渦輪增壓器
radial-piston hydraulic motor	径向柱塞式液压马达	徑向活塞液力馬達
radiant heat	辐射热	輻射熱
radioactive solid waste storage tank	放射性废物箱	固體輻射性廢棄物儲存箱櫃
radioactive substance	放射性物质	放射物質,放射性物質
radioactive waste water tank	放射性废水箱	輻射性廢水箱櫃
radio beacon	无线电信标	無線電指示標
radiobeacon station	无线电信标电台	無線電示標電台,無線電示台
radio bearing	无线电方位	無線電方位
radio bearing position line	无线电方位位置线	無線電方位位置線
radiocommunication	无线电通信	無線電通信

英 文 名	祖 国 大 陆 名	台 湾 地 区 名
radiocommunication service	无线电通信业务	無線電通信業務
radio determination	无线电测定	無線電測定術
radio determination-satellite service	卫星无线电测定业务	衛星無線電測定業務
radio determination service	无线电测定业务	無線電測定業務
radio determination station	无线电测定电台	無線電測定電台
radio direction finder	无线电测向仪	無線電測向儀
radio direction finder deviation	无线电测向仪自差	無線電測向儀自差
radio direction finding	无线电测向	無線電測向
radio direction-finding station	无线电测向电台	無線電探向電台
radiographic inspection	X 射线照相探伤	放射線檢查
radio great circle bearing	无线电大圆方位	無線電大圓方位
radio link	无线线路	無線電鏈
radiolocation	无线电定位［术］	無線電定位［術］
radiolocation land station	无线电定位陆地电台	陸上無線電定位電台
radiolocation mobile station	无线电定位移动电台	行動無線電定位電台
radio location service	无线电定位业务	無線電探向電台
radio log	无线电台日志	無線電日誌
radio maritime letter	海上无线电书信	海上無線電書信
radio navigation	无线电导航	無線電導航
radio navigational lattice	无线电导航图网	無線電導航網路圖
radio navigational warning	无线电航行警告	無線電航行警告
radio navigation land station	无线电导航陆地电台	陸上無線電導航電台
radio navigation mobile station	无线电导航移动电台	行動無線電導航電台
radio navigation service	无线电导航业务	無線電導航業務
radio officer	无线电报员	無線電人員
radio operator	无线电操作员	無線電操作人員
radio personnel	无线电人员	無線電人員
radio pratique message	无线电免检电报	無線電檢疫信文
radio regulation	无线电规则	無線電規則
radio relay system	无线电接力系统	無線電中繼系統
radio room	无线电室	無線電室
radio service	无线电业务	無線電業務
radio sextant	射电六分仪	無線電六分儀
radio station	无线电台	無線電台
radiotelegram	无线电报	無線電報
radiotelegraph auto-alarm	无线电报自动报警［器］	無線電報自動警報［器］
radiotelegraph installation	无线电报设备	無線電報裝置
radiotelegraph installation for lifeboat	救生艇无线电报设备	救生艇無線電報裝置

英　文　名	祖国大陆名	台湾地区名
radiotelephone alarm signal generator	无线电话报警信号发生器	無線電話警報信號産生器
radiotelephone call	无线电话呼叫	無線電話呼叫
radiotelephone distress frequency	无线电话遇险频率	無線電話遇險頻率
radiotelephone installation	无线电话设备	無線電話裝置
radiotelephone officer	无线电话员	無線電話務員
radiotelephone service	无线电话业务	無線電話業務
radio telephony	无线电话学	無線電話術
radio telex call	无线电用户电报呼叫	無線電傳呼叫
radio telex letter	无线电用户电报书信	無線電交換電報書信
radiotelexogram	无线电用户电报电文	無線電傳電報
radio telex service	无线电用户电报业务	無線電電報交換業務
radio theodolite	无线电经纬仪	無線電經緯儀
radio time signal	无线电时号	無線電對時信號
radio true bearing(RTB)	无线电真方位	無線電真方位
Radio Watch	无线电值守	無線電當值
radio weather service	无线电气象业务	無線電氣象業務
radius of the earth	地球半径	地球半徑
radome	天线罩	天線外罩
rafted ice	重叠冰	載冰
rags	破布	破布
rail	轨	軌
rainfall	雨量	雨量
raised quarter-deck vessel	艉升高甲板船	高艉主甲板船
raising by dewatering with compressed air	压气排水打捞	壓縮空氣排水浮升
raising by injection plastic foam	充塞泡沫塑料打捞	噴以泡沫塑膠浮升
raising by sealing patching and pumping	封舱抽水打捞	封艙抽水浮升
raising of a wreck	沉船打捞	打撈沉船
raising with salvage pontoons	浮筒打捞	用救難浮箱浮升
raked bow	前倾[型]艏	斜艏
raked stem(=raked bow)	前倾[型]艏	斜艏
ramark	雷达指向标	雷達航標
ram tensioner	撞锤张力器	撞鎚張力機
ram-wing craft	①气翼艇 ②冲翼艇	①氣翼艇 ②衝翼艇
random error	随机误差	偶發誤差
range finder	测距仪	測距儀
range marker	固定距标	距離指標
range of audibility	可听距离	聞距

英　文　名	祖国大陆名	台湾地区名
range of visibility	能见距	能見距
range rate	潮差比	潮距率
range resolution	距离分辨力	距離分解［度］
rapid stream	急流	急流
RAS(=replenishment-at-sea)	海上补给	海上整補
Rasalhague	候(蛇夫α)	候(蛇夫α)
ratchet gear	棘轮装置	棘輪裝置
rated breaking capacity	标定断开容量	額定斷電容量
rated engine speed	标定转速	額定引擎轉速
rated load weight	额定起重量	額定載量
rated making capacity	标定接通容量	額定接續容量
rated output	标定功率	額定功率
rated power(=rated output)	标定功率	額定功率
rated stock torque	标定转舵扭矩	額定舵桿扭矩
rated working condition	标定工况	額定工作狀況
rate of bearing variation	位变率	方位變動率
rate of distance variation	距变率	距離變動率
rate of transverse motion	横移率	橫移率
rat guard	防鼠挡	防鼠板,防鼠罩
rating	普通船员	乙級船員
rating corrections	标定功率修正	定額修正
ratio of broken space	亏舱率	堆貨餘隙率
ratline hitch	梯级结	梯級結
RCC(=Rescue Coordinator Center)	搜救协调中心	搜救協調中心
reach	横风航驶	橫餓
reaction formulas	化学方程式	反應式(化學)
reaction stage	反动级	反應級
reaction steam turbine	反动式汽轮机	反動式汽輪機
reactive power(=wattless power)	无功功率	無功功率
reactor control system	核反应堆控制系统	核反應器控制系統
reactor period	反应堆周期	反應器週期
reactor protective system	核反应堆保护系统	核反應器保護系統
realtime mode	实时方式	實時模式
rear ship	殿后舰	殿後船艦
rear view	后视图	後視圖
reasonable despatch	合理速遣	合理派遣
received for shipment bill of lading	收货待运提单	候裝載貨證券
receiver	①接收机 ②储液器	①接收機 ②受液器

英 文 名	祖国大陆名	台湾地区名
receiving antenna	接收天线	接收天線
receiving point	受理点	接收點
receiving ship	接受船	受補船
receiving station	接收站	接收電台,接收站
reception facilities	收受设施	收受設施
reciprocating engine	往复式发动机	往復機
reciprocating pump	往复泵	往復泵
reciprocating pump in series	串联往复泵	串聯往復泵
reciprocating refrigeration compressor	往复式制冷压缩机	往復冷凍壓縮機
reciprocating type steering gear	往复式转舵机构	往復式操舵裝置
recommended route	推荐航线	推薦航路
recording tachometer	自记式转速表	紀錄轉速計
record on spot	现场纪录	現場紀錄
recovering repair	基本恢复修理	修復
recovery line	回收索	回收索
rectilinear current(=reciprocating pump)	交流电	交流電,直線流
red bill of lading	红色提单	附帶保險載貨證券
redelivery of vessel	还船	還船
red lead paint	红丹	紅丹漆
red star-signal	红色信号	紅星信號
red tide	赤潮	赤潮
reduced carrier emission	减载波发射	減載波發射
reduced-voltage starting	降压起动	降壓起動
reduction gear	减速齿轮	減速齒輪
reduction ratio	减速比	減速比,縮減比
reduction to the meridian	近中天高度改正	折合中天高度
red water (=red tide)	赤潮	赤潮
Redwood viscosity	雷氏黏度	雷氏黏度
reef	缩帆	摺帆葉
reefer container	冷藏箱	冷凍貨櫃
reef knot	平结	平結
reel	绳车	捲盤
refined products	炼制品	精煉品
refining plant	净化装置	淨化裝置
reflector	反射器	反射器
reflector compass	反射罗经	反射羅經
refloat	脱浅	浮飄
refloating force	出浅力	再浮力

英　文　名	祖国大陆名	台湾地区名
refraction	蒙气差,折光差	折光差,折射
refractory material	耐火材料	耐火材料
refrigerant	制冷剂	冷凍劑,冷媒
refrigerant metering device	制冷剂计量装置	冷凍劑計算裝置
refrigerated cargo clause	冷藏货条款	冷凍貨條款
refrigerated cargo hold	冷藏货舱	冷凍貨艙
refrigerated chamber	冷冻间	冷凍室
refrigerated room	冷藏间	冷凍室
refrigerated space(=refrigerated room)	冷藏间	冷凍室
refrigerating capacity	制冷量	冷凍能量
refrigerating effect per brake horse power	单位轴马力制冷量	單位制動馬力冷凍效果
refrigerating effect per unit swept volume	单位容积制冷量	單位氣缸容積冷凍效果
refrigerating medium pump	冷剂泵	冷媒泵
refrigerating ton	制冷吨	冷凍噸
refrigeration agent(=refrigerant)	制冷剂	冷凍劑,冷媒
refrigeration cargo	冷藏货	冷藏貨
refrigeration cycle	制冷循环	冷凍循環
refrigeration system	制冷系统	冷凍系統
refrigerator oil	冷冻机油	冷凍油
refrigerator ship	冷藏船	冷凍船,冷藏船
regenerative braking	再生制动	再生制動
regenerative cycle gas turbine	回热循环燃气轮机	回熱循環燃氣渦輪機
regenerative steam turbine	回热式汽轮机	回熱式蒸汽渦輪機
register	注册	註冊,登記
registered breadth	登记宽度	登記寬度
registered depth	登记深度	登記深度
registered length	登记长度	登記長度
register of cargo handling gear of ship	船舶起货设备检验簿	船舶貨物裝卸設備登記簿
register of shipping	船级社	船級機構
registration of alteration	变更登记	變更登記
registration of withdrawal	注销登记	撤銷登記
Registro Italiano(RI)	意大利船级社	意大利驗船協會
regular maintenance	定期维修	定期保養
regular repair	期修	定期檢修
regular wave	规则波	規則波浪
regulating lock light	节制闸灯	調整閘燈
regulating rod	调节棒	調整桿

英　文　名	祖国大陆名	台湾地区名
regulations for ship formation movement	舰艇编队运动规则	船艦編隊運動規則
regulations of fishery harbor	渔港规章	漁港規則
Regulus	轩辕十四(狮子α)	軒轅十四(獅子α)
reheater	再热器	重熱器,再熱器
reheat factor	重热系数,再热系数	重熱因素
reheat steam turbine	再热式汽轮机	再熱式蒸汽渦輪機
Reid vapor pressure	雷德蒸汽压力	瑞德蒸氣壓
reinforced concrete vessel	钢筋水泥船	鋼筋混凝土船
reinforced rib	加强肋	加強肋
rejection risks	货物拒收险	貨物拒收險
relative bearing	舷角,相对方位	相對方位
relative bearing of radio	无线电舷角	無線電相對方位
relative density	相对密度	相對密度
relative efficiency	相对效率	相對效率
relative humidity	相对湿度	相對濕度
relative log	相对计程仪	相對計程儀
relative motion display	相对[运动]显示	相對運動顯示
relative motion radar(RM radar)	相对运动雷达	相對運動雷達
relative positioning	相对定位	相對定位
relative rotor displacement	转子相对位移	轉子相對位移
relative speed	相对速度	相對速[率]
relative speed of wind	相对风速	相對風速
relay	继电器	繼電器
release	释放	釋放
reliability	可靠性	可靠性
relief valve	卸压阀	洩壓閥,保險閥
remaining deviation	剩余自差	剩餘自差
remark list	批注清单	批註清單
remote control abnormal alarm	遥控异常报警	遙控異常警報
remote control mine-sweeping navigation service	遥控扫雷航海勤务	遙控掃雷航海勤務
remote water level indicator	远距离水位指示计	遠隔水位指示器
removing mud around wreck	船外除泥	船外除泥
removing rust	除锈	除銹
renewal survey	换证检验	換證檢驗
Renold's number	雷诺数	雷諾數
repair list	修理单	修理單
repair port	修理港	修理港

英　文　名	祖国大陆名	台湾地区名
repeat(RPT)	重复	重複
repeated starting sequence	重复起动程序	重複起動程序
repeater mode	重发器模式	重發器模式
rephasing	重新定相	重行定相
replenishing ship	补给舰	補給艦
replenishment at sea(RAS)	海上补给	海上整補
replenishment course	补给航向	整補航向
replenishment distance abeam	补给横距	整補橫距
replenishment distance astern	补给纵距	整補縱距
replenishment speed	补给航速	整補航速
replenishment station	补给阵位	整補站
report	报告	報告
reporting point(RP)	报告点	報告點
reporting procedures	报告程序	報告程序
report of clearance	出口报告书	結關報告
re-radiation	二次辐射	二次輻射
rescue basket	救生吊蓝	救生吊籃
rescue boat	救助艇	救難艇
Rescue Coordination Center (RCC)	搜救协调中心	搜救協調中心
rescue coordinator center	救助协调中心	救難協調中心
rescue litter	救生担架	救生擔架
rescue net	救生网	救生網
rescue seat	救生吊座	救生吊座
rescue sling	救生吊带	救生吊帶
rescue sub-center (RSC)	救助分中心	救助分中心
rescue unit(RU)	救助单位	救助單位
reserve	储备	備用
reserve antenna	备用天线	備用天線
reserve buoyancy	储备浮力	預[留]浮力
reserve receiver	备用收信机	備用接收機
reserve transmitter	备用发信机	備用發射機
reset	复位,重调	重置,復歸
residual current	余流	剩餘電流
residual dynamical stability	剩余动稳性	剩餘動穩度
residual fuel oil	残渣油	殘油
residual gas	残余废气	殘留氣體
residual oil standard discharge connection	残油标准排放接头	殘油標準排洩接頭
residual righting lever	复原稳性力臂	剩餘扶正力臂

英　文　名	祖国大陆名	台湾地区名
residue cargo	不在第一港卸的货	留船未卸貨
resistant to liquid (=liquid-tight)	液密	液密
resistant to oil (=oil-tight)	油密	油密
resolved position	确信位置	經確定之位置
resonance	共振,谐振	共振,諧振
respondentia	货抵押贷款	船貨押貸
responder beacon	转发信标	轉發信標
response lag	响应滞后	反應落後
responsible operator position	负责操作人员的职务	負責作業人員之職位
restoration	恢复	復原,回復
restricted landing area mark	限制降落区标志	限制降落區域標誌
restricted operator's certificate (ROC)	限用操作员证书	限用值機員證書
restriction	限制	限制
resultant	合力	合力
retractable fin stabilizer	伸缩式减摇鳍装置	伸縮式鰭板穩定器
retrieval	收回	救回
retro-reflective material	反光材料	反光材料
return air	回风	回風
return cargo	回程货	回程貨
revenue cutter	辑私船	缉私船,缉私艇
reverse	反向	反向,回動
reverse current test	逆电流试验	逆流試驗
reverse osmosis method	反渗透法	反滲透法
reverse power protection	逆功率保护	逆功率保護
reverse power test	逆功率试验	逆功率試驗
reverse spiral test	逆螺旋试验	逆蝸旋試驗
reverse starting sequence	换向起动程序	換向起動程序
reverse steam brake	回汽刹车	回汽煞俥
reverse stopping distance	倒车冲程	倒俥停俥距離
reverse towing	倒拖	倒拖
reversible diesel engine	可倒转柴油机	可逆轉柴油機
reversible pump	变向泵	可逆泵
reversing	换向	逆轉
reversing arrangement	换向装置	逆轉裝置
reversing interlock	换向联锁	逆轉聯鎖
reversing servomotor	换向伺服器	回動伺服電動機
reversing time	换向时间	換向時間
revolution counter	转数计数器	轉數計

英　文　名	祖国大陆名	台湾地区名
revolution per minute	每分钟转数	每分鐘轉數
revolution speed of propeller	螺旋桨转速	螺槳轉速
reward	报酬	報酬
RF(=running fix)	移线船位	航進定位
rheotaxis	向流性	向流性
rhumb line	恒向线	恒向線
rhumb line bearing(RLB)	恒向线方位	恒向線方位
rhumb line sailing	恒向线航线算法	恒向線航法
RI(=Registro Italiano)	意大利船级社	意大利驗船協會
ridge	高压脊,脊	隆起緣,脊
ridge line	脊线	脊線
riding anchor	力锚	著力錨
riding one point anchors	一点锚,平行锚	一字力錨
riding to single anchor	单锚泊	單錨泊
riding to two anchors	八字锚泊	[八字]雙錨泊
Rigel	参宿七(猎户β)	参宿七(獵戶β)
rigging	索具	索具
rigging screw	松紧螺旋扣	伸縮螺絲,緊索螺絲
right ascension(RA)	赤经	赤經
right cabotage	沿海航运权	沿海航運權
right for sailing	开航权	航行權
right handed propeller	右旋螺旋桨	右旋螺槳
right-handed turning	右旋	右轉
right-hand engine	右转发动机	右轉動力機
right-hand rotation diesel engine	右旋柴油机	右轉柴油機
right-hand screw	右旋螺旋杆	右旋螺釘,右旋螺桿
righting arm	复原力臂	扶正力臂
righting arm curve	复原力臂曲线	扶正力臂曲線
righting couple	复原力偶	扶正力偶
righting lever(=righting arm)	复原力臂	扶正力臂
righting lever curve(=righting arm curve)	复原力臂曲线	扶正力臂曲線
righting moment	复原力矩	扶正力矩
right of angary	非常征用权	擄用破壞中立國船舶非常徵用權
right of claim	请求权	請求權,求償權
right of fishery	捕鱼权	捕魚權
right of hot pursuit	紧追权	緊追權

英　文　名	祖国大陆名	台湾地区名
right of innocent passage	无害通过权	無害通過權
right of lien(=lien)	留置权	留置權
right of management of coastal strip	海岸带管理权	海岸帶管理權
right of navigation	航行权	航行權
right of notary	公证权	公證權
right of passage	通行权	通行權
right of passage between archipelagoes	群岛通过权	群島間通過權
right of passenger	旅客权利	旅客權利
right of priority	优先权	優先權
right of protection	保护权	保護權
right of retention	保留权	拘留權,保留權
right of search	检查权	搜索權
right of seizure	拘留权	扣押權
right of subrogation	代位求偿权	代位求償權
right of visit and search	临检权	臨檢搜索權
rigid construction	刚性结构	剛性結構
rigid helmet	硬盔	硬盔
rigidity	刚度	剛性,剛度
rigid liferaft	刚性救生筏	硬式救生筏
rigid replica of the International Code flag "A"	国际信号旗"A"字硬质复制品	複製硬質國際代碼信號"A"旗
rigid rotor	刚性转子	堅固轉子
Rigil Kent	南门二(半人马 α)	南門二(人馬 α)
rim	轮缘,齿圈	輪緣
rip current	离岸流	急浪流
rips	花水	急浪
rise and set of celestial body	天体出没	天體出沒
risk of collision	碰撞危险	碰撞危機
river boat	内河船	内河船
river mouth(=estuary)	河口	河口
rivet	铆钉	鉚釘
RLB(=rhumb line bearing)	恒向线方位	恒向線方位
RM radar(=relative motion radar)	相对运动雷达	相對運動雷達
road stead	泊船处	泊地,錨泊處
road tank vehicle	公路罐车	公路液罐車
ROC(=restricted operator's certificate)	限用操作员证书	限用值機員證書
rock awash	适淹礁	平水礁
rocket	火箭	火箭

英　文　名	祖国大陆名	台湾地区名
rocket parachute flare	火箭降落伞信号	火箭式降落伞照明弹
rock uncovered	明礁	露礁
Rockwell hardness	洛氏硬度	洛氏硬度
rod	竿钓	桿
roller bearing	圆柱滚子轴承	滚子軸承
roller fairleader	滚轮导缆器	滚子導索器
roller tube expander	滚柱式扩管器	擴管器
rolling	横摇	横摇
rolling bearing	滚动轴承	滚動軸承
rolling error	摇摆误差	搖擺誤差
rolling hatch cover	滚动式舱盖	滚動式艙蓋
rolling period	横摇周期	横摇週期
roll on/roll off	滚装	車輛駛上駛下船,車輛運輸艦
roll on/roll off ship	滚装船	滚装船
roll stowing hatch cover	滚卷式舱盖	滚動收放式艙蓋
rope	绳	索,纜,繩
rope ladder	绳梯	繩梯
rope reel	卷纲机	捲繩機
rope socket	索头环	鋼索接眼
rope stopper	制动索	制索繩
ro/ro cargo	滚装货	滚装貨
rotary current	回转流	旋轉流
rotary loop antenna	旋转环形天线	旋轉環形天線
rotary scavenging valve	回转式扫气阀	轉動驅氣閥
rotary sliding-vane refrigerating compressor	回转叶片式制冷压缩机	轉動滑葉冷凍壓縮機
rotary starting air distributor	回转式起动空气分配器	旋轉式起動空氣分配器
rotary vane pump(=vane pump)	叶片泵	轉葉泵
rotary vane steering gear	转叶式转舵机构	轉葉式操舵裝置
rotating magnetic field	旋转磁场	旋轉磁場
rotating stall	旋转失速	旋轉失速
rough cargo	粗糙货	粗貨
roughness	粗糙度	粗糙度
rough sea	大浪(5级)	洶濤
rough sea resistance	汹涛阻力	狂浪阻力
round haul net	旋网	旋網
roundness	圆度	圆度
round stern	圆型尾	圆形艉

英　文　名	祖国大陆名	台湾地区名
round turn	绕一道(圈)	繞轉
round turn and two half hitches	旋圆双半结	繞轉加雙半套結
rouse in	拉进	拉進錨鏈
rouse out	拉出	拉出錨鏈
route	航路	航線,航路
routine calls	例行呼叫	例行呼叫
routine maintenance	日常例行维修	例行保養
routine organization	执行机构	執行機構
routine priority	日常优先等级	日常優先順序
routine repetition	例行复述	例行複述
routing chart	航路设计图	航路圖
routing for storm avoidance	避风航路	避風航路
row boat	划桨船	划槳船
RP(=reporting point)	报告点	報告點
RPT(=repeat)	重复	重複
RSC(=rescue sub-center)	救助分中心	救助分中心
RTB(=radio true bearing)	无线电真方位	無線電真方位
RU(=rescue unit)	救助单位	救助單位
rubber bearing	橡胶轴承	橡膠軸承
rubber boat	橡皮艇	橡皮艇
rubber mat	橡胶垫	橡皮墊
rudder	舵	舵
rudder angle(=helm angle)	舵角	舵角
rudder angle indicator	舵角指示器	舵角指示器
rudder area ratio	舵面积比	舵面積比
rudder axle	舵轴	舵軸
rudder blade	舵叶	舵葉
rudder brake	舵掣	舵靭,舵制動器
rudder carrier	舵承	舵承
rudder circling stop	左右舵停船,蛇航制动	循環操舵停船法
rudder coupling	舵杆接头	舵頸接頭
rudder effect	舵效	舵效
rudder force	舵力	舵力
rudder plate	舵板	舵板
rudder post	舵柱	舵柱
rudder pressure	舵压力	舵壓力
rudder pressure center	舵压力中心	舵壓力中心
rudder quadrant	舵扇	扇形舵柄

英　文　名	祖国大陆名	台湾地区名
rudder stock	舵杆	舵桿
rudder stopper	制舵器	制舵器
rudder yoke	横舵柄	舵軛
rule base	规则库	規則庫
Rules for Classification and Construction of Ships	船舶入级和建造规范	船舶入級與建造規範
running bowline	活套结	套馬扣
running days	连续日	連續自然日
running fit	转动配合	轉動配合
running fix	移线船位	航進定位
running fixing(RF)	移线定位	航進定位
running free	顺风航驶	順風航駛
running-in	磨合	適配運轉
running rigging	动索	動索
running with the sea	顺浪航行	順浪航行
rust	锈	銹
rust preventive oil	防锈油	防銹油
rust proof	防锈	防銹
rustproof graduated drinking vessel	刻度防锈饮水杯	有刻度之防銹飲器

S

英　文　名	祖国大陆名	台湾地区名
SA(=selective availability)	选择可用性	選擇可用性
Sabbath cycle	萨巴蒂循环	定壓定容混式循環(柴油機)
sacrificial anode protection	牺牲阳极保护	耗蝕性陽極防護
SAE Viscosity Classification	美国汽车工程师协会黏度分级	美國汽車工程師學會黏度分級
safe carrying capacity	安全载流量	安全載流量
safety and environmental protection policy	安全与环保政策	安全與環保政策
safety and interlock device	安全与联锁装置	安全與聯鎖設施
safety arrival insurance	安全到达保险	安全到達保險
safety call	安全呼叫	安全呼叫
safety call format	安全呼叫格式	安全呼叫格式
safety communication	安全通信	安全通信
safety communication procedure	安全通信程序	安全通信程序
safety factor(SF)	安全系数	安全因素

英　文　名	祖国大陆名	台湾地区名
safety fairway	安全航路	安全主航道
safety injection system	安全注射系统	安全噴射系統
Safety Management Certificate	安全管理证书	安全管理證書
safety management manual	安全管理手册	安全管理手冊
safety management system(SMS)	安全管理制度	安全管理制度
safety message	安全报告	安全信文
safety NET	安全通信网	安全通信網
safety priority	安全优先等级	安全優先順序
safety rod	安全棒	安全棒
safety service	安全业务	安全業務
safety signal	安全信号	安全信號
safety speed	安全航速	安全速度
safety system	安全系统	安全系統
safety valve setting pressure	安全阀调定压力	安全閥調定壓力
safety water mark	安全水域标志	安全水域標誌
safe working load(SWL)	安全工作负荷	安全工作負荷
sagging	中垂	舯垂[现象]
sailer	帆船	帆船
sailing directions	航路指南	航行指南
sailing plan	航行计划	航行計劃
sailing plan report	航行计划报告	航行計劃報告
sailing schedule	船期表	船期表
sailing ship(=sailer)	帆船	帆船
sailmaker's tool	缝帆工具	縫帆工具
salinometer	盐度计	鹽份計,鹽度計
saloon	客厅	大餐廳,客廳
salvage and rescue ship	海难救助船	救難船
salvage at sea	海上救助	海上救助
salvage bond	海难救助合同	海難救助契約
salvage contract (=salvage bond)	海难救助合同	海難救助契約
salvage operation	海难救助作业	海難救助作業
salvage pump	救助泵	救助泵
salvage remuneration	救助报酬	救助報酬
salvage ship	救捞船	救撈船
salvor	救助人	施救者
sample	样品,试样	樣品,試樣
sample feed pump	样品供给泵	樣品供給泵
sampling point	采样点	抽樣點

英　文　名	祖国大陆名	台湾地区名
sampling probe	取样器	取樣管,取樣器
SAN（ =strong acid number）	强酸值	強酸值
sand blast	喷砂(除锈)	噴砂
sand-blocking dam	防淤堤	攔沙壩
sand ridge	沙脊	沙畦
sandstorms	沙暴	沙暴
sanitary pressure tank	卫生水压力柜	壓力衛生水櫃
sanitary pump	卫生泵	衛生泵
sanitary system	卫生水系统	衛生系統
SAR coordinating communications(=search and rescue coordinating communication）	搜救协调通信	搜救協調通信
sargasso sea	藻海	藻海
SAR point of contact(=search and rescue service）	搜救联络点	搜救聯絡點
SARR receive antenna	搜救重发器接收天线	搜救重發器接收天線
SAR service	搜救业务	搜救業務
[satellite] almanac	[卫星]历书	[衛星]曆書
satellite based navigational system	星基导航系统	衛星基導航系統
satellite cloud picture	卫星云图	衛星雲圖
satellite communication	卫星通信	衛星通信
satellite constellation	卫星星座	衛星星座
satellite coverage	卫星覆盖区	衛星覆蓋區
satellite disturbed orbit	卫星摄动轨道	衛星攪動軌道
satellite Doppler positioning	卫星多普勒定位	衛星都卜勒定位
satellite emergency position-indicating radiobeacon	卫星紧急无线电示位标	衛星應急指位無線電示標
[satellite] ephemeris	[卫星]星历	衛星曆表
satellite fix	卫星船位	衛星定位
satellite link	卫星链路	衛星鏈路
satellite message	卫星电文	衛星信文
satellite navigation system	卫星导航系统	衛星導航系統
satellite navigator	卫星导航仪	衛星導航儀
satellite network	卫星网络	衛星網路
satellite orbit	卫星轨道	衛星軌道
satellite system	卫星系统	衛星系統
saturated steam	饱和蒸汽	飽和蒸汽
saturated vapor pressure	饱和蒸汽压力	飽和蒸汽壓力

英 文 名	祖 国 大 陆 名	台 湾 地 区 名
Saturn	土星	土星
Saybolt viscosity	赛氏黏度	色博黏度
S/B (= stroke-bore ratio)	行程缸径比	衝程口徑比
SBM (= shore-based maintenance)	岸上维修	岸上維修
SBRS(= selective broadcast receiving station)	选择性广播接收台	選擇性廣播接收台
SBSS(= selective broadcast sending station)	选择性广播发射台	選擇性廣播發射台
SBT(= segregated ballast tank)	专用压载舱	隔離壓載水艙
SC(= special charge)	特别业务费	特別費用
scale deposits	水垢沉淀物	水垢沈積
scale effect	尺度效应	尺度效應
scale plate	刻度板	刻度板
scaler	除鳞机	除鱗機
scavenging air manifold	扫气箱	驅氣歧管
scavenging air port	扫气口	驅氣孔
scavenging air pressure fuel limiter	扫气压力燃油限制器	驅氣空氣壓力燃油限制器
scavenging box fire	扫气箱着火	驅氣箱著火
scavenging-compression stroke	换气-压缩行程	驅氣-壓縮衝程
scavenging efficiency	扫气效率	驅氣效率
Schedar	王良四(仙后 α)	王良四(仙后 α)
schooner	双桅帆船	雙桅帆船
Scleroscope hardness	肖氏硬度	反跳硬度
scraper	刮刀	刮刀
scraper ring	刮油环	刮油張圈
scraping of bearing	轴承刮削	軸承刮削
scratch	划痕	刮痕,割痕
screen of sidelight	舷灯遮板	舷燈遮光板
screw current	螺旋桨流	螺槳流
screw propeller	螺旋桨	螺[旋]槳
screw propeller ship	螺旋推进器船	螺槳推進船
screw pump	螺杆泵	螺泵
screw type refrigerating compressor	螺杆式制冷压缩机	螺桿式冷凍壓縮機
scrubber	洗涤器	清洗器,擦洗器,洗氣器,清除器
scudding	顺风航行	乘風航行
scuttle	舷窗	舷窗

英　文　名	祖国大陆名	台湾地区名
SD（＝semidiameter）	半径差	半徑差
SDR（＝special drawing right）	特别提款权	特別提款權
sea accident	海上事故	海上事故
sea anchor	海锚	海錨
sea area A1	A1 海区	A1 海域
sea area A2	A2 海区	A2 海域
sea area A3	A3 海区	A3 海域
sea area A4	A4 海区	A4 海域
sea areas under national jurisdiction	国家管辖海域	國家管轄海域
sea bank（＝sea wall）	海堤	海岸
sea berth	海上泊位	海上泊位
sea-borne cargo	海运货	海運貨
sea breeze	海风	海風
sea chest	通海阀箱	海底門
sea condition	海况	海面狀況
sea connection	通海接头	通海裝置
sea damage	水损	海[水漬]損
sea echo	海浪回波	海浪回波
seafarers	海员	航海人員
seagoing qualities	航海性能	航海性能
sea-going vessel	海船	海船,海輪
sea ice	海冰	海冰
sea ice concentration	海冰密集度	海冰密集度
seakeeping quality	耐波性	凌海性
seakindliness	凌波性	凌波性
sea letter telegram（SLT）	海上书信电报	海上書信電報
sea-level pressure	海平面气压	海平面氣壓
sealing steam system	密封蒸汽系统	封閉蒸汽系統
seal water system	密封水系统	水封系統
seamanship	船艺	船藝
seaman's book	海员证	船員證
sea mark	海上航路标志	海上航標
seaplane	水上飞机	水上飛機
search and rescue coordinating communication（SAR coordinating communication）	搜救协调通信	搜救協調通信
search and rescue mission coordinator（SMC）	搜救任务协调员	搜救任務協調人

英　文　名	祖国大陆名	台湾地区名
search and rescue procedure	搜救程序	搜救程序
search and rescue radar transponder	搜救雷达应答器	搜救雷達詢答器
search and rescue region(SRR)	搜救区	搜救區
search and rescue repeater	搜救中继器	搜救重發器,搜救中繼器
search and rescue satellite aided tracking	①搜救卫星辅助跟踪 ②卫星搜救跟踪系统	①搜救衛星輔助追蹤 ②衛星輔助搜救追蹤系統
search and rescue satellite system	搜救卫星系统	搜救衛星系統
search and rescue service(SAR service)	搜救业务	搜救業務
search and rescue unit	搜救单元	搜救單位
search datum	搜寻基点	搜索基點
searching	抄关	抄關
search light	探照灯	探照燈
search maneuvre	搜索机动	搜索策略
search pattern	搜寻方式	搜索方式
search radius	搜寻半径	搜索半徑
search track	搜寻航线	搜索航跡
search warrant	搜查证	搜查令狀
sea service	海上资历	海勤資歷
seasickness bag	呕吐袋	嘔吐袋
sea smoke	海上蒸汽雾	海面蒸汽霧
seasonal change in mean sea level	平均海面季节改正	平均海平面季節性變更
seasonal periods	季节期	季節期間
seasonal route	季节航路	季節性航路
seasonal tropical area	季节性热带区域	季節性熱帶區域
sea speed	海上速度	海流速率
sea surface temperature	海面温度	海面溫度
sea temperature	海水温度	海水溫度
sea tourist area	海上旅游区	海上旅遊區
sea trial	航行试验	[海上]試航
sea trial condition	试航条件	試航條件
sea valve	通海阀	海水閥
sea wall	海堤	海堤
sea water circulating pump	海水循环泵	海水循環泵
seawater color	海水水色	海水水色
seawater density	海水密度	海水密度
sea water desalting plant	海水淡化装置	海水淡化裝置

英 文 名	祖 国 大 陆 名	台 湾 地 区 名
sea water evaporator	海水蒸发器	海水蒸發器
sea water pump	海水泵	海水泵
seawater salinity	海水盐度	海水鹽度
sea water service system	海水系统	海水系統
seawater transparency	海水透明度	海水透明度
seaway bill	海运单	海運單
seaworthiness	适航性	適航性
seaworthy trim	适航吃水差	適航俯仰差
secondary circuit	二回路	副電路,二次電路,二次迴路
secondary distribution system	二次配电系统	二次配電系統
secondary field	二次场	二次場
secondary injection	二次喷射	二次噴射
secondary phase factor(SPF)	二次相位因子	二次相位因數
secondary station	副台	副台
second-class radioelectronic certificate	二级无线电电子证书	第二級無線電電子員證書
second engineer	大管轮,二车	二管輪
second mate	二副	二副
second officer (=second mote)	二副	二副
second-trace echo	二次行程回波	二次回波
secret language	密语	密語
sectional navigation	分段航行	分段航行
sectional pilotage	分段引航	分段引水
sectional view	剖视图	剖視圖
section board	区配电板	分段配電板
sector	扇形	扇形
sector of light	光弧	號燈光弧
sector search	扇形搜寻	扇形搜索
secure the anchor	收好锚	收錨固定
securing fitting	固定件	固定裝具
securing to buoy	系浮筒	繫浮筒
security	安全	安全保證
sediment	①沉淀 ②渣滓	①沈澱 ②渣滓
segment signal	段信号	段信號
segment synchronization	段同步	段同步
segregated ballast system	专用压载系统	隔離壓載系統
segregated ballast tank(SBT)	专用压载舱	隔離壓載水艙

英　文　名	祖国大陆名	台湾地区名
segregation	①隔票 ②离析	①隔離 ②析離
segregation table	隔离表	隔離表
seizing	缠扎	紮縛
selective availability(SA)	选择可用性	選擇可用性
selective broadcast receiving station (SBRS)	选择性广播接收台	選擇性廣播接收台
selective broadcast sending station(SBSS)	选择性广播发射台	選擇性廣播發射台
selective calling	选择呼叫	選擇呼叫
selective calling number	选择呼叫号码	選擇呼叫號碼
selective error correcting mode	选择前向纠错方式	選擇前向偵錯模式
selective forward error correction	选择前向纠错	選擇前向偵錯
selectivity of a receiver	接收机选择性	接收機選擇性
selectivity protection	选择性保护	選擇性保護
self-activating smoke signal	自燃烟雾信号	自動煙號
self-aligning bearing	自整位式轴承	自動對正軸承
self-checking function	自检功能	自核功能
self-cleaning separator	自清洗分油机	自清分離器
self-cleaning strainer	①自清洗式滤器 ②自清洗粗滤器	①自清洗濾器 ②自淨過濾器
self-contained air conditioner	立柜式空气调节器	自給式空調
self-contained air support system	自给空气支持系统	自供空氣支援系統
self-contained navigational aids	自主式导航设备	自備助航設備
self-excitation	自励,自激	自激,自勵
self-excited AC generator	自励交流发电机	自激交流發電機
self-identification	自身标识	自船識別碼
self-igniting light	自亮浮灯	自燃信號燈
self-ignition engine	压燃式发动机	壓燃式引擎
self-priming centrifugal pump	自吸式离心泵	自注離心泵
self propulsion test	自航试验	自[行]推[動]試驗
self repair	自修	自行檢修
self-right partially enclosed lifeboat	自动扶正部分封闭救生艇	自行扶正部分圍蔽救生艇
self stripping unit	自扫舱装置	自行收艙裝置
self trimming collier	自动平舱煤船	自動均載煤船
semaphore signaling	手旗通信	手旗通信
semianalytic inertial navigation system	半解析式惯性导航系统	半解析慣性導航系統
semi-balanced rudder	半平衡舵	半平衡舵
semicircular deviation	半圆自差	半圓自差

英　文　名	祖国大陆名	台湾地区名
semicircular method	半圆法	半圓法
semiconductor	半导体	半導體
semi conductor air conditioner	热电式空气调节器	半導體空調
semiconductor refrigeration	半导体制冷	半導體冷凍
semi-container ship	半集装箱船	半貨櫃船
semidiameter(SD)	半径差	半徑差
semidiesel	烧球式柴油机	半柴油機
semi-diurnal tide	半日潮	半日週潮
semi-duplex operation	半双工操作	半雙工作業
semi-hermetic refrigerating compressor unit	半封闭式制冷压缩机	半封密冷凍壓縮機裝置
semi-major axis of ellipse	轨道长半径	軌道半長徑
semisubmerged ship	半潜船	半潛船
senhouse slip shot	脱钩链段	鵜嘴形滑脱鉤鏈段
sense determination	定边	定向
sensibility	灵敏度	靈敏度
sensible horizon	地面真地平	感觀水平面
sensitive element	灵敏部分	敏感元件
sensitivity of a receiver	接收机灵敏度	接收機靈敏度
sensitivity of follow-up system	随动系统灵敏度	隨動系統靈敏度
sensor	传感器	察覺器,測感子,測感器
separate channel mark	左右通航标	分道航行標
separated by a complete compartment or hold from	货舱隔离	全艙隔離
separated from	隔离	隔離
separating bowl	分离盆	分離盆
separating disc	分离盘	分離盤
separation	隔票	分離
separation line	分隔线	分道線
separation zone	分隔带	分道區
sequence valve	顺序阀	順序閥
sequential single frequency code	顺序单频编码	順序單頻編碼
service advice	业务公电	業務通知
service area	服务区	服務區
service code	业务代码	業務代碼
service signal	业务信号	業務信號
service space	服务处所	服務空間
service tank(=daily tank)	日用柜	日用櫃
service telegram	公务电报	公務電報

英　文　名	祖国大陆名	台湾地区名
servo	伺服,随动	伺服
servo control unit	伺服控制单元	伺服控制系統
SES(=ship earth station)	船舶地球站	船舶地面台,船舶衛星電台
servo-motor	伺服电动机,伺服马达	伺服電動機
SES commissioning test	船舶地球站启用试验	船舶衛星電台啓用試驗
SES ID(=ship earth station identification)	船舶地球站识别码	船舶衛星電台識別碼
set the anchor ready for letting go	备锚	備便拋錨
settling position	稳定位置	調定位置
settling tank	沉淀柜	澄清櫃
settling time	稳定时间	安定時間
set up rigging	拉紧索具	拉緊索具
set value	给定值	調定值
sewage	生活污水	污水,穢水
sewage piping system	生活污水排泄系统	穢水管路系統
sewage pump	污水泵	污水泵
sewage standard discharge connection	生活污水标准排放接头	穢水標準排洩接頭
sewage tank	生活污水柜	穢水櫃
sewage treatment plant	污水处理设备	污水處理設備
sewage treatment unit	生活污水处理装置	穢水處理裝置
sextant	六分仪	六分儀
sextant adjustment	六分仪校正	六分儀校正
sextant altitude	六分仪高度	六分儀高度
sextant error	六分仪误差	六分儀誤差
SF(=safety factor)	安全系数	安全因素
SHA(=sidereal hour angle)	共轭赤经	恒星時角
shackle	卸扣	卸扣
shadow diagram	阴影图	陰影圖
shadow sector	阴影扇形	扇形陰影
shaft alley	轴墜	軸道
shaft bossing	轴毂	軸轂
shaft bracket	艉轴架	艉軸架
shaft brake	轴制动器	軸韌,軸制動器
shaft disengaged test	脱轴试验	脱軸試驗
shaft-driven generator	轴带发电机	軸驅動發電機
shafting alignment	轴系校中	軸系校準
shafting brake	轴系制动器	軸系制動器

英　文　名	祖国大陆名	台湾地区名
shaft locked test	锁轴试验	鎖軸試驗
shaft power	轴功率	軸功率
shaft tunnel	轴隧	軸道
shallow and narrow channel navigation operating mode management	浅水与窄航道航行工况管理	淺窄水道航行操作模式管理
shallow water effect	浅水效应	淺水效應
shallow water tide	浅水潮	淺水潮
shape	号型	號標,型材
shape for crossing ferry	横江轮渡号型	橫江渡輪號標
shear line	切变线	剪切線
sheep shank	缩结	縮短結
sheer	航弧,偏摆	船身迴擺(指錨泊急流中時)
sheer strake	舷顶列板	舷側厚板列
sheetbend	单编结	魯班扣,魯班單扣
shell and tube condenser	壳管式冷凝器	殼管式冷凝器
shell-and-tube heat exchanger	壳管式热交换器	殼管式交換器
shell expansion plan	外板展开图	外板展開圖
shelter	避风锚地	遮蓋,遮蔽
shelter deck	遮蔽甲板	遮蔽甲板
sheltered deck vessel	遮蔽甲板船	遮蔽甲板船
Sheratan	娄宿一(白羊 β)	婁宿一(白羊 β)
shielded thermocouple	屏蔽式电热偶	遮蔽式電熱鍋
shifting	移泊	移位
shifting angle of grain	谷物移动角	穀類移動角
shifting beam	活动梁	艙口活動梁
shifting board	止移板	防動板
shifting weight method	移载法	移重法
shim rod	补偿棒	填隙棒
ship	船舶	船,艦
ship/aircraft coordinated search	船舶/飞机协作搜寻	船舶/飛機協作搜尋
shipboard management	船上管理	船上管理
shipboard oil pollution emergency plan	船上油污染应急计划	船上油污染應急計劃
ship bottom alignment check	船底验平	船底校準
shipbroker	船舶经纪人	船舶經紀人
shipbrokerage	船舶经纪人佣金	船舶經紀人傭金
ship building berth	船台	造船台
ship business	船舶[通信]业务	船舶[通信]業務

英　文　名	祖国大陆名	台湾地区名
ship carriage requirements	船舶配备要求	船舶配備要求
ship certificate	船舶证书	船舶證書
shipchandler	船舶供应商	船舶供應商
ship classification society(=register of shipping)	船级社	船级機構
ship class mark	船级标记	船级標誌
ship class symbol notation	船级符号	船级符號
ship clearance	两船间距	兩船間距
ship collision	船舶碰撞	船舶碰撞
ship collision prevention	船舶避碰	船舶避碰
ship domain	船舶领域	船舶領域
ship earth station(SES)	船舶地球站	船舶地面台,船舶衛星電台
ship earth station identification(SES ID)	船舶地球站识别码	船舶衛星電台識別碼
ship elevator(=ship lift)	升船机	船舶升降機
ship formation course alteration	舰艇编队转向	艦艇編隊轉向
ship formation movement	舰艇编队运动	艦艇編隊運動
ship formation pattern	舰艇编队队形	艦艇編隊隊形
shiphandling	船舶操纵	船舶操縱
shiphandling in heavy weather	大风浪中船舶操纵	大風浪中操船
shiphandling in ice	冰中操船	冰中操船
ship induced magnetism	感应船磁	船體感應磁
ship lane	船舶保向宽度	船舶巷道
ship lift	升船机	船舶升降機
ship magnetism	船磁	船磁
ship manager	船舶经理人	船舶經理人
ship mechanical ventilation	船舶机械通风	船舶機械通風
ship model	船模	船模
ship mortgage	船舶抵押权	船舶抵押權
ship movement service	船舶动态业务	船舶移動業務
ship natural ventilation	船舶自然通风	船舶自然通風
ship operator	船舶经营人	船舶營運人
shipowner	船舶所有人,船东	船舶所有人,船東
shipped bill of lading	已装船提单	裝船載貨證券
shipper	托运人	託運人
ship permanent magnetism	永久船磁	永久船磁
shipping-bill	装货准单(海关)	載貨單
shipping business	航运业务	航運業務

英　文　名	祖国大陆名	台湾地区名
shipping economics	水运经济学	海運經濟學
shipping law	航运法	航業法
shipping note	装船通知单	下貨單
shipping order	装货单	裝貨通知單
shipping route	协定航线	航路
ship position(=fix)	船位	船位
ship power station	船舶电站	船舶發電所
ship radio licence	船舶无线电执照	船舶無線電台執照
ship registration	船舶登记	船舶登記
ship safety inspection	船舶安全检查	船舶安全檢查
ship speed	船速	船速
ship speed distribution	船速分布	航速分配
ship stabilizing gear	船舶减摇装置	船舶穩定裝置
ship station	船舶电台	船舶電台
ship station identity	船舶电台识别	船舶電台識別
ship survey	船舶检验	船舶檢驗
ship system(=marine system)	船舶系统	船舶系統
ship-to-ship distress alerting	船到船遇险报警	船與船間遇險警報
ship-to-shore distress alerting	船到岸遇险报警	船與岸間遇險警報
ship traffic simulation	船舶交通模拟	船舶交通模擬
ship ventilation	船舶通风	船舶通風
ship wave	航行波	船行浪
ship weather report	船舶气象报告	船舶氣象報告
ship wind	航行风	船行風
ship yard	造船厂	造船廠
ship's agent	船舶代理	船舶代理
ship's call sign	船舶呼号	船舶呼號
ship's certificates surveying record book	船舶证书检验簿	船舶證書檢驗紀錄簿
ship's constant	船舶常数	船舶常數
ship's doctor	船医	船醫
ship's fire fighting	船舶消防	船舶滅火
ship's husband	船舶管理人	船舶管理人
ship's meeting circle	舰艇相遇圆	船艦相遇圓
ship's passenger	船舶旅客	船舶旅客
ship's routing	船舶定线制	船舶定航線制
shooting net	放网	投網
shore	撑柱	撑柱
shore-based maintenance(SBM)	岸上维修	岸上維修

英　文　名	祖国大陆名	台湾地区名
shore based management	岸上管理	岸上管理
shore ice	岸冰	岸冰
shore pass	登岸证	登岸證
shore power	岸电	岸電
shoring	支撑	支撑
short blast	短声	短[笛]聲
short circuit	短路	短路
short circuit current	短路电流	短路電流
short-cut route	捷水道	捷徑航路
short-delivery	短卸	短卸
short form bill of lading	简式提单	簡式載貨證券
short handed	缺员	缺額
short international voyage	短程国际航行	短程國際航程
short-landed(=short-delivery)	短卸	短卸
shortlanded cargo	短卸货	短卸貨
short-line connection	短缆系结	短纜繫結
short period accident	短周期事故	短週期事故
short sea	碎浪海面	三角波近海
short-shipped	短装	短裝
short splice	短[插]接	短編接
short stowage cargo	填空货	填空貨
short ton	短吨	短噸
shot	链节	(錨鏈)節
shot line return bag	抛绳回收袋	射索回收袋
shower	阵雨	陣雨
shrinkage fit	冷缩配合	收縮配合
shrink-on	红套	紅套
shrouded impeller	闭式叶轮	罩筒葉輪
shutdown	停车	停轉
shut-down depth	停堆深度	
shut out	退关	退貨
shut-out cargo	退关货	退關貨
shuttle tanker	穿梭油轮	梭運油輪
side error	定镜差	水平鏡誤差
side girder	旁桁材	側縱梁
sidelight	舷灯	舷燈
sidelights combined in one lantern	合座舷灯	合併燈
side-lobe echo	旁瓣回波	旁波瓣回波

英 文 名	祖 国 大 陆 名	台 湾 地 区 名
side plate	舷侧板	側邊板
side port	舷门	舷門
side ramp	舷门跳板	舷門跳板
side reach	横向冲距	橫向衝距
sidereal day	恒星日	恒星日
sidereal hour angle(SHA)	共轭赤经	恒星時角
sidereal month	恒星月	恒星月
sidereal time(ST)	恒星时	恒星時
side rolling hatch cover	侧移式舱盖	側滾式艙蓋
side scan sonar	侧扫声呐	水平掃描聲納
side scuttle(=scuttle)	舷窗	舷窗
side stream(=tributary)	支流	支流
side stringer	舷侧纵桁	側縱材
side thruster	侧推器	側推器
sidewise force of propeller	螺旋桨横向力	螺槳橫向力
sight-feed lubricator	明给注油器	顯給滑潤器
sighting port	观察孔	窺孔
signal control	信号控制	信號控制
signal flag	信号旗	信號旗
signaling appliance	信号设备	信號設備
signaling by hand flags(=semaphore sig-naling)	手旗通信	手旗通信
signaling packet	信令分组	信號分組
signal light	信号灯	號誌柱
signal line	信号绳	信號繩
signal mark	信号标志	信號標誌
signal mast	信号桅	信號桅
signal shell	信号弹	信號彈
signal to attract attention	招引注意信号	引起注意信號
significant wave height	有效波高	有義波高
silence period(SP)	静默时间	靜默時間
similar stage of construction	建造相应阶段	建造達類似階段
simple cycle	简单循环	簡單循環
simplex	单工	單工,單式
simplex operation	单工操作	單工作業
Simpson's rules	辛氏法则	辛普生法則
simulation	模拟	模擬
simulation test	模拟试验	模擬試驗

英　文　名	祖国大陆名	台湾地区名
simulator	模拟器,仿真器	模擬機設施
simultaneous disengaging gear	联动脱钩装置	聯動脫鉤裝置
simultaneous disengaging unit	连动脱钩装置	連動脫鉤裝置
single acting cylinder	单作用油缸	單動式氣缸
single-acting diesel engine	单作用式柴油机	單動式柴油機
single anchor leg mooring	单锚腿系泊	單錨腿繫泊
single boat purse seine	单船围网	單船圍網
single boom system	单杆作业	單吊桿系統
single cam reversing	单凸轮换向	單凸輪換向
single column	单纵队	單縱隊
single difference	单差	單差
single drum winch	单卷筒绞车	單筒絞車
single effect evaporation	单效蒸发	單效蒸發
single expansion steam engine	单胀式蒸汽机	單脹式蒸汽機
single gyro pendulous gyrocompass	单转子摆式罗经	單轉子擺式電羅經
single-hose rigs	单管设备	單管設備
single letter signal code	单字母信号码	單字母信號碼
single line abreast	单横队	單橫隊
single line ahead (= single column)	单纵队	單縱隊
single plate rudder	单板舵	單板舵,平板舵
single point mooring (SPM)	单点系泊	單點繫泊
single probe	单油头	單油頭
single rope	单头缆	單頭纜
single-row ball bearing	单排球轴承	單排滾珠軸承
single shafting	单轴系	單軸系
single sideband emission (SSB emission)	单边带发射	單邊帶發射
single sideband radiotelephone (SSB RT)	单边带无线电话	單邊帶無線電話
single stage compressor	单级压缩机	單級壓縮機
single stage flash evaporation	单级闪发	單級閃蒸發
single-stage steam turbine	单级汽轮机	單級蒸汽輪機
single-stage turbocharging	单级废气涡轮增压	單級渦輪增壓
single strainer	单联滤器	單過濾器
single turn	单向旋回法	單迴旋法
single whip	单绞辘	單滑車
sinker	沉子	沈錘,沈子
Sirius	天狼(大犬 α)	天狼(大犬 α)
site error	场地误差	場地誤差
skeletal container	框架箱	框架貨櫃

英　文　名	祖国大陆名	台湾地区名
sketch of barge train formation	驳船队形图	駁船隊形圖
sky condition	天空状况	天空狀況
sky diagram	天象图,星图	天象圖
skylight	天窗	天窗
sky wave	天波	天波
sky wave correction	天波改正量	天波修正量
sky wave delay	天波延迟	天波延遲
sky wave delay curves	天波延迟曲线	天波延遲曲線
slack away	放松	鬆弛
slack current channel	缓流航道	憩流航道
slacken speed(= slow down)	减速	减速
slack stream	缓流	憩流
slack water	憩流,平流	憩流,憩潮
slack water area	平流区域	憩流區域
slamming	砰击	波繫
slave signal	副台信号	副台信號
slewing angle	吊杆偏角	吊桿偏角
slewing winch	回转吊杆绞车	[吊桿]迴旋絞車
slicing machine	切片机	切片機
slide fit	滑动配合	滑動配合
sliding bearing housing	滑动轴承箱	滑動軸承殼
sliding door	滑动门	滑[拉式]門
slight sea	轻浪(3 级)	微波
slip	滑失	滑流
slipper	滑块	滑塊
slippery hitch	小艇结	鬆套結
slip racking	系缆活结	繫纜活結
slip rope	回头缆	滑索
slipway	船排	船台,滑道(船台),斜道
slip wire(= slip rope)	回头缆	滑索
slop	污油水	污油水
slop tank	污油水舱	污油[水]櫃
sloshing	晃击	沖激,晃擊
slot	列位	槽
slot antenna	缝隙天线	槽孔天線
slot in flange	法兰键槽	凸緣上之槽孔
slotted waveguide antenna	隙缝波导天线	槽嵌波導天線

英　文　名	祖 国 大 陆 名	台 湾 地 区 名
slow down	减速	減速
slow flash light	慢闪光	慢閃光
slow turning starting sequence	慢转起动程序	慢轉起動程序
slow turning valve	慢转阀	慢轉閥
SLT(=sea letter telegram)	海上书信电报	海上書信電報
sludge	污泥,油泥	油泥
sludge boat	运泥船	運泥船
sludge incinerator	油泥焚化炉	油泥焚化爐
sludge pump	污油泵	污水泵,排渣泵
sludge tank	油泥柜	油泥櫃,油泥艙櫃
small correction	小改正	小修正
small water plane twin hull ship (SWATH)	小水线面双体船	小水線面雙體船
SMC(=search and rescue mission coordinator)	搜救任务协调员	搜救任務協調人
SMNV(=standard marine navigational vocabulary)	标准航海用语	標準航海用語
smoke helmet	烟盔	防煙盔
smoke mask	防烟面具	防煙面具
smoke screen laying maneuvre	施放烟幕机动	施放煙幕操縱
smooth sea	小浪(二级)	微浪
smothering method	窒息法	窒息法
SMS(=safety management system)	安全管理制度	安全管理制度
smuggling	走私	走私
snatch block	开口滑车	活口滑車,開口滑車
snow-covered ice	雪盖冰	雪蓋冰
sodium and vanadium content	钠和钒含量	鈉釩含量
sodium grease	钠基润滑脂	鈉基滑脂
soft-iron sphere	软铁球	軟鐵球
soil discharging facility	排泥机具	排泥設施
solar annual [apparent] motion	太阳周年视运动	太陽週年[視]運動
solar day	太阳日	太陽日
solar eclipse	日蚀	日食
solar proton event	太阳质子效应	太陽質子效應
solar spot	太阳黑点	太陽黑點
SOLAS(=International Convention for Safety of Life at Sea)	国际海上人命安全公约	海上人命安全國際公約
sole arbitrator	单独仲裁员	單獨仲裁人

英 文 名	祖 国 大 陆 名	台 湾 地 区 名
solenoid directional control valve	电磁换向阀	電磁方向控制閥
solenoid valve	电磁阀	電磁閥
solicitation	揽货	攬貨
solid	固体	固體,固態
solid bulk cargo	固体散货	散裝固體貨物
solid cargo station	固体货站	乾貨整補站
solid crankshaft	整锻曲轴	實心曲柄軸
solidification point(=freezing point)	凝点	凝固點
solid injection	机械喷射	無氣噴射
solid-injection diesel	机械[无气]喷射柴油机	無氣噴射柴油機
solid state circuit	固态电路	固態電路
solitary wave	独立波	孤立波
sonar	声呐	聲納
sonar dome coating	声呐罩涂料	聲納罩塗料
soot blower	吹灰器	吹灰器
soot deposits	烟灰沉积物	煙灰沈積物
soot fires	烟灰着火	煙灰著火
sound channel	声道	聲道
sounding	测深	測深
sounding lead	测深锤	測深錘
sounding line	测深线(测深绳)	測深索,測深繩
sounding pipe	测深管	測深管
sounding scale	测深尺	測深標尺
sound-level meter	声级计	音強度計
sound powered telephone	声力电话	聲力電話
sound-proof chamber	隔音室	隔音室
sound signal	声响信号	音響信號
sound signaling	声号通信	音響通信
sound signal shell	音响榴弹	音響信號彈
sound velocity calibration	声速校准	音速校準
sound velocity error	声速误差	音速誤差
sous-palan clause	船边交货条款	船邊交貨條款
south celestial pole	南天极	南天極
south mark	南方标	南方標
south pole	南极	南極
SP(=silence period)	静默时间	靜默時間
SPA(=sudden phase anomaly)	相位突然异常	相位突然異常

英 文 名	祖国大陆名	台湾地区名
space book	舱位登记簿	艙位簿
spacer	隔片	間隔物
space radiocommunication	空间无线电通信	太空無線電通信
space segment	空间段	太空段
space segment provider	空间段提供者	太空部分提供者
space station	空间站	太空站
space system	空间系统	太空系統
space system for search of distress vessels	搜寻遇险船舶空间系统	搜索遇險船舶太空系統
space telecommand	空间遥令	太空遙令
space tracking	空间跟踪	太空追蹤
spacing of lights	号灯间距	號燈間隔
span wire fuel rig	张索加油装置	張索加油裝置
spanwire storage padeye	跨索系固眼板	跨索儲放眼板
spanwire winch	跨索绞车	跨索絞機
spar buoy	杆状浮标	椿標
spare	备件	備品
spare anchor	备用锚	備用錨
spark ignition engine	点燃式发动机	火花點火引擎
spawning ground	产卵场	産卵場
special area	特殊区域	特別海域
special cargo	特种货	特種貨
special category space	特种处所	特種空間
special certificate	特种证书	特種證書
special charge(SC)	特别业务费	特別費用
special circumstances	特殊情况	特殊狀況
special drawing right(SDR)	特别提款权	特別提款權
specialized salvage service	专业救助	專業救助業務
special mandatory method	特殊强制办法	特殊強制方法
special mark	专用标志	特殊標誌
special purpose channel	专用航道	專用航道
special service	特种业务	特種業務
special survey	特别检验	特別檢驗
special trade passenger	特种业务旅客	特殊貿易客船
Special Trade Passenger Ships Agreement	特种业务客船协定	特殊貿易客船協約
special training	特别培训	特別訓練
specific air consumption	空气消耗率	單位耗氣量
specific cylinder oil consumption	气缸油注油率	單位氣缸耗油量
specific fuel consumption	[有效]燃油消耗率	單位燃料消耗量

英　文　名	祖国大陆名	台湾地区名
specific lubricating oil consumption	滑油消耗率	單位滑油消耗量
specific power	比功率	比功率
specific repetition frequency	特殊重复频率	特殊重複頻率
specific speed	比转数	比速
speed	航速	速率,航速
speedability	快速性	快速性
speed by log	计程仪航速	計程儀航速
speed claim	航速索赔	航速索賠
speed control assembly	速度控制器	速度控制器
speed drop	速度降	速度降
speed drop knob	速度降旋钮	降速鈕
speed error	速度误差	速度誤差
speed error corrector	速度误差校正器	速度誤差校正器
speed error table	速度误差表	速度誤差表
speed length ratio	速长比	速長比
speed-limiting governor	限速器	限速器
speed loss	失速	失速
speed made good	推算航速	實在速率,終結速率
speed of advance	计划航速	前進速
speed of flooding	进水速度	泛水速度
speed over ground	实际航速	對地速度
speed regulating characteristic	调速特性	調速特性
speed regulating valve	调速阀	調速閥
speed regulation by cascade control	串级调速	梯列控制調速
speed regulation by constant power	恒功率调速	恒功率調速
speed regulation by constant torque	恒转矩调速	恒轉矩調速
speed regulation by field control	磁场调速	磁場控制調速
speed regulation by frequency variation	变频调速	變頻調速
speed regulation by pole changing	变极调速	變極調速
speed setting value	速度设定值	速度設定值
speed trial	测速	速率試驗,速率試航
speed trial ground	测速场	測速場
SPF(=secondary phase factor)	二次相位因子	二次相位因數
spherical buoy	球形浮标	球形浮標
spherical roller bearing	球面滚柱轴承	球面滾子軸承
spheroid of earth	地球椭球体	地球橢球體
Spica	角宿一(室女 α)	角宿一(室女 α)
spilling	溢出	溢出

英　文　名	祖国大陆名	台湾地区名
spill-over signal	交会信号	信號漏失
spill-valve injection pump	回油阀式喷油泵	溢流閥式噴射泵
spindle buoy	纺锤形浮标	紡錘形浮標
spiral	螺旋线	蝸線
spiral test	螺旋试验	蝸旋試驗
splash plate	防溅挡板	防濺板
splicing	插接	疊接(木材)
split pin	开口销	開口銷
split-shaft gas turbine	拼合轴燃气轮机	拼合軸燃氣渦輪機
splitting	[天波]分裂	脫裂
split type	拼合式	拼合式
SPM(=single point mooring)	单点系泊	單點繫泊
sponge	海绵	海綿
spontaneous combustion point	自燃点	自燃點
spray evaporative condenser	喷淋蒸发式冷凝器	噴灑蒸發冷凝器
spray type heat exchanger	喷淋式热交换器	噴霧式熱交換器
spread spectrum signal	扩频信号	擴頻譜信號
spring	倒缆	倒纜
spring layer	跃层	銳變層
spring lay rope	钢麻绳	鋼麻合燃索
spring rise(SR)	大潮升	大潮升
spring tide	大潮	大潮
sprinkler	喷洒装置	噴灑裝置
sprinkler installation	洒水装置	噴水裝置
sprinkler system	洒水系统	噴水系統
sprocket	链轮	鏈輪
SPS(=standard positioning service)	标准定位业务	標準定位業務
spurious emission	杂散发射	混附發射
squall line	飑线	颮線
square cut stern	方[型]艉	方型艉
square-rigged vessel	横帆船	橫帆船
square sail	横帆	橫帆,方帆
squat	船体下坐	艉坐
squelch	静噪	消雜音(静音)
squid angling machine	鱿鱼钓机	釣魷魚機
squirrel-cage motor	鼠笼式电动机	鼠籠式電動機
SR(=spring rise)	大潮升	大潮升
SRR(=search and rescue region)	搜救区	搜救區

英 文 名	祖国大陆名	台湾地区名
SS(=steam ship)	蒸汽机船	汽船,輪船
SSB emission(=single sideband emission)	单边带发射	單邊帶發射
SSB RT (=single sideband radiotele- phone)	单边带无线电话	單邊帶無線電話
ST(=sidereal time)	恒星时	恒星時
stability	稳定性	穩度
stability at large angle of inclination	大倾角稳性	大傾側角時之穩度
stability criterion numeral	稳性衡准数	穩度基準數
stability lever	稳性力臂	穩度力臂
stability moment	稳性力矩	穩度力矩
stability of motion(=course stability)	航向稳定性	航向穩定性
stabilization control	定值控制	穩定控制
stabilized loop	稳定回路	穩定迴路
stabilizer control gear	减摇控制设备	穩定器控制裝置
stabilizer equipment room	减摇设备舱	穩定裝備室
stacker	堆垛机	承座接
stacking cone	堆积锥	承接錐
stacking fitting	堆码件	堆積裝具
stacking test	堆装试验	堆積試驗
stagnation	滞止	停滯
stagnation steam parameter	滞止蒸汽参数	停滯蒸汽參變數
stairway	梯道	梯道
stale bill of lading	过期提单	過期載貨證券
stalling angle	失速角	失速角
standard	标准,规范	標準
standard atmosphere	标准大气	標準大氣
standard compass	标准罗经	標準羅經
standard compass course	标准罗航向	標準羅經航向
standard dimension of channel	航道标准尺度	航道標準尺度
standard discharge connection	标准排泄接头	標準排洩接頭
standard error	标准[误]差	標準誤差
standard form	标准格式	標準格式
standard frequency and time signal satel- lite service	卫星标准频率和时间信号业务	衛星標準頻時信號業務
standard frequency and time signal service	标准频率和时间信号业务	標準頻時信號業務
standard frequency and time signal station	标准频率和时间信号台	標準頻時信號台

英　文　名	祖国大陆名	台湾地区名
standard interface description	标准界面说明	標準介面說明
standard maneuvering test	Z形试验,标准操纵性试验	蛇航試驗,標準操縱性試驗
standard marine navigational vocabulary	标准航海用语	標準航海用語
standard parallel	基准纬度	基準緯度
standard positioning service(SPS)	标准定位业务	標準定位業務
standard reporting format	标准报告格式	標準報告格式
standard slide valve diagram	标准滑阀图	標準滑閥圖
standard tensioned replenishment along-side method	标准横向张索补给法	標準強力法輸送設備
standard time	标准时	標準時
standard void depth	标准空档深度	標準空隙深度
stand-by	备用	待命,預備
stand-by generator	备用发电机	備用發電機
stand-by ship	守护船	待命船
standing operating procedure	固定作业程序	現行作業程序
standing rigging	静索	固定索具,静索
standing wave	驻波	定波,駐波
stand into land	驶向海岸	駛向海岸
stand-on vessel	直航船	直航船舶
stand out to sea	离港出海	出海
star	恒星	星,恒星
star apparent place	恒星视位置	恒星視位置
star atlas	恒星图	星座圖
starboard	右舷	右舷
starboard engine	右舷发动机	右舷引擎
starboard hand buoy	右侧浮标	右舷通過浮標
starboard side(=starboard)	右舷	右舷
star catalogue	星表	星表
star chart(=star atlas)	恒星图	星座圖
star-delta connection	星形–三角形接法	星形三角連接
star-delta starting	星–三角起动	星角起動
star finder	索星卡	尋星盤,辨星儀
star finding	索星	尋星
star globe	星球仪	示星球
star identification	认星	識星
star identifier(=star finder)	索星卡	辨星儀,尋星盤
star number	星号	星號

英　文　名	祖国大陆名	台湾地区名
start blocking control	［电动机］起动阻塞控制	起動阻塞控制
starter	起动器	起動機,起動器
start failure alarm	起动故障报警	起動故障警報
start finished signal	起动结束信号	起動結束信號
starting	起动	起動,開動
starting air cut off	起动空气切断	起動空氣切斷
starting air distributor	起动空气分配器	起動空氣分配器
starting air manifold	起动空气总管	起動空氣歧管
starting air reservoir	起动空气瓶	起動氣瓶
starting air system	起动空气系统	起動空氣系統
starting and accelerating operating mode management	起航与加速工况管理	起動與加速操作型式管理
starting cam	起动凸轮	起動凸輪
starting control valve	起动控制阀	起動控制閥
starting device	起动装置	起動裝置
starting engine speed	起动转速	引擎起動轉速
starting interlock	起动联锁	起動聯鎖
starting performance	起动性能	起動性能
start-up accident	起动事故	起動事故
start-up blind-zone	起动盲区	起動盲區
state room	预备间	特別房艙
states recording system	状态记录系统	衛星報況系統
statical stability	静稳性	靜穩度
static balance	静平衡	靜力平衡
static course stability	静航向稳定性	靜航向穩定性
static head	静压头	靜落差
static heeling angle	静横倾角	靜橫傾角
static positioning	静态定位	靜態定位
static pressure regulator	静压调节器	靜壓力調整器
static state	静态	靜態
station	电台	電台
stationary	固定的	固定的
stationary blade	静叶片	固定葉片
stationary fishing gear	定置渔具	定置漁具
stationary fishing net	定置网	定置漁網
stationary fit	静配合	靜配合
stationary front	静止锋	滯留鋒

英 文 名	祖 国 大 陆 名	台 湾 地 区 名
stationary satellite	静止卫星	定置衛星
station bill	应变部署表	部署表
station call	叫号电话	叫號電話
station light	队形灯	部位燈
station marker light box	队形标志灯箱	信號箱
station of destination	收报台	收信台
station of origin	发报台	發信台
station pointer	三杆分度器	三臂定位器
status enquiry	状态询问	狀況詢問
statutory survey	法定检验	法定檢驗
stay	支索,拉条	牽索,支桿,拉條
stay bolt	牵条螺栓	牽條螺栓
steady	稳定的	穩定的
steadying lines	稳定索	穩定索
steady [state]	稳态	穩定狀態
steam	蒸汽	蒸汽
steam bleeding system	抽汽系统	抽汽系統,分汽系統
steam cargo winch	蒸汽起货机	蒸汽吊貨絞機
steam distribution adjustment	配汽调整	配汽調整
steam distribution device	配汽机构	配汽設施
steam drum	汽机鼓	蒸汽鼓,汽鼓
steam ejector gas-freeing system	蒸汽喷射油气抽除装置	噴汽清除[有害]氣體系統
steam heating system	蒸汽供暖系统	暖汽系統
steam jet refrigeration	蒸汽喷射制冷	蒸汽噴射冷凍
steam rate	耗汽率	耗汽率
steam ship(SS)	蒸汽机船	汽船,輪船
steam steering engine	蒸汽舵机	蒸汽舵機
steam stop valve	蒸汽截止阀	停汽閥
steam trap	阻汽器	蒸汽除水閘
steam turbine ship	汽轮机船	蒸汽渦輪機船
steam turbine single-cylinder operation	汽轮机单缸运行	蒸汽渦輪機單缸運轉
steam turbine stage	汽轮机级	蒸汽渦輪機級
steel	钢	鋼
steel-cored copper wire	钢芯铜线	鋼心銅線
steel ship	钢船	鋼船
steerage	①舵效 ②统舱	①舵效 ②統艙
steering	操舵	操舵

英　文　名	祖国大陆名	台湾地区名
steering and sailing rules	驾驶和航行规则	操舵及航行規則
steering compass	操舵罗经	駕駛羅經
steering compass course	操舵罗航向	駕駛羅經航向
steering engine room	舵机舱	舵機艙
steering gear	操舵装置	操舵機,操舵裝置
steering gear room	舵机间	舵機室
steering hunting gear	舵机追随机构	舵機從動裝置
steering line	操纵缆	導引線
steering order(=helm order)	舵令	舵令
steering telemotor	操舵遥控传动装置	操舵液壓遙控裝置
steering test	操舵试验	操舵試驗
steering wheel	舵轮	舵輪
Stellite	司太立合金	史斗鉻鈷,銘鉻鎢合金
stem	艏柱	艏柱,艏材
stem the tide	顶潮流航行	頂潮航行
step-by-step	逐级,步进式	逐步
step control	分级控制	分級控制
step input	阶跃输入	分段輸入
stepless	无级的	無段
stepless speed regulation	无级调速	無段調速
stepped piston	阶梯形活塞	階段活塞,塔形活塞
step steering	阶跃操舵	階躍操舵
stern	艉	船尾
stern aft spring	艉向后倒缆	艉向後倒纜
stern all	齐退	全體倒划
stern anchor(=poop anchor)	艉锚	艉錨
stern bearing	艉轴承	艉軸承
stern breast	艉横缆	艉橫纜
stern door	尾门	艉門
stern engined ship	艉机型船	艉機艙船
stern forward spring	艉向前倒缆	艉向前倒纜
sternlight	艉灯	艉燈
stern line	艉缆	艉纜
stern post	艉柱	艉柱
stern ramp	艉门跳板	艉門跳板
stern shaft(=tail shaft)	艉轴	艉軸
stern trawl	尾拖网	尾拖網
stern tube	艉轴管	艉軸套

英　文　名	祖国大陆名	台湾地区名
stern tube lubricating oil	艉轴管滑油	艉軸管潤滑油
stern tube sealing	艉轴管密封装置	艉軸管密封裝置
stern tube sealing oil pump	艉轴管轴封泵	艉軸管軸封油泵
stern tube stuffing box	艉轴管填料函	艉軸管填料函
stevedore	装卸工,码头工人	裝卸工,碼頭裝卸工人
stiffener	扶强材	加強肋
stiffening	加强,加固	加強,補強
stiffness	刚性	剛性,抗撓性,勁度
stiff shafting	刚性轴系	剛性軸系
stiff ship	过稳船	高穩度船
still water bending moment	静水弯矩	静水彎曲力矩
stirrer	搅拌机	攪拌器
stirrup	蹬索	鐙,繫索
stock anchor	有杆锚	有桿錨,普通錨
stockless anchor	无杆锚	無桿錨,山字錨
stopper	制索器	制鏈器
stopper hitch	制索结	絆索套結
stopping ability	停船性能	停船性能
stopping distance	冲程	衝止距
stopping test	停船试验	停船試驗
stop posts for towline	拖缆限位器	拖纜限位器
stop valve	截止阀	停止閥
storage	储藏	貯藏室,倉庫
storage charge	仓储费	倉儲費
storage life before unpack	油封期	拆封前儲存壽命
store and forward	存储转发	存儲轉發
store and forward unit	存储转发单元	儲存前管器
stores list	物料单	物料單
storm	10 级风	狂風(10级)
storm tide	风暴潮	暴風潮
storm valve	防浪阀	止浪閥
storm warning	暴风警报	暴風警報
stowage	堆码	裝載
stowage category	配装类	裝載種類
stowage factor	积载因数	裝載因數
stowage plan	积载图	貨物裝貨圖,裝載圖
stowage reel	卷绳车	儲放捲盤
stowaway	偷渡	偷渡者

英　文　名	祖国大陆名	台湾地区名
straddle carrier	跨运车	跨載機
straight bill of lading	记名提单	記名載貨證券
straight blade	直叶片	直葉片
straightened up in place	现场校正	現場矯直
straightening vane	整流叶片	整流葉片
straight mineral oil	直馏矿物油	純礦油
straight stem(=vertical bow)	直立[型]艏	直立艏
straight-through boiler	直流式锅炉	直通式鍋爐
straight type engine	直列式发动机	直列型動力機
strainer	[粗]过滤器	濾器
strain-gauge transducer	应变仪传感器	應變計測感器
straining pulley	张力轮	拉緊帶輪
strait	海峡	海峽
strand	索股	索股
strapping	捆扎	捆紮
stratosphere	平流层	平流層
stratus	层云	層雲
stratus-cumulus	层积云	層積雲
stream anchor	中锚	中錨,流錨
stream ice	狭长流冰区	狹長流冰
streamline	流线	流線
streamline rudder	流线型舵	流線型舵
strength	强度	強度
stress concentration	应力集中	應力集中
stress corrosion cracking	应力腐蚀裂纹	應力腐蝕龜裂
stress fatigue	应力疲劳	應力疲勞
strike clause	罢工条款	罷工保險條款
strike on a rock	触礁	觸礁
stringer angle	舷边角钢	舷緣板角鐵
stripping pump	扫舱泵	殘油泵,收艙泵
strobe light	频闪灯	閃光燈
stroke-bore ratio(S/B)	行程缸径比	衝程口徑比
stroke volume	气缸工作容积	衝程容積
strong acid number	强酸值	強酸值
strong breeze	6 级风	強風(6級)
strong gale	9 级风	烈風(9級)
structural member	构件	構件
structure	结构	結構,構造

英　文　名	祖国大陆名	台湾地区名
stud	双头螺栓	螺栓
sub-area coordinator	分区协调人	分區協調人
sub-channel	副航道	副航道
subchartering	转租	轉租
subdivision	［水密］分舱	艙區劃分
subdivision loadline	分舱载重线	艙間吃水線
subject indication type	主题表示类型	主題顯示之形式
subject indicator character	主题标识符	主題標識符
subletting（＝subchartering）	转租	轉租
submarine	潜艇	潛［水］艇
submarine cable	海底电缆	水底電纜
submarine chaser	猎潜艇	驅潛艇
submarine depot ship	潜艇母舰	潛艇母艦
submarine diving	潜艇下潜	潛艇下潛
submarine handling	潜艇操纵	潛艇操縱
submarine pipeline	海底管道	水底管線
submarine quick diving	潜艇速潜	潛艇急潛
submarine quick surfacing	潜艇速浮	潛艇急浮
submarine sound signal	水下音响信号	水中音響信號
submarine surfacing	潜艇起浮	潛艇上浮
submarine survey work	水下测量作业	水下測量作業
submarine trimming	潜艇均衡	潛艇均衡
submarine's adverse maneuverability	潜艇反操纵性	潛艇反操縱性
submarine's cruising state	潜艇巡航状态	潛艇巡航狀態
submarine's maneuverability	潜艇操纵性	潛艇操縱性
submarine's maneuvering strength	潜艇操纵强度	潛艇操縱強度
submarine's proceeding state	潜艇航行状态	潛艇航行狀態
submarine's proceeding state underwater	潜艇水下航行状态	潛艇水下航行狀態
submarine's proceeding state with snorkel	潜艇通气管航行状态	潛艇通氣管航行狀態
submarine's relative diving	潜艇相对下潜	潛艇相對下潛
submarine's relative surfacing	潜艇相对上浮	潛艇相對上浮
submarine's surface proceeding state	潜艇水面航行状态	潛艇水面航行狀態
submarine's trimmed diving	潜艇平行下潜	潛艇平行下潛
submarine's trimmed surfacing	潜艇平行上浮	潛艇平行上浮
submerged anchor dropping	水下抛锚	水下抛錨
submerged anchor weighing	水下起锚	水下起錨
submerged diving chamber	潜水舱	潛水艙
submerged running astern	水下倒车	水下倒俥

英　文　名	祖国大陆名	台湾地区名
submersible platform	下潜平台	下潛平台
submersible pump	潜水泵	潛水泵
sub-program	子程序	子程序
sub-refraction	欠折射	次折射
subrogation	代位	代位
subscriber number	用户码	用戶碼
subsonic velocity	亚音速	次音速
substandard ship	低标准船	次標準船
substation	分站	配電所,變電所
substitute flag	代旗	代旗
sub-telegraph	副车钟	副俥鐘
subtropical high	副热带高压	副熱帶高壓
subzero temperature	零下温度	零下溫度
suction current	吸入流	吸流
suction head	吸入压头	吸入高度
suction loss	失吸现象	失吸現象
suction stroke	吸气行程	吸入衝程
suction well	吸井	吸引井
sudden phase anomaly	相位突然异常	相位突然異常
sulfur content	硫分	硫含量
summer load line	夏季载重线	夏期載重線
summer oil	夏季用润滑油	夏季用潤滑油
summer solstice	夏至点	夏至
summer time	夏令时	夏令時
summer zone	夏季区带	夏期地帶
sun and planet gear	行星型齿轮转动机构	太陽行星齒輪
sun compass	太阳罗经,天体罗经	日規
sunken danger	水下障碍物	水下障礙物
sunken rock	暗礁	暗礁
sunk object	沉没　物	沈沒物
sun rise	日出	日出
sun set	日没	日沒
sun's azimuth table	太阳方位表	太陽方位表
supercharge	增压	增壓
supercharged engine	增压式发动机	增壓發動機
superconducting generator	超导发电机	超導發電機
superconductivity	超导电性	超導電性
superconductor electric propulsion plant	超导电力推进装置	超導電力推進裝置

英　文　名	祖 国 大 陆 名	台 湾 地 区 名
supercritical velocity	超临界速度	超臨界速度
superheater	过热器	過熱器
super large engine	超大型发动机	超大型引擎
super-long stroke diesel engine	超长行程柴油机	超長衝程柴油機
super-power accident	超功率事故	超功率事故
super-refraction	超折射	超折射
supersonic method	超声波探伤法	超音波探傷法
superstructure	上层建筑	船艛建築,上層建築
superstructure deck	上层建筑甲板	船艛甲板
supplemental instruction	补充说明	補充說明
supplementary report	补充报告	補充報告
supplementary ［weather］forecast	补充[天气]预报	補充[天氣]預報
supplement of sailing directions	航路指南补篇	航行指南增補篇
supplying helicopter	补给直升机	補給直升機
supplying station	补给站	補給站
supply network	供电网	供電網絡
support	支架	支座
supporting liquid	支承液体	支持液
support level	支持级	助理級
suppressed carrier emission	抑制载波发射	遏止載波發射
surf	拍岸浪	潑岸浪,拍岸浪
surface	表面	表面
surface air cooler	表面式空气冷却器	表面空氣冷卻器
surface current	表层流	表面流
surface navigation	水面航行	水面航行
surface search coordinator	海面搜寻协调船	海面搜索協調船
surface tension	表面张力	張力
surface ［weather］chart	地面[天气]图	地面天氣圖
surge	喘振	顫動,激度
surge-preventing system	防喘系统	防波動系統
surging	纵荡	激變
surging limit	喘振限	喘振限,波振限
surplus power	剩余功率	剩餘動力
surrounding fishing	围网作业	旋網漁業
surveying	测量	測量
surveying ship	测量船	測量船,測量艦
survey of anchor and chain gear	锚设备检验	錨與錨設備檢驗
survey of refrigerated cargo installation	货物冷藏装置检验	貨物冷凍裝置檢驗

英 文 名	祖国大陆名	台湾地区名
survey on damage to cargo	残损鉴定	貨損檢驗
surveyor	验船师	驗船師,公證人,檢驗師
survey report	检验报告	檢驗報告
survival at sea	海上求生	海上求生
survival craft	救生艇筏	救生艇筏
survival craft recovery arrangement	救生艇筏回收装置	救生艇筏回收裝置
survival manual	救生手册	求生手冊
survival station	营救器电台	營救器電台
suspension height boom length ratio	悬高杆长比	懸高桿長比
swage	铁模	縱移
swage block	型砧	型鐵砧,花砧
swash bulkhead	制荡舱壁	制水艙壁
swash plate	制荡板	制水板
SWATH(=small water plane twin hull ship)	小水线面双体船	小水線面雙體船
sway	横荡	横移
sweep	搜索线	搜索線
sweeping	扫海	掃艙地腳
sweep line	手纲	手網
swell	涌浪	湧
swell compensator	波浪补偿器	波浪補償器
swell height	涌浪高度	湧高
swell scale	涌级	湧級
swinging area	掉头区	迴旋區
swinging ground	自差校正场	迴旋場所
swirl-chamber diesel engine	涡流室式柴油机	渦流室柴油機
swirl type atomizer	旋涡式雾化器	漩渦式霧化器
switch	开关	開關
switching over to heavy oil	切换到使用重油位置	切換使用重油
swivel	转环	轉環
SWL(=safe working load)	安全[工作]负荷	安全工作負荷
synchro light	同步指示灯	同步指示燈
synchronism	谐摇	同步
synchronizer	同步器	同步器,協調器
synchronizer knob	同步器旋钮	同步旋鈕
synchronous generator	同步发电机	同步發電機
synchronous impedance	同步阻抗	同步阻抗
synchronous induction motor	同步感应电动机	同步感應電動機

英　文　名	祖国大陆名	台湾地区名
synchronous rolling(= synchronism)	谐摇	同步
synchronous satellite	同步卫星	同步衛星
synchroscope	同步指示器	同步儀
synodical month	朔望月	朔望月
synoptic chart	天气图	天氣圖
synoptic process	天气过程	天氣過程
synoptic situation	天气形势	綜觀[天氣]大勢
synthetical decision making of collision avoidance	避碰综合决策	綜合避碰決策
synthetic fiber rope	化学纤维绳	合成纖維索
synthetic fishing net	人造纤维渔网	人造纖維漁網
synthetic oil	合成油	合成油
synthetic rope	人造纤维绳	人造纖維繩索
system	系统	系統,制度
systematic error	系统误差	系統誤差
systematic observation	系统观察	系統觀測
systematic test of ship model	船模系列试验	船模系列試驗
system fail	系统故障	系統故障
system flow chart	系统操作图	系統流程圖
system information	系统信息	系統資訊
system operability test	系统操作测试	系統操作測試
system response	系统响应	系統回應

T

英　文　名	祖国大陆名	台湾地区名
tabling	校直	鑿榫
tabulated altitude	表列高度	表列高度
tachogenerator	测速发电机	測速發電機
tachometer	转速表	轉速計
tacking	迎风换抢	迎風棹餿
tackle	滑车组	滑俥轆轤
tackline	隔绳	間索
tactical diameter	旋回初径	迴旋直徑
tail shaft	艉轴	艉軸
tail wind	顺风	順風
tally	理货	理貨
tallyman	理货员	理貨員,記算員

英 文 名	祖 国 大 陆 名	台 湾 地 区 名
TAN(=total acid number)	总酸值	總酸值
tandem loading system	串联系泊装油系统	串列裝卸系統
tandem propeller	串列螺旋桨	串聯螺槳
tandem propulsion system	串联式推进装置	串聯推進系統
tangential component	切向分量	切線方向分量
tangential stress	切向应力	切線應力
tank bottom water	垫水	艙底水
tank cleaning facilities	清舱设备	清艙設施
tank-cleaning plant	货油舱清洗装置	洗艙裝置
tank cleaning pump(=Butterworth pump)	洗舱泵	洗艙泵,巴特華斯泵（洗艙用）
tank container	罐式箱	槽[货]櫃
tanker	液货船	液貨船
tanker piping system	液货船管系	液貨船管路系統
tank experiment	船模试验	船模試驗
tank steaming-out piping system	蒸汽熏舱管系	蒸汽窒火管路系統
tank washing machine	洗舱机	洗艙機
tank washing opening	洗舱口	洗艙開口
tank washing water	洗舱水	洗艙水
taper	锥形	楔形,推拔
taper gauge	锥度规	斜度計,推拔規
taper roller bearing	圆锥滚子轴承	滾錐軸承
target acquisition	目标录取	目標獲得
tariff	①资费表 ②费率本,运价本	①價目表 ②費率規章,收費制
task light	专用灯,特别工作灯	整補作業燈
TB(=true bearing)	真方位	真方位
TBN(=total base number)	总碱值	總鹼值
TC(=true course)	真航向	真航向
TCT(=time charter on trip basis)	航次期租合同	航次計時租船
TCPA(=time to closest point of approach)	最近会遇时间	最接近點時間
TD(=time difference)	时间差	時差
TDC(=top dead center)	上止点	上死點
TDM (=time division multiplexing)	时分复用	割時多工制
technical bulletin	技术通报	技術通報
technical specification	技术规范	技術規範
telecommand	遥控指令	遙控指令

英 文 名	祖国大陆名	台湾地区名
telecommunication	电信	電信
telecontrol	遥控	遙控
telegraph book(=bell book)	车钟记录簿	俥鐘紀錄簿
telegraph cable beacon	电报线立标	電報線標桿
telegraph logger	车钟记录仪	俥鐘紀錄儀
telegraph receiver	车钟接收器	俥鐘接收器
telegraph transmitter	车钟发送器	俥鐘發送器
telegraphy	电报学	電報術
telephony	电话	電話
teleprinter exchange	电传	電傳
telescope pipe	伸缩套管	套筒伸縮管
telescopic alidade	望远镜方位仪	望遠鏡照準儀
telex letter service	用户电报书信业务	電報交換書信業務
telex over radio(TOR)	无线电用户电报	無線電電報交換
telex service	用户电报业务	電報交換業務
telex telephony(TEXTEL)	用户电报电话	用戶電報電話
temperature compensation	温度补偿	溫度補償
temperature controlling system	温控系统	控溫系統
temperature lapse rate	气温直减率	氣溫遞減率
temperature switch	温度继电器	溫度開關
tempering	回火[处理]	回火
temporary notice	临时通告	臨時通告
temporary radio beacon service	临时无线电信标业务	臨時無線電示標業務
temporary repair	临时修理	臨時修理
temporary replenishing rig	简易补给装置	暫時補給裝置
temporary unlighted buoy	临时无灯浮标	臨時無燈浮標
tender	供应船	勤務艦,補給船
tender ship	欠稳船	低穩度船
tensile strength	抗拉强度	抗拉強度
tension control assembly	张力控制器	張力控制器
tension-leg platform	张力腿平台	張力腳式鑽油台
tension transducer	张力传感器	張力轉換器
terminal	终端	終端
terminal charge	港口使费	貨櫃碼頭費用
terminal reheat air conditioning system	末端再加热空气调节系统	終端再加熱空調系統
term of voyage	航行期间	航程條款
terrestrial communication network	地面通信网络	地面通信網絡

英 文 名	祖 国 大 陆 名	台 湾 地 区 名
terrestrial communication system	地面通信系统	地面通信系統
terrestrial fix(TF)	陆标船位	地文定位
terrestrial pole(=earth pole)	地极	地極
terrestrial radiocommunication	地面无线电通信	地面無線電通信
terrestrial sphere	地球圆球体	地球球體
terrestrial station	地面电台	地面電台
territorial sea	领海	領海
territorial water(=territorial sea)	领海	領海
test for lifeboat	救生艇试验	救生艇試驗
testing-bed test	台架试验	試驗台試驗
test report	试验报告	試驗報告
test the steering gear	试舵	操舵裝置試驗
TEU(=twenty equivalent unit)	标准箱(20 英尺集装箱)	20 呎貨櫃［相］當［數］量
text	文本	原文,本文
TEXTEL(=telex telephony)	用户电报电话	用戶電報電話
TF(=terrestrial fix)	陆标船位	地文定位
the first order inertia force	一次惯性力	
the first order moment compensator	一次力矩补偿器	一次力矩補償器
theoretical heat cycle	理论热循环	理論熱循環
theory of errors	误差理论	誤差理論
thermal capacity	热容量	熱容量
thermal container	保温箱	保溫［貨］櫃
thermal cracking	热裂	熱裂煉,熱分解
thermal creep	热蠕变	熱潛變
thermal electric type air conditioner	热电式空气调节器	熱電式空調器
thermal fatigue(=heat fatigue)	热疲劳	熱疲勞
thermal protective aid	保温用具	保溫用具
thermal shielding	热屏蔽	熱遮蔽
thermistor	热敏电阻	熱阻［半導］體
thermocouple	热电偶	熱電偶
thermocurrent	热电流	熱電流
thermodynamic temperature scale	热力学温标	熱力學溫度標
thermo electromotive force	热电动势	熱電動勢
thermoregulator	温度调节器	溫度調節器
thermostability	热稳定性	熱穩定性
thermostat	恒温器	恒溫器
thermostatic expansion valve	热力膨胀阀	熱力膨脹閥

英　文　名	祖国大陆名	台湾地区名
thermostat regulator	恒温调节器	溫度調節器
the second order moment compensator	二次力矩补偿器	二次力矩補償器
thickness gauge	厚度规	厚度規
thickness measuring	测厚	測厚
thick steel shell type bearing	厚壁轴承	厚鋼殼型,軸承
thick wall	厚壁	厚壁
thimble	心环	繩眼襯環
thin film evaporation	薄膜蒸发	薄膜蒸發
thin fog	薄雾(4级)	薄霧(4級)
thin fuel oil	薄燃料油	稀燃油
thin wall	薄壁	薄壁
thin wall bearing	薄壁轴承	薄壁軸承
thin-walled structure	薄壁结构	薄壁結構
third engineer	二管轮	三管輪
third generation engine	第三代中速柴油机	第三代柴油機
third mate	三副	三副
third officer (=third mate)	三副	三副
thread	螺纹	螺紋
threading die	螺纹板牙	螺紋模
thread pitch gauge	螺距规	螺距規
thread plug gauge	螺纹塞规	螺紋塞規
three-arm protractor(=station pointer)	三杆分度器	三臂定位器
three-dimensional flow	三元流动	三維流動
three island vessel	三岛型船	三島型船
three letter signal code	三字母信号码	三字母信號碼
three-phase asynchronous motor	三相异步电动机	三相異步電動機
three-phase four-wire system	三相四线制	三相四線制
three-phase inverter	三相反向变流机	三相變流機
three-pin plug	三角插塞	三腳插塞
three-position four way directional control valve	三位四通换向阀	三位四通方向控制閥
three-thrown crank shaft	三拐曲轴	三拐曲軸
three-way valve	三通阀	三通閥
three-winding transformer	三绕组变压器	三卷線變壓器
threshold factor	阈限系数	界限係數
threshold limit value	阈限值	低限值
throttle	节流	節流
throttle governing	节流调节	節流調速

英　文　名	祖国大陆名	台湾地区名
throttle valve	节流阀	節流閥
through bill of lading	联运提单	聯運載貨證券
through bolt	贯穿螺栓	貫穿螺栓
through cargo	联运货	聯運貨
through transport cargo(=through cargo)	联运货	聯運貨
throw	行程,摆幅	動程,曲柄臂
thrust	推力	推力
thrust bearing	推力轴承	推力軸承
thrust collar	推力环	推力軸環
thrust of propeller	螺旋桨推力	螺槳推力
thrust shaft	推力轴	推力軸
thrust shoe	推力块	推力履片
thunderlight	雷光	雷光
thunder-shower	雷阵雨	雷陣雨
thunder storm	雷雨	雷雨
thyristor converter set	可控硅变流机组	閘流體換流器/組
thyristor excited system	可控硅励磁系统	閘流體勵磁系統
tidal age	潮龄	潮齡
tidal bore	涌潮	怒潮
tidal current	潮流	潮流
tidal datum	潮高基准面	潮汐基準面
tidal flat	干出滩	潮灘
tidal harmonic constant	潮汐调和常数	潮汐調和常數
tidal period	潮汐周期	潮汐週期
tidal range	潮差	潮差
tidal reduction for soundings	测深潮汐订正	測深訂正潮汐
tidal river	感潮河	潮河
tidal stream(=tidal current)	潮流	潮流
tidal stream table	潮流表	潮流表
tidal wave	潮波	潮波,潮浪
tide	潮汐	潮汐
tide-generating force	引潮力	生潮力,引潮力
tide level	潮面	潮面
tide reaching zone	感潮河段	潮達區
tide table	潮汐表	潮汐表
tier	层位	層列
tiering	排链	錨艙排列錨鏈
tight fit	紧配合	緊配合

英 文 名	祖 国 大 陆 名	台 湾 地 区 名
tightness test for hull	船体密性试验	船體密性試驗
tiller	舵柄	舵柄
tilting	倾斜	傾斜
timber and half hitch	拖材结	曳索套結
timber clause	木材条款	木材條款
timber hitch	圆材结	木材結
timber load line	木材载重线	[装]载木[材]载重線
time base	时基	時基
time-base generator	时基[信号]发生器	時基[信號]發生器
time calibration	时间校正	時間校正
time charter	定期租船	計時雇船,計時租賃
time charter on trip basis（TCT）	航次期租合同	航次計時租船
time difference(TD)	时间差	時差
time difference of tide	潮时差	潮時差
time division	时间分隔	時間分割
time division multiplexing(TDM)	时分复用	劃時多工制
time division system	时间分隔制	時分制
time insurance	定期保险	定期保險
time-lag relay	时限继电器	緩動繼電器
time limitation	时效	時限
time limit of general average	共同海损时限	共同海損時限
time of closet approach(TCPA)	最接近时间	最接近之時刻
time of GEOSAR alert notification	静止卫星报警通报时间	定置衛星警報通報時間
time of rudder movement	转舵时间	轉舵時間
time of week（TOW）	周计时	以週計時
time signal	时号	對時信號,報時信號
time to closest point of approach	最近会遇时间	最接近點時間
time zone chart	时区图	時區圖
timing diagram	正时图	正時圖
timing mark	时标	定時記號
tin-base white-metal linings	锡基白合金衬层	錫基白金襯層
tin opener	开罐器	開罐器
tin sheet	马口铁,镀锡铁皮	鍍錫鐵片,馬口鐵皮
tip-leakage loss	叶尖漏泄损失	葉梢漏洩損失
title	标题	標題,名稱
t-junction	三通接头	三通接頭
TK（=track）	航迹	航跡
TM radar（=true motion radar）	真运动雷达	真運動雷達

英　文　名	祖 国 大 陆 名	台 湾 地 区 名
toilet	盥洗室	盥洗室
tolerance	公差	公差
tolerance error(=admissible error)	容许误差	容許誤差
tonnage	吨位	噸位
tonnage dues	吨税	噸税
tons of cargo handled	货物操作吨	貨物操作噸
tons per centimeter immersion	每厘米吃水吨数	每公分吃水噸數
tooth	齿	齒
top dead center(TDC)	上止点	上死點
topmark	顶标	頂[上]標[誌]
top of the flood	涨潮顶点	最高潮
topographic and coastal survey	地形岸线测量	地形岸線測量
topping lift	千斤索	吊桿頂索
topping lift winch	千斤索绞车	吊桿頂索絞車
topsail halyard bend	扬帆结	上桅揚帆結
top-side paint	干舷漆	乾舷漆
top side tank	顶边舱	舷側氣櫃
TOR(=telex over radio)	无线电用户电报	無線電電報交換
torque	转矩,扭矩	扭矩
torque of propeller	螺旋桨转矩	螺槳轉矩
torquer	力矩器	扭矩器
torque/speed limit	等转矩限	轉矩転速限
Torremolinos International Convention for Safety of Fishing Vessels	托列莫利诺斯国际渔船安全公约	托里莫列路斯漁船安全國際公約
torsion	①扭转 ②扭力	①扭轉 ②扭力
torsional fatigue cracks	扭转疲劳裂纹	扭轉疲勞裂痕
torsional meter	扭力计	扭力計,轉矩計
torsional resonance	扭[振]共振	扭轉共振
torsional strength	扭转强度	抗扭強度
torsional vibration	扭转振动	扭轉振動
torsional vibration damper	扭振减振器	扭轉振動減振器
toss oars	立桨	舉槳
total	全体	全部,全體,總計
total acid number(TAN)	总酸值	總酸值
total amount of general average	共同海损总额	共同海損總額
total base number(TBN)	总碱值	總鹼值
total efficiency	总效率	總效率
total head	总压头	總落差

英　文　名	祖国大陆名	台湾地区名
total loss	全损	全损
total loss only	船舶全损险	全损险
totally enclosed lifeboat	全封闭救生艇	全圍蔽救生艇
total resistance	总阻力	總阻力
tourist ship	旅游船	遊覽船
tourist submersible	水下游览艇	觀光潛艇
TOW（＝time of week）	周计时	以週計時
towage contract	拖航合同	拖船合約
towed vessel	被拖船	被拖船
towing	拖曳,牵引	拖曳,牽引
towing alongside	傍拖	傍拖
towing arch	拖缆弓架	拖纜拱架
towing beam	拖缆承架	拖索承梁
towing bitt	拖缆桩	拖纜柱,拖纜椿
towing gear	拖曳设备	拖曳設備
towing hook	拖钩	拖纜鉤
towing in ice	冰区拖航	冰區拖纜航
towing light	拖带灯	拖航號燈
towing line	拖缆	拖纜
towing operating mode management	拖曳作业工况管理	拖曳作業方式管理
towing post	拖桩	拖纜椿
towing side light	偏缆灯	拖帶邊燈
towing train	拖曳船队	拖曳船隊
towing vessel（＝tug）	拖船	拖船
towing winch	拖缆机	拖纜絞機
tow worthiness	适拖	適拖
TPC（＝moment to change trim per centi-meter）	每厘米吃水吨数	每公分吃水噸數
trace	①痕迹 ②微量	①跡 ②微量
traces of oil	油迹	油跡
track（TK）	①航迹 ②轨迹	①航跡 ②軌道
track calculating	航迹计算	航跡計算
track distribution	航迹分布	航跡分佈
tracking	跟踪	目標追蹤
track made good	航迹推算	實際航跡
track reach	航迹冲程	航跡衝距
track spacing	航线间隔	航路間隔
trade route	贸易航线	貿易航線

英 文 名	祖 国 大 陆 名	台 湾 地 区 名
trade wind	信风	信風
traffic	通信业务量	業務通報量
traffic boat	交通艇	交通艇
traffic capacity	交通容量	交通容量
traffic control area	交通控制区	交通控制區
traffic control zone	交通管制区	交通管制區
traffic density	交通密度	交通密度
traffic enquiry	通信询问	通信詢問
traffic flow	交通流	交通流量
traffic lane	通航分道	航行巷道
traffic list	通报表	通報表
traffic mark	通行信号标	通航標誌
traffic separation schemes(TSS)	分道通航制	分道通航制
traffic volume	交通量	交通量
trailing edge	随边	殿緣(螺槳)
trailing wake	尾流	航跡流
train ferry	火车轮渡	火車渡船
training ship	实习船	訓練船
tramp service	不定期船运输	不定期船業務
transducer	换能器	轉換器,轉發器
transducer directivity	换能器指向性	轉換器指向性
transfer	横距	迴轉橫距
transfer of bill of lading	提单转让	轉讓載貨證券
transfer of control station	操纵部位转换	控制位置轉換
transfer pump	驳运泵	轉駁泵
transient working condition	过渡工况	過渡工作情況
transistor	晶体管	電晶體
transit(=meridian passage)	中天	中天
transit beacon	叠标	疊標桿
transit bill of lading	过境提单,转口提单	接運載貨證券
transit cargo	过境货,转口货	過路貨,轉口貨,接運貨
transition period	过度阶段	過渡階段
transit leading mark	过渡导标	疊導標
transit route	传递路由	中轉途徑
Transit System	海军导航卫星系统,子午仪系统	子午儀系統
transmission	传动,传输	傳動,傳輸
transmission shafting	传动轴系	傳動軸系

英　文　名	祖国大陆名	台湾地区名
transmission system	传向系统	傳動系統
transmitter identification character	发信台识别符	發射台識別符
transmitting antenna	发射天线	發射天線
transmitting compass	复示磁罗经	電導羅經
transportable moisture limit	适运水分限	適運水分限
transport ship	运输船	運輸艦
transshipment bill of lading	转船提单	轉口載貨證券
transshipment cargo	转船货	轉船貨
transshipment clause	转船条款	轉船條款
transtainer	门式起重机	門式起重機
transverse	横向的	橫向的
transverse force of discharge current	排出流横向力	排流橫向力
transverse force of propeller submergence	螺旋桨浸深横向力	螺槳浸深橫向力
transverse force of wake current	伴流横向力	伴流橫向力
transverse frame system	横骨架式	橫肋系統
transverse section plan	横剖面图	橫剖面圖
transverse stability	横稳性	橫向穩度
transverse strength	横强度	橫向強度
trapezoidal board	梯形牌	梯形板
travel	移动	衝程
traveller	移动式起重机,滑马	吊運車
traverse survey	导线测量	折航測量
trawl	拖网	拖網
trawl buoy	拖网浮标	拖網浮標
trawler	拖网渔船	拖網漁船
trawl fishing(=trawling)	拖网作业	拖網作業
trawl gallows	网板架	網板架
trawling	拖网作业	拖網作業
treaty of commerce and navigation	通商航海条约	通商航海條約
trend analysis	趋势分析	趨向分析
trial maneuvering	[雷达避碰]试操纵	試操縱
tributary	支流	支流
trigger	触发器	觸發器
trim	吃水差	俯仰差
trim by bow	艏倾	艏俯
trim by stern	艉倾	艉俯
trim diagram	吃水差曲线图	俯仰差圖
tri-metal bearing	三[层]金属轴承	三[層]金屬軸承

英 文 名	祖国大陆名	台湾地区名
trimming	平舱	扒平,平艙
trimming angle	纵倾角	俯仰角
trimming moment	纵倾力距	俯仰力矩
trimming table	吃水差比尺	俯仰表
Trinity House London	伦敦引航公会	英國導航協會
trip	脱扣	跳脱
trip account	航次结账单	航次帳單
trip charter	航次租船	航次傭船
trip device	脱扣装置	跳脱設施
triple difference	三差	三差
triple expansion steam engine	三胀式蒸汽机	三段膨脹蒸汽機
triple point	三相点	三相點
tripod mast	三脚桅	三腳桅
tripping coil	脱扣线圈	跳脱線圈
trolley	架空吊车	吊運車
trolling gurdy	曳绳钓起线机	曳繩釣起繩機
troll line	曳绳钓	曳繩釣
troopship	运兵船	運兵船
tropical cyclone	热带气旋	熱帶氣旋
tropical depression	热带低压	熱帶低壓
tropical disturbance	热带扰动	熱帶擾動
tropical fresh water load line	热带淡水载重线	熱帶淡水載重線
tropical load line	热带载重线	熱帶載重線
tropical storm	热带风暴	熱帶風暴
tropical zone	热带区带	熱帶地帶
tropic tide	回归潮	回歸潮
troposphere	对流层	對流層
tropospheric refraction correction	对流层折射改正	對流層折射修正
trouble	①故障 ②干扰	①故障 ②干擾
trough	低压槽	槽形低壓
trough line	槽线	槽線
T-R switch	收发开关	收發開關
truck light	桅冠灯	桅冠燈
true altitude	真高度	真高度
true anomaly	真近点角	真近點角
true bearing(TB)	真方位	真方位
true course(TC)	真航向	真航向
true density-temperature correction coeffi-	密度温度系数	實密度–溫度修正係數

英　文　名	祖国大陆名	台湾地区名
cient		
true error	真误差	真誤差
true horizon	真地平	真地平線,真水平線
true motion display	真[运动]显示	真運動顯示
true motion radar(TM radar)	真运动雷达	真運動雷達
true north	真北	真北
true rise and set	真出没	真出沒
true sun	真太阳	真太陽
true vapour pressure	真蒸汽压力	真蒸汽壓力
true wind	真风	真風
trunk	①围井筒 ②筒	①圍壁 ②筒
trunk line routes(=main channel)	主航道	主航道
trunk piston type diesel engine	筒形活塞式柴油机	筒狀活塞型柴油機
trunk stream	干流	幹流
TSS(=traffic separation schemes)	分道通航制	分道通航制
tsunami	海啸	海嘯
tubular combustor	管形燃烧室	管形燃燒室
tug	拖船	拖船
tug hire	拖轮费	拖船費
tuna long liner	鲔钓船	鮪釣船
tuna mother ship	鲔钓母船	鮪釣母船
tune	调谐	調諧
turbine	涡轮机	渦輪機
turbine oil	汽轮机油	輪機滑油
turbo-	涡轮驱动	渦輪驅動
turboblower	涡轮增压器	渦輪增壓器
turbocharger(=turboblowen)	涡轮增压器	渦輪增壓器
turbocharger surge	增压器喘振	渦輪增壓器波振
turbocharger turbine characteristics	增压器涡轮特性	增壓器渦輪特性
turbocharging	废气涡轮增压	渦輪增壓
turbocharging auxiliary blower	增压系统辅助鼓风机	渦輪增壓器輔助鼓風機
turbocharging emergency blower	增压系统应急鼓风机	渦輪增壓器應急鼓風機
turbogenerator	涡轮发电机	渦輪發電機
turbulence	紊流,湍流	擾動,擾流
turn	转动	迴轉
turning ability	旋回性	迴轉能力
turning buoy	转向点浮标	轉向點浮標
turning circle	旋回圈	旋回圈,迴轉圈,迴旋圈

英　文　名	祖国大陆名	台湾地区名
turning circle trial(=turning trial)	旋回试验	迴旋試驗
turning gear	转车机	迴轉裝置,盤俥裝置
turning gear interlocking device	盘车联锁装置	盤俥裝置聯鎖裝置
turning indices	旋回性指数	迴旋指數
turning operating mode management	转向工况管理	迴轉操作形式管理
turning period	旋回周期	迴旋週期
turning point	转向点	轉向點
turning radius	旋回半径	迴旋半徑
turning rate	旋转角速度	迴轉速率
turning short round	掉头	短迴轉
turning short round by ahead and astern engine	进倒车掉头	進退俥短迴轉掉頭
turning short round by one end touch the shoal	触浅掉头	觸淺短迴轉掉頭
turning short round with anchor	抛锚掉头	抛錨短迴轉掉頭
turning short round with the aid of current	顺流掉头	藉流短迴轉掉頭
turning trail	旋回试验	迴旋試驗
turn of tidal current	转流	潮流旋轉
turret mooring system	转塔式系泊系统	轉塔式繫泊系統
TW(=typhoon warning)	台风警报	颱風警報
tween deck	二层甲板	中甲板
twenty equivalent unit （TEU）	标准箱(20 英尺集装箱)	20 呎貨櫃[相]當[數]量
twin-engine single-shaft system	双机单轴式	雙機單軸系統
twin gyro pendulous gyrocompass	双转子摆式罗经	
twin propellers	双推进器	雙螺槳
twin shafting	双轴系	雙軸系
twin trawling	对拖	雙拖
twisted blade	扭叶片	扭葉片
twist lock	扭锁	扭轉鎖定器
two course beacon	双向信标	雙向示標
two-cycle engine	二冲程发动机	二衝程引擎
two-degree of freedom gyroscope	二自由度陀螺仪	二自由度迴轉儀
two-edged	双刃的	雙刃的
two-element air ejector	双组空气抽逐器	雙組空氣抽射器
two half hitches	双半结	雙半套結
two lane canal	双航道运河	雙向道運河
two letter signal code	双字母信号码	雙字母信號碼

英　文　名	祖国大陆名	台湾地区名
two-noded vibration	双节点振动	雙節點振動
two-pass superheater	双流程过热器	雙通道過熱器
two-pin plug	两脚插头	兩腳插塞
two-ply	双层的	雙層
two-position	双位	雙位
two-position three way directional control valve	二位三通换向阀	二位三通方向控制閥
two-row impulse wheel	二列冲动叶轮	雙排衝動葉輪
two sector system	双扇形系统	雙扇形系統
two-speed motor	双速电动机	雙速電動機
two-stage supercharging	二级增压	二級增壓
two-start screw	双头螺纹	雙頭螺紋
two-step injection system	两级喷射系统	兩級噴射系統
two stroke diesel engine	二冲程柴油机	二衝程柴油機
two-throw crank	双联曲柄	二拐曲柄
two-unit	双机组	雙機組
two-way make-before break contact	双向先合后断触点	雙向先合後斷觸點
two-way route	双向航路	雙向航路
two-way VHF radiotelephone apparatus	双向甚高频无线电话设备	雙向特高頻無線電話
two-wire system	双线制	雙線制
type	型式	型,式
typhoon	台风	颱風
typhoon anchorage	防台锚地	防颱錨地
typhoon eye	台风眼	颱風眼
typhoon track	台风路径	颱風路徑
typhoon warning(TW)	台风警报	颱風警報
typical insulating flange joint	标准绝缘法兰接头	典型絕緣凸緣接頭

U

英　文　名	祖国大陆名	台湾地区名
UERE (=user equivalent range error)	用户等效距离误差	用戶等效距離誤差
UHF communication	特高频通信	超高頻通信
UKC(=under keel clearance)	富余水深	餘裕水深
ULCC(=ultra large crude carrier)	超大型油船	超級油輪
ullage	空距	隙尺
ullage port	舱顶空档测量孔	油面測距孔

英 文 名	祖 国 大 陆 名	台 湾 地 区 名
ullage scale	空距尺	油面計
ultimate	①极限的 ②最后的	①極限的 ②最後的
ultimate life	极限使用寿命	極限使用壽命
ultimate strength	极限强度	極限強度
ultra large crude carrier(ULCC)	超大型油轮	超級油輪
ultrasonic	超声波	超音波
ultrasonic detector	超声波探伤器	超音波探傷器
ultrasonic examination	超声波探伤	超音波檢查
unattended	无人值班的	無人當值
unattended machinery space	无人[值班]机舱,无人 机舱	無人當值機艙,無人化 機艙
unavailable energy	无用能	無用能
unbalanced force	不平衡力	不均衡力
unbalanced rudder	不平衡舵	不平衡舵
unberthing	离泊	離泊
uncertainty phase	不明阶段	不明階段
unclassified	未入级的	未入級
undamped	无阻尼的	無阻尼
undercooling	过度冷却	過冷
under-deck cargo	舱内货	艙內貨
under-frequency	欠频	欠頻
under hung rudder(=hanging rudder)	悬挂舵	懸舵
under keel clearance(UKC)	富余水深	餘裕水深
under-voltage	欠压	欠壓
under-voltage test	欠压试验	欠壓試驗
underwater explosive cutting	水下爆破切割	水下爆切
underwater habitat	水下居住舱	水下起居艙
underwater hovering	水下悬浮	水下懸浮
underwater mooring device	水下系泊装置	水下繫泊設施
underwater navigation	水下航行	水下航行
underwater oil storage tank	水下储油罐	水下儲油櫃
underwater operation ship	水下作业船	水下作業船
underwater robot	水下机器人	水下機器人
underwater ship	全潜船	全潛船
underwater sightseeing boat	水下游览船	水下觀光船
underwater sound projector	水下声标	水下聲標
underwater stage decompression	水下阶段减压法	水下階段減壓分
underwater turning	水下旋回	水下迴旋

英　文　名	祖国大陆名	台湾地区名
underway	在航	航行中
undocking	出坞	出坞
uniflow scavenging	直流扫气	單向流驅氣
uniflow steam engine	单流式蒸汽机	單流蒸汽機
uniform	①均匀的 ②同一的	①均匀的 ②同一的
uniform liability system	同一责任制	同一責任制
union	管节	管套節
union crane service	双吊联合作业	雙吊桿作業
union purchase system	双杆作业	雙桿固定合吊裝置
unit	①单位 ②机组	①單位 ②組
United Nations Convention on Conditions for Registration of Ships	联合国船舶登记条件公约	船舶登記條件聯合國公約
United Nations Convention on International Multimodal Transport of Goods	联合国国际货物多式联运公约	國際貨物多式聯運聯合國公約
United Nations Convention on the Law of the Sea	联合国海洋法公约	聯合國海洋法公約
United Nations Convention on the Liability of Operators of Terminals in International Trade	联合国国际贸易运输港站经营人赔偿责任公约	國際貿易終端站營運人責任聯合國公約
unit effective efficiency	机组有效效率	機組有效效率
unitization	成组	單元化
unitized cargo	成组货	單元化貨物
unit power plant	成套动力装置	整套動力設備
unit size	粒度	單位尺度
universal	通用的,万用的	通用的,萬用的
universal fairleader	万向导缆器	萬向導纜器
universal joint	万向接头	萬向接頭
universal time(GMT)	世界时	世界時
unknown clause	不知条款	不知條款
unlit mark	无灯标志	無燈標誌
unloading	卸载,卸货	卸貨
unloading valve	卸荷阀	卸載閥
unlocated alert	未定位警报	未能定位之警報
unmanned	无人管理的	無人當值
unmanned machinery space	无人机舱	無人化機艙
UN number	联合国编号	聯合國編號
unravel water tank	整理水池	整理水池
unsupercharged	非增压的	非增壓的

英　文　名	祖国大陆名	台湾地区名
unsymmetrical flooding	不对称浸水	不對稱泛水
unwanted emission	无用发射	無用發射
up-bound vessel	上行船	上行船
u-pipe	U 型管	U 型管
upper	上部的	上部
upper branch of meridian	午圈	上子午線
upper deck	上甲板	上甲板
upper-level［weather］chart	高空[天气]图	高空[天氣]圖
upper meridian passage(=upper transit)	上中天	上中天
upper reach	上游	上游
upper transit	上中天	上中天
upright	①直立 ②正浮	①直柱 ②正浮
upright position	正浮位置	正浮位置
up stream	逆水	逆流
upstream vessel	逆流船	逆流船
upward stroke	上行行程	上行衝程
upwelling	上升流	湧升流
urgency	紧急	緊急
urgency call	紧急呼叫	緊急呼叫
urgency call format	紧急呼叫格式	緊急呼叫格式
urgency communication	紧急通信	緊急通信
urgency communication procedure	紧急通信程序	緊急通信程序
urgency message	紧急报告	緊急信文
urgency priority	紧急优先等级	緊急優先順序
urgency signal	紧急信号	緊急信號
urgent	紧急的	緊急的
urgent meteorological danger report	紧急气象危险报告	緊急氣象危險報告
urgent navigational danger report	紧急航行危险报告	緊急航行危險報告
useful	①有用的 ②有效的	①有用的 ②有效的
user equivalent range error（UERE）	用户等效距离误差	用戶等效距離誤差
UTC(=coordinated universal time)	协调世界时	協調世界時

V

英　文　名	祖国大陆名	台湾地区名
vaccination certificate	预防接种证书	預防接種證書
vacuum	真空	真空
vacuum distillation	真空蒸馏	真空蒸餾

英　文　名	祖国大陆名	台湾地区名
vacuum manometer	真空压力表	真空壓力器
vacuum pump	真空泵	真空泵
valuable cargo	贵重货	高值貨
value of property salved	获救财产价值	獲救財産價值
valve	①阀 ②气门 ③电子管	①閥 ②氣門 ③真空管
valve chest	阀箱	閥櫃,閥箱
valve clearance	气阀间隙	閥餘隙
valve-in head type engine	顶置气阀式发动机	頂置閥式引擎
valve lift	气阀升程	閥升程
valve lift diagram	气阀升程图	閥升圖
valve oscillator	电子管振荡器	真空管振盪器
valve rotating mechanism	气阀旋转机构	閥旋轉機構
valve spindle guide	阀杆导承	閥桿導件
valve timing	气阀正时	閥動定時
vane	桨叶	輪葉
vaneless diffuser	无叶式扩压器	無葉擴散器
vane pump	叶片泵	輪葉泵
vanishing angle of stability	稳性消失角	穩度消失角
vapor compression distillation plant	压汽式蒸馏装置	蒸汽壓縮蒸餾裝置
vapor compression refrigeration	蒸发压缩制冷	蒸汽壓縮冷凍
Var	磁差	地磁差
variable	可变的	變數
variable capacity pump	变量泵	變量泵
variable compression ratio	可变压缩比	可變壓縮比
variable delivery pump (=variable capacity pump)	变量泵	變量泵
variable-displacement oil motor	变量油马达	變量油馬達
variable exhaust valve closing device	可变排气阀关闭机构	可變排氣閥關閉機構
variable geometric turbine nozzle device	涡轮可变几何形状喷嘴装置	渦輪可變幾何形狀喷嘴裝置
variable injection timing mechanism	可变喷油正时机构	可變噴油定時機構
variable range marker	活动距标	可變距指標
variable working condition	变工况	可變工作情況
v-beet	三角皮带	三角皮帶
vector	矢量	向量,矢量
vector display	矢量显示	向量顯示
vee-type diesel engine	V 型柴油机	V 型柴油機
Vega	织女一(天琴 α)	織女一(天琴 α)

英　文　名	祖国大陆名	台湾地区名
vehicle	运载体	载具
velocity	速度	速度
velocity of geostrophic	地转风速	地轉風速
velocity of ship wind	船行风速,视风速	船行風速
velocity stage	速度级	速度級(輪機)
vent	通风口	通氣孔
ventilating cowl	通风帽	通風帽
ventilation and air-conditioning room	通风与空调机间	通風與空氣調節機室
ventilator	通风筒	通風筒,通風器
vent pipe	透气管	通風管
Venus	金星	金星
verification and control	查核与管制	查核與管制
vernal equinox	春分点	春分點
vertex	大圆顶点	頂點
vertical	①垂直 ②立式的	①垂直 ②立式
vertical antenna	垂直天线	垂直天線
vertical beam width	垂直波束宽度	垂直波束寬度
vertical bow	直立[型]艏	直立艏
vertical circle	垂直圈	垂直大圓
vertical danger angle	垂直危险角	垂直危險角
vertical line	垂直线	垂直線,豎線
vertical magnet	垂直磁棒	上升用磁鐵
vertical positioning of lights	号灯垂直位置	號燈之垂直位置
vertical replenishment	垂直补给	垂直整補
vertical sector	垂直光弧	垂直弧區
vertical working range	垂直工作范围	垂直工作範圍
very high altitude	特大高度	特高高度
very high frequency radio direction finder (VHF RDF)	甚高频无线电测向仪	特高頻無線電測向儀
very high sea	狂涛(8级)	浪甚高
very large crude carrier(VLCC)	大型油船	巨型油輪
very rough sea	巨浪(6级)	極大浪(6级)
vessel(=ship)	船舶	船,艦
vessel aground	搁浅船	擱淺船
vessel constrained by her draught	限于吃水船	吃水受限船
vessel engaged in fishing	从事捕鱼的船舶	從事捕魚中船舶
vessel engaged in mineclearance operation	清除水雷船	掃雷船
vessel in sight of one another	互见中的船舶	船舶互見

英 文 名	祖 国 大 陆 名	台 湾 地 区 名
vessel not under command	失控船	操縱失靈船
vessel reporting system	船舶报告系统	船舶報告系統
vessel restricted in her ability to maneuver	操纵能力受限船	操縱能力受限船
vessel safety engineering	船舶安全学	船舶安全工程
vessel to leeward	下风船	下風船
vessel to windward	上风船	上風船
vessel traffic engineering	船舶交通工程	船舶交通工程
vessel traffic investigation	船舶交通调查	船舶交通調查
vessel traffic service(VTS)	船舶交通管理	船舶交通服務
vessel traffic service center(VTS center)	船舶交通管理中心	船舶交通服務中心
vessel traffic survey (=vessel traffic in-vestigation)	船舶交通调查	船舶交通調查
vessel under sail	驶帆船	揚帆行駛之船舶
vessel's speed and fuel consumption clause	航速燃油消耗量条款	船速與耗油量條款
vestigial-sideband emission	残余边带发射	殘邊帶發射
VHF communication	甚高频通信	特高頻通信
VHF emergency position-indicating ra-diobeacon	甚高频紧急无线电示位标	特高頻應急指位無線電示標
VHF radio installation	甚高频无线电设备	特高頻無線電裝置
VHF radiotelephone installation	甚高频无线电话设备	特高頻無線電話裝置
VHF RDF(=very high frequency radio direction finder)	甚高频无线电测向仪	特高頻無線電測向儀
viaduct	高架桥	高架道路,陸橋
vibration	振动	振動
vibration-absorbing coupling	减振联轴器	减振聯軸節
vibration mode	振型	振動模式
victualler	运粮船	糧食船
view	对景图	視圖
violation	违反	違反
violent storm	11级风	暴風(11级)
virtual height	有效高度	實際高度
Visby Rules	维斯比规则	威斯比規則
viscometer	黏度计	黏度計
viscosimeter (=viscometer)	黏度计	黏度計
viscosity	黏度	黏度,黏性
viscosity automatic control system	黏度自动控制系统	黏度自動控制系統
viscosity classification	黏度分级	黏度分级
viscosity index	黏度指数	黏性指數

英 文 名	祖国大陆名	台湾地区名
viscosity index improver	增黏剂	增黏劑
viscosity-temperature	黏度–温度图	黏度–溫度[曲線]圖
visibility	能见度	能見度
visibility excellent	极好能见度(9级)	能見度極佳(9级)
visibility good	好能见度(7级)	能見度良好(7级)
visibility moderate	中能见度(6级)	能見度中等(6级)
visibility of light	号灯能见距	號燈能見距
visibility poor	能见度不良(5级)	能見度不良(5级)
visibility very good	良好能见度(8级)	能見度甚佳(8级)
visible	可见的	可見的
visible horizon	能见地平	視水平線
visible range	测者能见地平距离	視程
visible trace	可见痕迹	可見痕迹
visual bearing	目测方位	目測方位
visual identification	视力识别	視力識別
visual signal	视觉信号	視覺信號
visual signaling	视觉通信	視覺通信
vital	极重要	極重要
VLCC(=very large crude carrier)	大型油船	巨型油輪
VLF communication	甚低频通信	特低頻通信
v-mode chain	V 型链	V 型鏈
voice/data group call	语音/数据群呼	語音/數據群呼
voice messaging service	话传电报业务	話傳電報業務
void space	空位	空艙
Voith Schneider Propeller（VSP）	平旋推进器	擺線推進器
voltage build-up	起压	電壓建起
voltage-current transducer	电压电流变换器	電壓電流轉變器
volume	容积	容積
volume conversion coefficient	体积系数	容積換算係數
volume of GEOSAR traffic	静止卫星通信容量	定置衛星通訊之容量
volume of non-GEOSAR traffic	非静止卫星通信容量	非定置衛星通訊之容量
volume-temperature correction coefficient	体积温度系数	容積穩定修正係數
volumetric efficiency	容积效率,充气系数	容積效率,體積效率
voluntary stranding	有意搁浅	故意擱淺
vortex	旋涡	旋渦,渦動,渦流
voyage	航次	航次
voyage charter(=trip charter)	航次租船	航次備船
voyage insurance	航次保险	航次保險

英 文 名	祖国大陆名	台湾地区名
voyage repair	航修	航修
voyage report	航次报告	航次報告
voyage specified	特定航次	指定航程
v-shaped formation	人字队	V 形隊
VSP(=Voith Schneider propeller)	平旋推进器	擺線推進器
VTS(=vessel traffic service)	船舶交通管理	船舶交通服務
VTS center(=vessel traffic service center)	船舶交通管理中心	船舶交通服務中心
VTS station	船舶交通管理站	船舶交通服務站
v-type derrick post	V 型起重柱	V 型吊桿柱
Vulkan flexible coupling	伏尔肯弹性联轴器(一种橡胶弹性元件联轴器)	福爾幹撓性聯軸節

W

英 文 名	祖国大陆名	台湾地区名
WA(=with average)	货物水渍险	水漬險
WAAS(=wide area augmentation system)	广域增强系统	廣域系統
wagon deck	汽车甲板	車輛甲板
waist breast	腰横缆	舯横纜
waiting	等待	等候
wake	①伴流 ②迹流	①伴流 ②跡流
wake fraction	伴流系数	跡流因數
wake light	航迹灯	接近受補船艉燈
walkaway	出走(背绳走)	背著走
wall	壁,墙	壁,牆
wall plug	壁式插座	壁式插座
warm	温暖	溫暖
warm advection	暖平流	暖平流
warm air mass	暖气团	暖氣團
warm current	暖流	暖流
warm front	暖锋	暖鋒
warming-up	暖机	暖機
warming-up steam system	暖机蒸汽系统	暖機蒸汽系統
warning	警告	警告
warning device	报警装置	警報設施
warning signal	警告信号	警告信號

英　文　名	祖国大陆名	台湾地区名
warp	①曳纲 ②绞船索	①曳網,曳繩 ②拖索
warping end	绞缆筒	捲索筒
warping head(＝warping end)	绞缆筒	捲索筒
warping the berth	绞缆移船	絞纜移船
warranted period	保用期	保固期
war risk clause	战争条款	戰時險條款
warship	军舰	軍艦
wash	冲洗	沖洗
washed overboard	浪击落水	浪擊落水
wash port	排水口	排水口
waste	废弃物	廢棄物
waste disposal system	废物处理系统	廢物處理系統
waste heat recovery	废热回收	廢熱回收
waste-heat recovery plant	废热回收装置	廢熱回收裝置
watch buoy	守位浮标	示位置浮標
watch keeping	值班	當值
water	水	水
water charging system	补水系统	加水系統
water circulation system	水循环系统	水循環系統
water content	水分	含水量
water discharge capacity	排水能力	排水能量
water drum	水鼓	水鼓
water filling test	灌水试验	注水試驗
water filter tank	滤水柜	濾水櫃
water fire extinguishing system	水灭火系统	噴水滅火系統
water hammer	水击	水鎚
water head test	压水试验	壓水試驗
waterjet vessel(＝hydrojet boat)	喷水推进船	噴水推進船
water level	水位	水位
water-level indicator	水位表	水位計
waterline	水线	水線
water lubricating	水润滑	水潤滑
waterplane coefficient	水线面系数	水線面[積]係數
water pressure tank	压力水柜	壓力水櫃
water-proof	防水的	不透水
waterproof electric torch	防水手电筒	防水手電筒
water proof type	防水型	防水型
water regulating valve	水量调节阀	水量調整閥

英 文 名	祖国大陆名	台湾地区名
water-soluble acids	水溶性酸(强酸)	水溶性酸(强酸)
water stand	停潮	平潮
water surface equilibrium vapor pressure	水面饱和水汽压	水面平衡蒸汽壓
water tank coating	水舱涂料	水艙塗料
watertight	水密	水密
watertight adhesive tape	防水胶带	防水膠帶
watertight bulkhead	水密舱壁	水密艙壁
watertight compartment	水密舱室	水密艙區
watertight door	水密门	水密門
watertight integrity	水密完整性	完整水密
watertight type	水密型	水密型
water treatment	水处理	水處理
water tube boiler	水管锅炉	水管鍋爐
water vapor pressure	水汽压	水汽壓
wattless load	无功负荷	無功負載
wattless power	无功功率	無功功率
wave	①波 ②波动	①波 ②波動
wave amplitude	波振幅	波幅
wave bending moment	波浪弯矩	波載彎曲力矩
wave crest	波峰	波峰
wave forecast	海浪预报	波浪預測
wave hollow	波谷	波谷
wave making resistance	兴波阻力	興波阻力
wave parameter	波浪要素	波浪要素
wave period	波浪周期	波浪週期,波週期
wave quelling oil	镇浪油	鎮浪油
wave ridge (=wave crest)	波峰	波峰
wave scale	浪级	波浪等級
wave spectrum	波浪谱	波譜
wave trough (=wave hollow)	波谷	波谷
way enough	停桨!	到了(操艇口令)
way point	航路点	航路點
WD(=working day)	工作日	工作日
weak	弱的	弱的
weak acid number	弱酸值	弱酸值
weak link	弱链环	弱鏈
weak spring diagram	弱弹簧示功图	弱彈簧示功圖
weak swell	弱涌	弱湧

英　文　名	祖国大陆名	台湾地区名
wear	磨损	磨耗
wearing	顺风换抢	轉向迎風行駛
wear rate	磨损率	磨耗率
wear ring	承磨环	耐磨環
weather	天气	天氣
weather anchor	上风锚	上風錨
weather bulletin	天气公报	天氣公報
weather facsimile receiver	气象传真接收机	氣象傳真接收機
weather forecast	天气预报	天氣預報
weather phenomena	天气现象	天氣現象
weather-proof	防风雨的	耐候
weather report	天气报告	氣象報告
weather routing	气象定线	天氣定航
weather shore	上风岸	上風岸
weather side	上风舷	上風舷
weather symbol	天气符号	天氣符號
weathertight	风雨密	風雨密
weather working day(WWD)	晴天工作日	晴天工作日
wedge	楔	楔
weekly inspection	每周检查	每週檢查
weigh	起锚	起錨
weight	重量	重量
weight cargo	计重货物	計重貨
weight governor	飞重式调速器	配重調速器
weld	焊接	熔接
welded type crankshaft	焊接型曲轴	焊接型曲柄軸
welding flux	焊剂	熔接劑
welding wire	焊条	焊條
westing	向西航行	往西航
west mark	西方标	西方標
wet	湿气	濕
wet cylinder liner	湿式缸套	濕式缸襯[套]
wet sand blasting	湿喷砂除锈	濕噴砂
wet sump lubrication	湿底润滑	濕油槽潤滑
whaler	捕鲸船	捕鯨船
whaling mother ship	捕鲸母船	捕鯨母船
wharf	码头	碼頭
wheel house	操舵室	駕駛室

英　文　名	祖国大陆名	台湾地区名
whipping	扎绳头	紮頭,繩頭紮束
whistle	号笛	汽笛,號笛
whistle-requesting mark	鸣笛标	鳴笛標誌
whistle signal	笛号	笛號
whistling buoy	装哨浮标	鳴笛浮標
white metal bearing	白合金轴承	白合金軸承
white star rocket	白星火箭	白星火箭
WHO (=World Health Organization)	世界卫生组织	世界衛生組織
wide area augmentation system （WAAS）	广域增强系统	廣域系統
width of formation	队形宽度	隊形寬度
wildcat	锚链轮	嵌鏈輪
Williamson turn	威廉逊旋回法	威廉生掉頭法
winching area mark	悬降区标志	懸降區域標誌
wind aft	艉风	艉風
wind direction	风向	風向
wind driven current	风生流	風生流
wind duration	风时	風時
wind heeling lever	风压横倾力臂	風壓傾側力臂
wind heeling moment	风压横倾力矩	風壓傾側力矩
windlass	起锚机	起錨機
windlass room	锚机舱	錨機室
window type air conditioner	窗式空气调节器	窗型空調
wind pressure	风压	風壓
wind resistance	风阻力	風阻力
wind resistance coefficient	风阻力系数	風阻力係數
wind rose	风花	風花圖
wind scale	风级	風級
wind scooper	风斗	風斗
wind speed	风速	風速
wind tunnel	风洞	風洞
wind vane	风向标	風向標
wind velocity (=wind speed)	风速	風速
windward side	受风舷	上風舷
wind wave	风浪	風成浪
wind wave height	风浪高度	風[成]浪高
wing tank	边舱	翼櫃,翼艙
winter load line	冬季载重线	冬期載重線
winter North Atlantic load line	北大西洋冬季载重线	冬期北大西洋載重線

英　文　名	祖国大陆名	台湾地区名
winter seasonal area	冬季季节区域	季節性冬期區域
winter seasonal zone	冬季季节区带	季節性冬期地帶
winter solstice	冬至点	冬至
Wireless telegraphy	无线电报学	無線電報術
wire rope	钢丝绳	鋼絲索
wire rope cutter	钢丝绳剪	切繩器
with average	货物水渍险	水漬險
withdrawal of ship	撤船	撤船
wooden ship	木船	木船
wood raft	木排	木排
working day（WD）	工作日	工作日
World administrative radio conference	世界无线电行政大会	世界行政無線電會議
World Health Organization（WHO）	世界卫生组织	世界衛生組織
world meteorological organization	世界气象组织	世界氣象組織
world wide navigational warning service（WWNWS）	全球航行警告业务	全球航行警告業務
wreck	沉船	沈船,船舶殘骸
wreckage	沉船残体	難船漂流物
wreck remains	沉船残留物	沈船殘留物
wreck surveying	沉船勘测	沈船勘測
WWD（=weather working day）	晴天工作日	晴天工作日
WWNWS（=World wide navigational warning service）	全球航行警告业务	全球航行警告業務

X

英　文　名	祖国大陆名	台湾地区名
x-axis	X轴	X軸線
x-component	X轴分量	X軸分量
X-ray	X射线	X射線
X-ray examination	X射线检查	X射線檢查

Y

英　文　名	祖国大陆名	台湾地区名
yacht	游艇	遊艇
yard	桁	橫桁
yard repair	厂修	廠修

英　文　名	祖国大陆名	台湾地区名
yaw	首摇	平擺
yaw angle	航摆角(**艏摇角**)	平擺角
yaw checking anchor	止荡锚	止擺錨
yaw checking test	抑制偏摆试验	止擺試驗
yawing	偏荡	縱橫搖盪,[艏艉]平擺
yawing in anchoring	锚泊偏荡	錨泊橫搖
y-axis	Y 轴	Y 軸線
y-engine	三缸星形发动机	三缸星形引擎
yield	屈服	降伏
yield limit	屈服极限	降伏極限
York-Antwerp Rules	约克-安特卫普规则	約克-安特衛普規則
Young's modulus	杨氏模数	楊氏模數,彈性模數

Z

英　文　名	祖国大陆名	台湾地区名
z-axis	Z 轴	Z 軸線
ZD(=zone description)	时区号	時區標號
Z drive	Z 型传动	Z 型傳動
zenith	天顶	天頂
zenithal chart	方位投影图	天頂投影圖
zenith distance	天顶距	天頂距
zenith projection	天顶投影	天頂投影法
zero	零	零
zero adjustment	零位调节	零位調整
zero error	零点误差	零點誤差
zero lift	零升力	無升力
zero-lift angle	零升力角	零升力攻角
zero-power experiment	零功率实验	零功率實驗
zero time reference	零时基点	零時基準點
ZG(=Zhongguo Shipping Register)	中国船级社	中國船級社
Zhongguo Shipping Register (ZG)	中国船级社	中國船級社
zigzag maneuvre	曲折机动	曲折操縱
zinc	锌	鋅
zinc-anode for protection	防腐锌阳极	鋅陽極防蝕
zinc-base alloy	锌基合金	鋅基合金
zodiac	黄道带	黃道帶
zone description(ZD)	时区号	時區標號

英 文 名	祖 国 大 陆 名	台 湾 地 区 名
zone letter	区号	區號
zone of contact	作用区	接觸區
zone of fracture	断裂区	破裂區,破裂帶
zone reheat air conditioning system	区域再加热空气调节系统	區域再熱空調系統
zone time(ZT)	区时	區域時,區時
Z propeller	全向推进器	全向推進器
Z transmission(=Z drive)	Z 型传动	Z 型傳動
ZT(=zone time)	区时	區域時,區時